# Environmental Health Criteria 1

# Mercury

Published under the joint
sponsorship of the United
Nations Environment
Programme and the World
Health Organization

WORLD HEALTH ORGANIZATION GENEVA 1976

This report contains the collective views of an international group of experts and does not necessarily
represent the decisions or the stated policy of either
the World Health Organization or the United
Nations Environment Programme

*Environmental Health Criteria 1*

# MERCURY

Published under the joint sponsorship of
the United Nations Environment Programme
and the World Health Organization

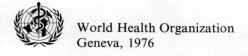

World Health Organization
Geneva, 1976

ISBN 92 4 154061 3

PRINTED IN THE UNITED KINGDOM

# CONTENTS

4

## ORIGIN AND OBJECTIVES OF THE PROGRAMME

During the last two decades, evaluation of the health hazards from chemical and other environmental agents has received considerable attention in several WHO programmes. High priority was given to drinking water quality (1), food additives (2), and pesticide residues (3), to occupational exposure (4), air quality in urban areas (5), and, more recently, to the carcinogenic risk of chemicals to man (6).

In most instances, man's *total* exposure to a given agent, from different media or conditions (air, water, food, work, home), was not considered. The inadequacy of this approach is obvious for pollutants that may reach man by several pathways, as is the case with lead, cadmium, and some other metals, and certain persistent organic compounds. In response to a number of World Health Assembly resolutions (WHA23.60, WHA24.47, WHA25.58, WHA26.68) and taking into consideration the relevant recommendations of the United Nations Conference on the Human Environment (7) held at Stockholm in 1972, and of the Governing Council of the United Nations Environment Programme (UNEP) (8), an integrated and expanded programme on the assessment of health effects of environmental conditions was initiated in 1973 under the title of: WHO Environmental Health Criteria Programme, with the following objectives:

(i) to assess existing information on the relationship between exposure to environmental pollutants (or other physical and chemical factors) and man's health, and to provide guidelines for setting exposure limits consistent with health protection, i.e., to compile environmental health criteria documents;

(ii) to identify new or potential pollutants by preparing preliminary reviews on the health effects of agents likely to be increasingly used in industry, agriculture, in the home or elsewhere.

(iii) to identify gaps in knowledge concerning the health effects of recognized or potential pollutants or other environmental factors, to stimulate and promote research in areas where information is inadequate, and

---

[a] Prepared by the WHO Secretariat. References are listed on page 14.

5

(iv) to promote the harmonization of toxicological and epidemiological methods in order to obtain research results that are internationally comparable.

The general framework of the Environmental Health Criteria Programme was formulated by a WHO meeting held in November 1972 (9), and further elaborated by a WHO Scientific Group that met in April 1973 (10).

# DEFINITIONS, TERMINOLOGY, AND UNITS

## Terminology

In the framework of the WHO Environmental Health Criteria Programme, it is understood that the term "criteria" designates the relationship between exposure to a pollutant or other factor and the risk or magnitude of undesirable effects under specified circumstances defined by environmental and target variables (9). This corresponds to the definition proposed by the Preparatory Committee for the United Nations Conference on the Human Environment (11). Other Preparatory Committee definitions of immediate interest to the criteria programme are:

— "*exposure*: the amount of a particular physical or chemical agent that reaches the target";

— "*target* (or *receptor*): the organism, population, or resource to be protected from specific risks";

— "*risk*: the expected frequency of undesirable effects arising from a given exposure to a pollutant".

The WHO Scientific Group on Environmental Health Criteria (10) accepted these definitions for the purposes of its discussions, but felt that they were not altogether satisfactory, and recommended that WHO, in collaboration with other international organizations, should reconsider them, along with other necessary definitions, at an appropriate international meeting. In accordance with this recommendation, the WHO Secretariat is preparing a list of basic terms to be used in the Environmental Health Criteria Programme that will be submitted to the national institutions and other international organizations for discussion.

The Scientific Group (10) found the definition of "exposure" particularly inadequate and considered that it should be expanded to include the concepts of concentration and length of exposure in addition to the amount of the agent.

The WHO Secretariat considers it useful to attach specific meanings to the terms "effect", "response" and "dose" as was done by the Sub-committee on the Toxicology of Metals of the Permanent Commission and International Association on Occupational Health at the Tokyo meeting (12). These terms will be used in the following sense unless indicated differently in specific criteria documents:

— "*effect*: a biological change caused by (or associated with)[a] an exposure";

— "*response*: the proportion of a population that demonstrates a specific effect";

— "*dose*: the amount or concentration of a given chemical at the site of the effect".

The concept of "response" as defined above is generally accepted but the terminology used to describe this concept varies widely. Many toxicologists use the terms "effect" and "response" interchangeably to denote a specific biological change associated with exposure, whereas different terms are used to indicate the proportion of a population affected (e.g., incidence, cumulative response frequency, response rate, etc.).

There is no general agreement as to the use of the term "dose" for chemical agents. Its common usage is to express the amount of substance administered, for instance, to an experimental animal (e.g., oral dose, injected dose, etc.). In most cases, the amount or concentration of a given agent at the site where its presence induces a given effect cannot be determined by direct measurement and has to be estimated from experimental, occupational, or general environmental exposure, or from measurements in biological indicator media such as blood, urine, faeces, sweat, or hair (12). To avoid misunderstanding, it is, therefore, necessary in each case to make as clear as possible the way in which the "dose" is measured or estimated, including the units used.

Because of the existing differences in the use of terms, no attempt has been made at this stage to impose a uniform terminology in all criteria documents. Until an internationally agreed terminology becomes available, the task groups on specific criteria documents are given freedom to choose their terminology, provided the terms are defined and used consistently throughout the document under consideration.

---

[a] Added by the WHO Secretariat.

**Units**

An attempt has been made to express all numerical values in a uniform fashion, for instance, the concentrations are always expressed as mass concentrations in units acceptable to the SI system (e.g. mg/litre or mg/kg) (*13*). Some departures from this are made where the introduction of new units would cause confusion, e.g., lead in blood is expressed in µg/ 100 ml and not in µg/litre.

**Priorities**

Considering the large number of environmental agents and factors that may adversely influence human health, a practical programme for the preparation of criteria documents must be based on clearly defined priorities. The list of priorities has been established by a WHO Scientific Group (*10*), and is based on the following considerations:

— "*Severity and frequency of observed or suspected adverse effects on human health.* Of importance are irreversible or chronic effects, such as genetic, neurotoxic, carcinogenic, and embryotoxic effects including teratogenicity. Continuous or repeated exposures generally merit a higher priority than isolated or accidental exposures.

— *Ubiquity and abundance of the agent in man's environment.* Of special concern are inadvertently produced agents, the levels of which may be expected to increase rapidly, and agents that add to a natural hazard.

— *Persistence in the environment.* Pollutants that resist environmental degradation and accumulate, in man, in the environment, or in food chains, deserve attention.

— *Environmental transformations or metabolic alterations.* Since these alterations may lead to the production of chemicals that have greater toxic potential, it may be more important to ascertain the distribution of the derivatives than that of the original pollutant.

— *Population exposed.* Attention should be paid to exposures involving a large portion of the general population, or occupational groups, and to selective exposures of highly vulnerable groups represented by pregnant women, the newborn, children, the infirm or the aged."

The full list contains some 70 chemicals and physical hazards, and it will be periodically reviewed. In preparing this list, it was realized that each country must assess environmental health problems in the light of its own national situation and establish its own priorities, which may not have been covered by this list.

# SCOPE AND CONTENT OF ENVIRONMENTAL HEALTH CRITERIA DOCUMENTS

## Scope

As stated on page 5, the purpose of the criteria documents is to compile, review, and evaluate available information on the biological effects of pollutants and other environmental factors that may influence man's health, and to provide a scientific basis for decisions aimed at protecting man from the adverse consequences of exposure to such environmental factors, both in the occupational and general environment. Although attainment of this objective entails consideration of a wide range of data, no attempt is made to include in the documents an exhaustive review of all published information on the environmental and health aspects of specific agents. In the process of collecting the required information, the available literature has been carefully evaluated and selected as to its validity and its relevance to the assessment of human exposure, to the understanding of the mechanism of biological effects, and to the establishment of dose–effect and dose–response relationships. Environmental considerations are limited to information that can help in understanding the pathways leading from the natural and man-made sources of pollutants to man. Non-human targets (e.g., plants, animals) are not considered unless the effects of their contamination are judged to be of direct relevance to human health. For similar reasons much of the published information on the effects of chemicals on experimental animals has been omitted.

## Content

The criteria documents consist of three parts:

(i) A summary, which highlights the major issues, followed by recommendations for research to fill existing gaps in knowledge;

(ii) The bulk of the report, which contains the findings on which the evaluation of the health risks is based. This part has a similar structure in all the criteria documents on chemical agents and contains the following chapters: chemical and physical properties and analytical methods; sources of environmental pollution; environmental transport, distribution and transformation; metabolism; experimental studies of effects; and epidemiological and clinical studies of the effects. The subdivision of these chapters differs from document to document.

(iii) Evaluation of health risks to man from exposure to the specific agent. This part of the criteria document states the considered opinion of the task group, which examined the findings contained in the second part (see (ii) above), and typically contains the following sections: relative contributions to the total dose from air, food, water, and other exposures; dose–effect relationships; dose–response relationships and, whenever possible, guidelines on exposure or dose limits.

## Chemical and physical data

The chemical and physical data included in the criteria documents are limited to the properties that are considered relevant to the assessment of exposure and to the understanding of the effects. Where applicable, the impurities that may occur in commercial products are examined. Analytical techniques are discussed only to the extent needed to understand and evaluate data on levels in the environment and biological samples. The methods described should not be considered as recommended procedures. Where feasible, information is included on the applicability of a given method for the analysis of different types of sample, on detection limits, precision, and accuracy. The detection limit represents the smallest total amount the method is able to determine. In most cases, the amount of sample is limited so that it is useful in practice to express the smallest concentration that can be determined by that method. Precision of a method is defined in terms of the standard deviation or the coefficient of variation of a number of analyses made on the sample. Accuracy denotes systematic deviation of the measured values from the true value. It is impossible to ascertain the accuracy with absolute certainty; the evidence for the accuracy of a method is often circumstantial and is based either on inter-laboratory data-quality control studies or on the agreement of results obtained with procedures using different approaches. The results of one "accurate" procedure should agree with those of another "accurate" procedure for a given set of samples.

## Production, use, and environmental levels

Data on the production, use, and levels in the environment of pollutants are reported only to illustrate the magnitude and extent of the problem and are not meant to represent an exhaustive and critical review. It is hoped that, in the future, better data will be available and that closer collaboration will be established with other governmental and non-governmental organizations qualified to supply such information.

**Biological data**

Although every effort is made to review the whole literature, it is possible that some publications have been overlooked. Some studies have purposely been omitted because the information contained therein was not considered valid or relevant to the scope of the criteria documents, or because they only confirmed findings already described. In general, the information is summarized as given by the author; however, certain shortcomings of reporting or of experimental design are also pointed out. The data on carcinogenicity have been examined and evaluated in consultation with the International Agency for Research on Cancer.

Whenever possible, the dose–effect and the dose–response relationships reported in the criteria documents are based on epidemiological and other human studies, and animal data are used, in general, as supporting evidence.

## ARRANGEMENTS FOR THE PREPARATION OF CRITERIA DOCUMENTS

In order to obtain balanced and unbiased information, the collection and evaluation of information is done in close collaboration with national scientific and health institutions. About 20 Member States of WHO have designated national focal points for collaboration in the WHO Environmental Health Criteria Programme. Without this collaboration no progress could have been made in its implementation.

In addition, a number of WHO collaborating centres on environmental health effects have been designated to extend and complement the expertise available in the WHO Secretariat.

Two procedures have been used in preparing the criteria documents. One is based on the consolidation of national contributions and the other on a draft criteria document prepared by consultants or the collaborating centres in association with the Secretariat.

### Procedure based on national contributions

Criteria documents are prepared in four stages: (1) the preparation of national contributions by focal points in the Member States reviewing all relevant research results obtained in these countries; (2) consolidation of the national contributions into a draft document, which is done on a contractual basis with individual experts or WHO collaborating centres; (3) the draft criteria documents are circulated to the national focal points for comments and additions, based on which a second draft is prepared, and (4) the second draft document is reviewed and the information assessed at a meeting of internationally recognized experts (the task group meetings).

National contributions to the criteria documents consist of a review of data on health effects of environmental agents, as revealed by experimental, clinical, and epidemiological studies, and of other relevant information on research carried out in each country and published in scientific journals or official publications. In order to facilitate the integration of national contributions into draft criteria documents, detailed outlines are prepared for each environmental agent considered, and the national focal points are requested to follow these outlines as closely as possible and to attach all publications referred to in the review in the form of reprints or microfiches.

**Procedure for drafts prepared by the Secretariat**

With the exception of steps 1 and 2 (which are replaced by the preparation of a draft criteria document by individual experts or WHO collaborating centres), the procedure is the same as described above. This procedure is applied in cases where much preparatory work has been done in Member States and where criteria-like documents (WHO or national) already exist.

**Task group meetings**

The task group meetings that are convened to complete the criteria documents have the following terms of reference:

(i) to verify, as far as possible, that all available data have been collected and examined;

(ii) to select those data relevant to the criteria documents;

(iii) to determine whether the data, as summarized in the draft criteria document, will enable the reader to make his own judgement concerning the adequacy of an experimental, epidemiological, or clinical study;

(iv) to judge the health significance of the information contained in the draft criteria document, and

(v) to make an evaluation of the dose–effect, dose–response relationships and of the health risks from exposure to the environmental agents under examination.

Members of task groups serve in a personal capacity, as experts and not as representatives of their governments or of any organization with which they are affiliated. In addition to the first and second draft criteria documents, the members of the task group are requested to refer to the original publications whenever they deem that necessary, and to review national and other comments on the first draft criteria document to make sure that no significant information is omitted and that the final document properly reflects the work done in different countries.

## Collaboration with the United Nations Environment Programme (UNEP) and other international organizations

The WHO Environmental Health Criteria Programme has received substantial financial assistance from UNEP which is acknowledged with appreciation. In addition, the programme has been planned from the outset in consultation with the UNEP Secretariat. The UNEP Secretariat receives all the drafts of criteria documents and their comments are carefully considered in the preparation of the final documents. UNEP is regularly invited to be represented at the task group meetings.

The United Nations, their subsidiary bodies and specialized agencies, and the IAEA are as a rule invited to provide comments on the draft criteria documents and to participate in the task group meetings. The same applies to selected nongovernmental organizations in official relationship with WHO.

## Note to readers of the criteria documents

While every effort has been made to present information in the criteria documents as accurately as possible without unduly delaying their publication, mistakes might have occurred and are likely to occur in the future. In the interest of all users of the environmental health criteria documents, readers are kindly requested to communicate any errors found to the Division of Environmental Health, World Health Organization, Geneva, Switzerland, in order that they may be included in corrigenda which will appear in subsequent volumes.

In addition, experts in any particular field dealt with in the criteria documents are kindly requested to make available to the WHO Secretariat any important published information that may have inadvertently been omitted and which may change the evaluation of health risks from exposure to the environmental agent under examination, so that the information may be considered in the event of updating and re-evaluation of the conclusions contained in the criteria documents.

# REFERENCES

1. *International Standards for Drinking Water*, third edition, Geneva, World Health Organization, 1971.
2. WHO Technical Report Series, Nos: 129 (1957), 228 (1962), 281 (1964), 309 (1965), 339 (1966), 373 (1967), 383 (1968), 430 (1969), 445 (1970), 462 (1971), 488 (1972), 505 (1972), 539 (1974).
3. WHO Technical Report Series, Nos: 370 (1967), 391 (1968), 417 (1969), 458 (1970), 474 (1971), 502 (1972), 525 (1973), 545 (1974), 574 (1975), 592 (1976).
4. WHO Technical Report Series, No.: 415 (1969).
5. WHO Technical Report Series, No.: 506 (1972).
6. INTERNATIONAL AGENCY FOR RESEARCH ON CANCER. *IARC Monographs on the Evaluation of Carcinogenic Risk of Chemicals to Man*, Vol. 1–11 (1972–76).
7. UNITED NATIONS GENERAL ASSEMBLY. *Report of the United Nations Conference on the Human Environment held at Stockholm, 5–16 June 1972* A/CONF.48/14, 3 July 1972.
8. UNITED NATIONS ENVIRONMENT PROGRAMME. *Report of the Governing Council of the United Nations Environment Programme (First session)* UNEP/GC/10, 3 July 1973.
9. *The WHO Environmental Health Criteria Programme* (unpublished WHO document EP/73.1).
10. *Environmental Health Criteria.* Report of a WHO Scientific Group (unpublished WHO document EP/73.2).
11. UNITED NATIONS GENERAL ASSEMBLY. *Report of the Preparatory Committee for the United Nations Conference on the Human Environment on its Third Session.* United Nations document A/CONF.48/PC/13, 30 September 1971.
12. NORDBERG, G. F., ed. *Effects and dose–response relationships of toxic metals, Proceedings from an international meeting organized by the Sub-committee on the Toxicology of Metals of the Permanent Commission and International Associations on Occupational Health, Tokyo, 18–23 November 1974.* Amsterdam, Oxford, New York, Elsevier Scientific Publishing Company, 1976.
13. LOWE, D. A. *A guide to international recommendations on names and symbols for quantities and on units of measurement.* Geneva, World Health Organization, 1975, 314 pp. (*Progress in Standardization No. 2.*)

# WHO TASK GROUP ON ENVIRONMENTAL HEALTH CRITERIA FOR MERCURY

*Geneva 4–10 February 1975*

## Participants:

*Members*

Professor T. Beritić, Institute for Medical Research and Occupational Medicine, Zagreb, Yugoslavia

Dr H. Blumenthal, Division of Toxicology, Bureau of Foods, Food and Drug Administration, Department of Health, Education and Welfare, Washington, DC, USA (*Rapporteur*)

Dr J. Bouquiaux, Department of the Environment, Institute of Hygiene and Epidemiology, Brussels, Belgium

Dr G. J. van Esch, Laboratory for Toxicology, National Institute of Public Health, Bilthoven, Netherlands

Professor L. Friberg, Department of Environmental Hygiene, Karolinska Institute, Stockholm, Sweden (*Chairman*)

Professor G. L. Gatti, Istituto Superio di Sanità, Rome, Italy

Dr L. Magos, Toxicology Research Unit, Medical Research Council Laboratories, Carshalton, Surrey, England

Dr J. Parizek, Institute of Physiology, Czechoslovak Academy of Sciences, Prague, Czechoslovakia

Dr J. K. Piotrowski, Department of Biochemistry, Institute of Environmental Research, Medical Academy in Łódź, Łódź, Poland (*Vice-Chairman*)

Dr E. Samuel, Health Protection Branch, Department of National Health and Welfare, Ottawa, Ontario, Canada

Dr S. Skerfving, Department of Internal Medicine, University Hospital, Lund, Sweden

Dr T. Tsubaki, Brain Research Insitiute, Niigata University, Niigata, Japan

Professor H. Valentin, Institute for Occupational and Social Medicine, Erlangen, Federal Republic of Germany

*Representatives from other organizations*

Dr A. Berlin, Health Protection Directorate, Commission of the European Communities, Luxembourg

Dr D. Djordjević, Occupational Health and Safety Branch, ILO, Geneva, Switzerland

Dr W. J. Hunter, Commission of the European Communities, Luxembourg

Dr G. D. Kapsiotis, Senior Officer, Food Policy and Nutrition Division, FAO, Rome, Italy

Dr E. Mastromatteo, Chief, Occupational Health and Safety Branch, ILO, Geneva, Switzerland

*Secretariat*

Dr T. Clarkson, University Center in Environmental Health Sciences, The University of Rochester, School of Medicine and Dentistry, Rochester NY, USA (*Temporary Adviser*)

Dr F. C. Lu, Chief, Food Additives, WHO, Geneva, Switzerland (*Secretary*)

Dr B. Marschall, Medical Officer, Occupational Health, WHO, Geneva, Switzerland

# ENVIRONMENTAL HEALTH CRITERIA FOR MERCURY

A WHO Task Group on Environmental Health Criteria for Mercury met in Geneva from 4–10 February 1975. Dr B. H. Dietrich, Director, Division of Environmental Health, opened the meeting on behalf of the Director-General. The Task Group reviewed and amended the second draft criteria document and made an evaluation of health risks from exposure to mercury and its compounds. The revised draft was sent for comments to all members of the Task Group.

A group of WHO temporary advisers (Dr T. Clarkson, Dr L. Friberg, Dr A. Jernelöv,[a] Dr L. Magos, and Dr G. Nordberg[b]) assisted the Secretariat in the final scientific editing of the document. They met in Geneva on 13 and 14 November 1975.

The first and second draft criteria documents were prepared by Dr T. Clarkson, Environmental Health Sciences Centre, the University of Rochester School of Medicine and Dentistry, Rochester, New York, USA. The comments on which the second draft was based were received from the national focal points for the WHO Environmental Health Criteria Programme in Bulgaria, Czechoslovakia, the Federal Republic of Germany, Italy, Japan, the Netherlands, New Zealand, Poland, Sweden, the USA, and the USSR; and from the United Nations Industrial Development Organization (UNIDO), Vienna, and the United Nations Scientific, Educational and Cultural Organization (UNESCO), Paris. Comments from the International Labour Organisation, Geneva, the United Nations Food and Agriculture Organization, Rome, and the Commission of the European Communities Health Protection Directorate, Luxembourg, were submitted at the task group meeting.

Comments were also received, at the request of the Secretariat, from Dr L. Amin-Zaki, Iraq, Dr G. J. van Esch, Netherlands, Dr K. Kojima, Japan, and Dr S. I. Shibko, USA.

The collaboration of these national institutions, international organizations, WHO collaborating centres and individual experts is gratefully acknowledged. Without their assistance the document could not have been completed. The Secretariat wishes to thank in particular Dr T. Clarkson for his help in all phases of the preparation of the document.

---

[a] Institute for Water and Air Pollution Research, Stockholm, Sweden.
[b] Department of Environmental Hygiene, Karolinska Institute, Stockholm, Sweden.

This document is based primarily on original publications listed in the reference section. However, several recent publications broadly reviewing health aspects of mercury and its compounds have also been used. These include reviews by the Swedish Expert Group (1971)., Hartung & Dinman (1972), IAEA (1972), and Wallace et al. (1971). Reviews devoted primarily to the biological effects of mercury have been published by Clarkson (1972a, 1972b) and Miller & Clarkson (1973). Furthermore, several recent symposia have provided extensive reviews of the environmental aspects of mercury (Bouquiaux, 1974; D'Itri, 1972; Krenkel, 1975). A systematic review of various environmental health aspects of mercury, including a broad review of the accessible literature up to 1971, has been presented by Friberg & Vostal (1972).

# 1. SUMMARY AND RECOMMENDATIONS FOR FURTHER RESEARCH

## 1.1 Some definitions

In order to clarify the meaning of certain terms used in the document, some definitions are given below. However, it should be noted that these definitions have not been formally adopted by WHO.

The terms *critical effects*, *critical organ*, and *critical organ concentration* have recently been defined by the Sub-Committee on Toxicology of Metals of the Permanent Commission and International Association of Occupational Health (Nordberg, 1976). The term "critical" as defined by the Committee differs from its usual meaning in clinical medicine, where it refers to a situation in which the patient's condition may deteriorate suddenly and dramatically. It also differs in meaning from that used in the field of radiation protection, where the "critical" organ is defined as the organ of the body whose damage by radiation results in the greatest injury to the individual. In this document, the term "critical" does not refer to a life-threatening situation, but to a key decision point for taking preventive action. For example, at some point in the dose–effect relationship, a critical effect can be identified. The appearance of an effect in an individual signals the point at which measures should be taken to reduce or prevent further exposure.

## 1.2 Summary

### 1.2.1 Analytical methods

The method of choice for determining total mercury in environmental and biological samples is flameless atomic absorption. The technique is rapid and sensitive and the procedure is technically simple. Neutron activation is now principally used as a reference method against which the accuracy of atomic absorption procedures may be checked. Gas–liquid chromatography combined with an electron-capture detector is the most widely used method for identifying methylmercury in the presence of other compounds of mercury.

The methods of sampling require careful consideration of the type of exposure to be monitored and the material to be analysed. Errors arising in collection, storage, and transportation of samples may be as important as instrument errors in contributing to the total error in the measurement of mercury in the sample. These include contamination of the sample, and the loss of mercury by adsorption on the walls of the container, and by volatilization. In estimating human exposure, special care should be taken to see that the sample is truly representative, e.g. the mercury vapour concentration in the breathing zone and the concentration of methylmercury in the daily diet.

### 1.2.2 Sources of environmental pollution

The major source of mercury is the natural degassing of the earth's crust and amounts to between 25 000 and 125 000 tonnes per year. Anthropogenic sources are probably less than natural sources. World production of mercury by mining and smelting was estimated at 10 000 tonnes per year in 1973 and has been increasing by an annual rate of about 2%. The chloralkali, electrical equipment, and paint industries are the largest consumers of mercury, accounting for about 55% of the total consumption. Mercury has a wide variety of other uses in industry, agriculture, military applications, medicine, and dentistry.

Several of man's activities not directly related to mercury account for substantial releases into the environment. These include the burning of fossil fuel, the production of steel, cement, and phosphate, and the smelting of metals from their sulfide ores. It was extimated that the total anthropogenic release of mercury would amount to 20 000 tonnes per year in 1975.

### 1.2.3 Environmental distribution and transport

Two cycles are believed to be involved in the environmental transport and distribution of mercury. One is global in scope and involves the atmospheric circulation of elemental mercury vapour from sources on land to the oceans. However, the mercury content of the oceans is so large, at least seventy million tonnes, that the yearly increases in concentration due to deposition from the global cycle are not detectable.

The other cycle is local in scope and depends upon the methylation of inorganic mercury mainly from anthropogenic sources. Many steps in this cycle are still poorly understood but it is believed to involve the atmospheric circulation of dimethylmercury formed by bacterial action.

The methylation of inorganic mercury in the sediment of lakes, rivers, and other waterways and in the oceans is a key step in the transport of mercury in aquatic food chains leading eventually to human consumption. Methylmercury accumulates in aquatic organisms according to the trophic level, the highest concentrations being found in the large carnivorous fish.

Alkylmercury fungicides used as seed dressings are important original sources of mercury in terrestrial food chains. Mercury is passed first to seed eating rodents and birds and subsequently to carnivorous birds.

Accumulation of methylmercury in aquatic and terrestrial food chains represents a potential hazard to man by consumption of certain species of oceanic fish, of fish or shellfish from contaminated waters, and of game birds in areas where methylmercury fungicides are used.

### 1.2.4 Environmental exposure levels

The concentration of mercury in the atmosphere is usually below 50 ng/m$^3$ and averages approximately 20 ng/m$^3$. A concentration of 50 ng/m$^3$ would lead to a daily intake of about 1 µg. "Hot spots" near mines, smelting works, and refineries require further investigation but could lead to daily intakes as high as 30 µg. Daily intakes would be higher for occupational exposures to mercury vapour. An average mercury concentration in air of 0.05 mg/m$^3$ would lead to an average daily intake via inhalation of about 480 µg. The highest occupational exposures usually occur in mining operations but over 50 specific occupations or trades involve frequent exposure to mercury vapour.

Mercury in drinking water would contribute less than 0.4 µg to the total daily intake. Bodies of fresh water for which there is no independent evidence of contamination contain mercury at less than 200 ng/litre. Oceanic mercury is usually less than 300 ng/litre.

Food is the main source of mercury in nonoccupationally exposed populations, and fish and fish products account for most of the methylmercury in food. Mercury in food other than fish is usually present at concentrations below 60 µg/kg. Mercury is present in freshwater fish from uncontaminated waters at concentrations of between 100 and 200 µg/kg wet weight. In contaminated areas of freshwater, mercury levels between 500 and 700 µg/kg wet weight are often described and in some cases, concentrations are even higher. Most species of oceanic fish have mercury levels of about 150 µg/kg. However, the large carnivorous species (e.g. swordfish and tuna) usually fall in the range of 200–1500 µg/kg. With few exceptions methylmercury accounts for virtually all the mercury in both freshwater and marine fish.

Intake of mercury from food is difficult to estimate with precision. Daily intake from food other than fish is estimated as 5 µg but the chemical form of mercury is not known. Most of the methylmercury in diet probably comes from fish and fish products. The median daily intake of methylmercury in Sweden has been estimated as 5 µg. In most countries the daily intake is less than 20 µg but in subgroups in certain countries where there is an unusually high fish intake (dieters) the daily intake may rise to 75 µg and may even be as high as 200–300 µg (in coastal villages dependent on large oceanic fish as the main source of protein). In areas of high local pollution, daily intakes could be well in excess of 300 µg and these levels have led to two recorded outbreaks of methylmercury poisoning.

## 1.2.5  Metabolism of mercury

Approximately 80% of inhaled mercury vapour is retained. Information on pulmonary retention of other forms of mercury in man is lacking. Absorption of inorganic mercury compounds from foods is about 7% of the ingested dose. In contrast, gastrointestinal absorption of methylmercury is practically complete. Little information is available on skin absorption although it is suspected that most forms of mercury can penetrate the skin to some extent. In the case of methylmercury, poisoning has resulted from skin application.

Animal data indicate that the kidneys accumulate the highest tissue concentrations no matter what form of mercury is administered. The distribution of mercury between red cells and plasma depends upon the form of mercury. The red cell to plasma ratio is highest for methylmercury (approximately 10) and lowest for inorganic mercury (approximately 1) in man.

The hair is a useful indicator medium for people exposed to methylmercury. The concentration of mercury in hair is proportional to the concentration in the blood at the time of formation of the hair. The relationship between hair and blood concentrations is not known for other forms of mercury.

Most forms of mercury are predominantly eliminated with urine and faeces. In workers exposed over a long period to mercury vapour, urinary excretion slightly exceeds faecal elimination. On a group basis, mercury excretion in urine is proportional to the time-weighted average air concentration. Large individual fluctuations are common in daily mercury excretion in urine in people under the same exposure conditions.

Faecal elimination accounted for approximately 90% of total mercury elimination in volunteers given a single dose of methylmercury. Urinary concentrations of total mercury do not correlate with blood levels after exposure to methylmercury.

Animal data indicate that elemental mercury vapour rapidly crosses the placenta. The transplacental transfer of methylmercury compounds is well documented in man. The mercury concentrations in plasma in the mother and the newborn infant are similar but the concentration in the fetal red blood cells is approximately 30% higher than in those of the mother.

Details on transmission into breast milk are available only for methylmercury. The concentration of mercury in breast milk is approximately 5% of the simultaneous mercury level in blood in the mother, and infants can accumulate dangerously high blood concentrations by suckling if their mothers are heavily exposed.

Tracer studies in volunteers and in exposed populations have established the main features of the metabolic model for methylmercury in man. Clearance half-times from the whole body and from blood are about 70 days. Daily intakes of methylmercury will lead to a steady-state balance in about one year, when the body burden will be approximately one hundred times the daily intake. In steady-state, the numerical value of the concentration of mercury in whole blood in µg/litre is virtually equal to the numerical value of the daily intake in µg/day/70 kg body weight. Considerable individual variation around these average values has been noted, which must be taken into account in the estimation of risk in exposed populations.

The metabolic models for other forms of mercury are less well developed.

### 1.2.6 Experimental studies on the effects of mercury

Reversible and irreversible toxic effects may be caused by mercury and its compounds, depending upon the dose and duration of exposure. Reversible behavioural changes may be produced in animals by exposure to mercury vapour.

Methylmercury compounds produce irreversible neurological damage in animals. Many of the neurological signs seen in man have been reproduced in animals. Methylmercury is equally toxic to animals whether it is given in the pure chemical state or in fish where it has accumulated naturally. A latent period lasting weeks or months is observed between cessation of exposure and onset of poisoning. Morphological changes have been seen in the brain before onset of signs. This phenomenon has been referred to as "silent damage". Animal data support epidemiological evidence from Japan, that the fetus is more sensitive than the adult.

Little is known about the physical and chemical factors affecting the toxicity of mercury. Selenium is believed to be protective against inorganic and methylmercury compounds.

### 1.2.7 Epidemiological and clinical studies

The classic symptoms of poisoning by mercury vapour are erethism (irritability, excitability, loss of memory, insomnia), intentional tremor, and gingivitis. Most effects of mercury vapour are reversible on cessation of exposure, although complete recovery from the psychological effects is difficult to determine. Recovery may be accelerated by treatment with penicillamine and unithiol (2,3,dimercaptopropansulfonate).

Studies of occupational exposure to mercury vapour reveal that the classic symptoms of mercurialism do not occur below a time-weighted average mercury concentration in air of 0.1 mg/m$^3$. Symptoms such as loss of appetite and psychological disturbance have been reported to occur at mercury levels below 0.1 mg/m$^3$.

The most common signs and symptoms of methylmercury poisoning are paraesthesia, constriction of the visual fields, impairment of hearing, and ataxia. The effects are usually irreversible but some improvement in motor coordination may occur. Complexing and chelating agents may be useful in prevention if given early enough after exposure but BAL is contraindicated in cases of methylmercury poisoning as it leads to increased brain levels of mercury.

Epidemiological investigations have been made on populations in whom the intensity and duration of exposure to methylmercury through diet differs, for example, a population in Iraq having high daily mercury intakes (as high as 200 μg/kg/day) for a brief period (about 2 months), populations in Japan having lower daily intakes with exposure for several months or years, and several fish-eating populations having daily intakes of mercury usually below 5 μg/kg but with exposure lasting for the lifetime of the individual. The results of these studies indicate that the effects of methylmercury in adults become detectable in the most sensitive individuals at blood levels of mercury of 20–50 μg/100 ml, hair levels from 50–120 mg/kg, and body burdens between about 0.5 and 0.8 mg/kg body weight.

Observations on the Minamata outbreak in Japan indicate that the fetus is more sensitive to methylmercury than the adult but the difference in degree of sensitivity has not yet been established.

### 1.2.8 Evaluation of health risks to man from exposure to mercury and its compounds

Adverse health effects have not yet been identified in workers occupationally exposed to a time-weighted average air concentration of mercury of 0.05 mg/m$^3$. This air concentration is equivalent to an average mercury concentration in blood of 3.5 µg/100 ml and an average mercury concentration in urine of 150 µg/litre on a group basis. The corresponding ambient air concentration of mercury for exposure of the general population would be 0.015 mg/m$^3$.

It is estimated that the first effects associated with long-term daily intake of methylmercury should occur at intake levels between 3 and 7 µg/kg/day. The probability of an effect (paraesthesia) at this intake level is about 5% or less in the general population. These figures apply only to adults. Prenatal life may be the most sensitive stage of the life cycle to methylmercury. Furthermore experiments on animals indicate a potential for genetic damage by methylmercury.


### 1.3 Recommendations for Further Research

#### 1.3.1 Environmental sources and pathways of mercury intake

More information is needed on the physical and chemical forms of mercury in air, food, and water. With the exception of fish tissue, little is known of the proportion of total mercury in the diet that is in the form of methylmercury.

The concentration of mercury in the air in "hot spots" near points of industrial release is not yet adequately documented. The few reports reviewed in this criteria document indicate that people living near points of emission may receive substantial exposure to airborne mercury. Levels of mercury in the oceans are still inadequately documented. The pathways of methylation of mercury in the ocean and its uptake by fish of different trophic levels are poorly understood.

Studies are needed to estimate quantitatively the dietary intake of methylmercury in populations dependent on fish for their main source of protein. Average dietary intakes for the populations of several industrialized countries have been reported. However, of much greater importance are the identification of those subgroups of the population having unusually high dietary intakes of methylmercury and the careful quantitative estimation of average daily intake in these groups.

### 1.3.2 Metabolic models in man

The kinetic parameters of uptake, distribution, and excretion of methylmercury in man are documented in much more detail than for other forms of mercury. However, questions still remain on the linearity of this metabolic model at high toxic doses of methylmercury. Specifically, the applicability of the metabolic model derived from human tracer-dose studies should be verified at higher dose levels. Information on this point would greatly facilitate the interpretation of results of epidemiological studies on heavily exposed populations.

Recent findings of large individual variations in clearance half-times of methylmercury from blood are of considerable importance in the estimation of risks from long-term dietary intake. Further studies are needed to establish the statistical parameters of the distribution of individual clearance half-times, and on the biological mechanisms underlying these differences.

A more complete metabolic model for inhaled mercury vapour in man is urgently needed. Despite the continuous occupational exposure of thousands of workers annually and the long history of man's exposure to this form of mercury, we still do not have sufficient information to relate mercury concentrations in air to accumulated body burdens and to identify the most appropriate indicator media for levels of mercury vapour in the target organ (the brain). Animal experiments have indicated the ability of the inhaled vapour to cross the placenta; no information is available on human subjects concerning this important question.

### 1.3.3 Epidemiological studies

Several types of epidemiological study are needed. Long-term studies on adults should concentrate on those areas of the dose–response relationship where the effects of methylmercury become just detectable. There are still uncertainties concerning the concentrations of total mercury in indicator media and the equivalent long-term daily intake of mercury as methylmercury associated with the earliest effects in the most sensitive group in the adult population.

So far, dose–response relationships in human populations have been based on outbreaks of poisoning in which daily exposure was high and limited to months or a few years at the most. To extrapolate these relationships to the general population, more information is needed on the potential influence of long-term exposure.

In addition to continuing studies on mature adults, groups of the population specially sensitive to methylmercury should be identified.

25

Special studies should be made on the relationship between the dose received by the expectant mother and the effect on her infant including the development and growth of the child.

Further epidemiological studies are needed on groups occupationally exposed to mercury vapour. Whenever possible, collaborative studies should be carried out in which cohorts should be followed in time and different groups related to each other.

### 1.3.4 Interaction of mercury with other environmental factors

The extrapolation to the general population of epidemiological data from outbreaks of methylmercury poisoning that have occurred in certain parts of the world is fraught with uncertainties, unless the possible interaction of local environmental factors can be taken into account. For example, the conditions under which selenium exerts antagonistic and synergistic effects and its mode of action should be studied. Alcohol influences the metabolism of mercury and may affect the toxicity of inhaled vapour in man. Genetic factors should also be considered. Acatalasaemic individuals may metabolize inhaled mercury vapour differently from normal individuals.

Mercury, along with other heavy metals, has the potential to alter the activity of drug metabolizing enzymes. Studies should be made on these potential effects with special emphasis on those individuals carrying high body burdens of mercury.

### 1.3.5 Biochemical and physiological mechanisms of toxicity

Long-term investigations of the mode of toxic action of mercury and its compounds are needed to give an insight into the causes of individual differences in sensitivity to mercury and into differences in metabolism such as clearance half-times. Methylmercury is known to produce "silent damage" in that morphological changes can be seen in the brains of experimental animals before functional disturbances are detectable. Biochemical disturbances such as inhibition of protein synthesis precede overt signs of damage. There is a great need to develop sensitive biochemical and physiological tests, especially in the case of methylmercury poisoning.

A deeper understanding of the toxic action of mercury should lead to the development of more effective means of treatment. Present methods depend mainly on prevention, using complexing and chelating agents to remove the metal from the body before serious damage has occurred.

# 2. PROPERTIES AND ANALYTICAL METHODS

## 2.1 Chemical and Physical Properties

Mercury can exist in a wide variety of physical and chemical states. This property presents special problems to those interested in assessing the possible risk to public health. The different chemical and physical forms of this element all have their intrinsic toxic properties and different applications in industry, agriculture, and medicine, and require a separate assessment of risk.

The chemistry of mercury and its compounds has been outlined in several standard chemistry texts (Rochow et al., 1957; Gould, 1962; Cotton & Wilkinson, 1972). Mercury, along with cadmium and zinc, falls into Group IIb of the Periodic Table. In addition to its elemental state, mercury exists in the $+1$ (mercury(I)) and $+2$ (mercury(II)) states in which the mercury atom has lost one and two electrons, respectively. The chemical compounds of mercury(II) are much more numerous than those of mercury(I).

In addition to simple salts, such as chloride, nitrate, and sulfate, mercury(II) forms an important class of organometallic compounds. These are characterized by the attachment of mercury to either one or two carbon atoms to form compounds of the type RHgX and RHgR′ where R and R′ represent the organic moiety. The most numerous are those of the type RHgX. X may be one of a variety of anions. The carbon–mercury bond is chemically stable. It is not split in water nor by weak acids or bases. The stability is not due to the high strength of the carbon–mercury bond (only 15–20 cal/mol and actually weaker than zinc and cadmium bonds) but to the very low affinity of mercury for oxygen. The organic moiety, R, takes a variety of forms, some of the most common being the alkyl, the phenyl, and the methoxyethyl radicals. If the anion X is nitrate or sulfate, the compound tends to be "salt like" having appreciable solubility in water; however, the chlorides are covalent non-polar compounds that are more soluble in organic solvents than in water. From the toxicological standpoint, the most important of these organometallic compounds is the subclass of short-chain alkylmercurials in which mercury is attached to the carbon atom of a methyl, ethyl, or propyl group.

An expert committee, considering occupational hazards of mercury compounds, distinguished two major classes of mercury compounds— "organic" and "inorganic" (MAC Committee, 1969). Inorganic mercury compounds included the metallic form, the salts of mercury(I) and mercury(II) ions, and those complexes in which mercury(II) was reversibly bound to such tissue ligands as thiol groups and protein. Compounds in which mercury was directly linked to a carbon atom by a covalent bond were classified as organic mercury compounds. This distinction is of limited value because the toxic properties of elemental mercury vapour differ from those of the inorganic salts and, furthermore, the short-chain alkylmercurials differ dramatically from other mercurials that fall within the definition of organic mercury. From the standpoint of risk to human health, the most important forms of mercury are elemental mercury vapour and the short-chain alkylmercurials.

Mercury in its metallic form is a liquid at room temperature. Its vapour pressure is sufficiently high to yield hazardous concentrations of vapour at temperatures normally encountered both indoors and outdoors under most climatic conditions. For example, at 24°C, a saturated atmosphere of mercury vapour would contain approximately 18 mg/m$^3$—a level of mercury 360 times greater than the average permissible concentration of 0.05 mg/m$^3$ recommended for occupational exposure by the National Institutes of Safety and Health, USA (NIOSH, 1973). Apart from the noble gases, mercury is the only element having a vapour which is monatomic at room temperature. However, little is known about the chemical and physical states of mercury found in the ambient air and in the air where occupational exposure occurs.

Elemental mercury vapour is generally regarded as insoluble. Nevertheless, small amounts dissolved in water and other solvents are important from the toxicological point of view. At room temperatures, in air-free water, its solubility is approximately 20 µg/litre. In the presence of oxygen, metallic mercury is rapidly oxidized to the ionic form—mercury(II)—and may attain concentrations in water as high as 40 µg/litre.

Calomel or mercury(I) chloride ($Hg_2Cl_2$) is the best known mercury(I) salt. Widely used in the first half of this century in teething powders and in anthelmintic preparations, the low toxicity of this compound is due principally to its very low solubility in water. Mercury(I) forms few complexes with biological molecules. However, in the presence of protein and other molecules containing SH groups, it gives one atom of metallic mercury and one mercury(II) ion. In general, an equilibrium is established between $Hg^0$, $Hg_2^{++}$ and $Hg^{++}$ in aqueous solution. The distribution of mercury between the three oxidation states is determined by the redox (oxidation–reduction) potential of the solution and the concentration of halide, thiol, and other groups that form complexes with $Hg^{++}$. The dissociation of mercury(I) chloride by thiol groups should be understood in this context. Extra halide and thiol compounds, added to solution, form complexes with mercury(II) ions and the mercury(I) chloride splits to restore the equilibrium between $Hg^0$, $Hg_2^{++}$ and $Hg^{++}$. The split results in the formation of one atom of mercury for every mercury(I) chloride molecule dissociated.

The mercury(II) ion, $Hg^{++}$, is able to form many stable complexes with biologically important molecules. Mercury(II) chloride (corrosive sublimate), a highly reactive compound, readily denatures proteins and was extensively used in the past century as a disinfectant. It is soluble in water and, in solution, forms four different complexes with chloride, $HgCl^+$, $HgCl_2$, $HgCl_3^-$ and $HgCl_4^=$. It has been suggested that the negatively charged chlorine complexes are present in sea water (see section 5).

Phenylmercury compounds have a low volatility. However, the halide salts of methyl-, ethyl-, and methoxyethylmercury can give rise, at 20°C, to saturated mercury vapour concentrations of the order of 90, 8, and 26 mg/m$^3$, respectively (Swensson & Ulfvarsson, 1968). In the case of methylmercury this saturated vapour concentration is several orders of magnitude greater than the maximum allowable concentration in the working atmosphere. This hazardous property of the halide salts of the short-chain alkylmercurials is not always fully appreciated in industrial and agricultural use and even in research laboratories (Klein & Hermen, 1971). In contrast, methylmercury dicyandiamide, previously widely used as a fungicide, has a much lower vapour pressure, being 340 times less volatile than the chloride salt.

Although the carbon–mercury bond is chemically stable, in the living animal, the bond is subject to cleavage (for review, see Clarkson, 1972a). The nature of the R radical is all important. If R is a phenyl or methoxy-alkyl group, rapid breakdown occurs in animal tissues so that most of the organic compound has disappeared within a few days. Enzymes that break the carbon–mercury bond have been discovered and isolated (Tonomura et al., 1968a, 1968b, 1968c). The short-chain alkylmercurials undergo the slowest breakdown *in vivo* with methylmercury being the most stable. Differences in the stability of the carbon–mercury bond play an important role in determining the toxicity and mode of action in man. The rapid breakdown of phenyl- and methoxymercury results in toxic effects similar to those of inorganic mercury salts. The relative stability of the alkylmercurials is one important factor in their unique position with regard to toxicity and risks to human health.

The organic and inorganic cations of mercury, in common with other heavy metal cations, will react reversibly with a variety of organic ligands[a] found in biologically important molecules. The chemical affinity of

---

[a] Ligands are chemical groups within a molecule that are capable of donating electrons to a metal cation to form a chemical bond. Examples of biologically important ligands are the carboxyl, and especially with regard to heavy metals, the sulfhydryl (SH) groups.

mercury(II) and of its monovalent alkylmercury cations for a variety of biologically occurring ligands is so great that free mercury would be present *in vivo* at concentrations so low as to be undetectable by present methods.

## 2.2 Purity of Compounds

Impurities in mercury and its compounds are not important in assessing the hazards to man. Those compounds of mercury used in industry and agriculture have impurities of less than 10%. Bakir et al. (1973) reported that a methylmercury fungicide responsible for an epidemic of poisoning in Iraq contained 10% or less of ethylmercury as an impurity. Inorganic mercury usually amounts to no more than 1% of the total mercury in organomercurial preparations and rarely exceeds 5%.

Impurities are of importance in the preparation of standard solutions for analytical procedures and in experimental research in animals where impurities in radioactive mercury may give misleading results. Preparations of methylmercury labelled with the isotope $^{203}$Hg are subject to radiolytic breakdown to inorganic compounds depending on the pH. This instability must be taken into account in the interpretation of some original reports in which the purity of the radioisotope was not checked properly.

## 2.3 Sampling and Analysis

Before reviewing various aspects of sample collection and analysis it may be worth taking an overview of the various sources of error in the determination of mercury content. Not only are there errors in the instrumental determination of mercury and in the laboratory procedures, but significant and often major errors occur during the collection, transportation, and storage of the samples. The accuracy of the determination of mercury in environmental samples should be assessed from this broad point of view. The error will be the sum of the errors in collection, storage, transportation and, in the instrumental determination. It is of the greatest importance to determine the greatest source of error in each particular case. This, in itself, may lead to considerable improvement in the overall accuracy of the determination. For example, the introduction of a new and more sensitive instrumental technique may allow the collection of smaller samples and thus facilitate storage and transport. On the other hand, there is little value in proceeding further with improvements in instrumental measurements if major errors remain at the collection, storage, or transport stages.

### 2.3.1 Sample collection

Methods of sample collection for the determination of mercury in air have recently been reviewed (NIOSH, 1973). A recommended method for the determination of total mercury in air is presented. Essentially the method consists of using two bubblers in series, containing sulfuric acid and potassium permanganate. The mercury in these traps is subsequently determined by atomic absorption procedures. Problems of the determination of mercury in air are critically evaluated. Included in these problems is the fact that numerous chemical and physical forms of mercury may exist in air and that these are subject to interconversion. The volatility of mercury and its compounds is a special problem in the determination of mercury bound to particles. The separation of particulates from air, such as by filtration, may result in the loss of mercury by volatilization from the particulate. Published methods of sample collection consist of removal of mercury from the air by passing it through scrubbing devices, or direct collection of the air sample, for example in a plastic bag or syringe. The scrubbing device may take the form of bubblers, filters, absorbants, or amalgam collectors. Unfortunately many of the published procedures do not report collection efficiency. Attention is drawn to the importance of the use of standard dust chambers to check the efficiency of absorption.

The procedure recommended by NIOSH (1973) has a collection efficiency for total mercury of more than 90%, when mercury is in the form of elemental vapour or inorganic salts. Organomercurials in air are collected with an efficiency of more than 80%, except in the case of the short-chain alkylmercurials. Bramen (1974) has described a procedure for separating and measuring different physical and chemical forms of mercury in air. Previous reports distinguishing between mercury vapour and particle-bound mercury have not reported the efficiency of collection.

An early method (Polešajev, 1936) for the determination of mercury in air involved absorption in iodine and subsequent determination of the coloured complex in the sediment. This method is still widely used in the Soviet Union and some countries of eastern Europe.

Commercially available portable monitoring devices are used to determine mercury directly in air. The air is pumped through an optical cell that measures the absorption of light emitted from a mercury vapour lamp. These units, although convenient, measure only elemental mercury vapour and are subject to a wide variety of interferences and interfering substances many of which are likely to be present in the working environment. These units should be calibrated each time before use. The commercial units also suffer from the deficiency that they sample only small volumes of air that may not give a representative picture of the working environment. Research should be directed towards the development of personal monitoring devices. These devices should be small and portable so that they can be carried by workmen throughout the working day and thereby give a cumulative picture of the exposure of each individual. In most cases it would be necessary only to devise systems for collecting total mercury.

The method of Wolf et al. (1974) allows the direct detection of mercury using reactive tubes (Draeger tubes) providing a simple screening method for determining mercury in working places at sporadic intervals.

The collection of samples for the determination of mercury in water must take into account the following factors; (a) the low concentration of mercury in water, normally of the order of 10 ng/litre; (b) the tendency of mercury to adsorb on to the surface of the collection vessel at these low concentrations; (c) the possibility, if not likelihood, of volatilization of mercury from the sample (Toribara et al., 1970) and (d) the type of collection vessel. Greenwood & Clarkson (1970) have reported on the rates of loss of mercury from containers made from ten different materials and suggested that Pyrex, polycarbonate, and Teflon are the best materials for storing and handling mercury. Further studies of possible losses of organo-mercurials through the walls of some plastic containers should, however, be studied. Losses due to volatilization may be reduced by the addition of oxidizing substances such as potassium permanganate (Toribara et al., 1970). Lamm & Ruzika (1972) have recommended that radioactive-tracer mercury be added to the sample to check the losses discussed above. They note that this procedure has rarely been adopted to date.

For the collection and storage of food samples, acceptable procedures are usually followed. The most important food items for determination of mercury are those containing fish and fish products. Mercury levels in other foodstuffs usually do not amount to a significant fraction of daily exposure unless the food has accidently been contaminated, such as by the use of pesticides. In the collection and storage of food samples prior to analysis, care should be taken to avoid bacterial growth leading either to the breakdown of organic mercury compounds or to the volatilization of mercury (Magos et al., 1964).

Samples of blood, hair, and urine have been used to monitor the exposure of human beings to mercury. The methods of collecting and storing these samples are of great importance. With respect to blood samples, care should be exercised to avoid any clot formation. If this does occur, the sample should be homogenized thoroughly before analysis. It is useful, in certain situations, to determine mercury in the red cells and plasma and it is thus important to avoid any haemolysis of the blood sample. The nature of the anticoagulants used does not affect the mercury determinations, of either the total mercury in whole blood or the distribution of mercury between plasma and red blood cells. "Vacutainers"[a] are convenient for blood collection and allow storage of the blood samples in Pyrex tubing under aseptic conditions. Blood samples that have been contaminated by microorganisms and stored in the refrigerator at 4°C for a month or more may give misleading results due to the breakdown of methylmercury and other organic mercury compounds (Clarkson, personal communication, 1974). The storage of blood samples in the frozen state or freeze-dried is suitable providing that mercury is determined only for whole blood. Significant losses of mercury do not occur during freeze-drying procedures (Albanus et al., 1972).

Measurement of mercury in urine samples has been used as a measure of exposure to mercury under industrial conditions. The popularity of this approach in early studies was mainly due to the ease of digestion of the urine sample. However, there are serious problems in the collection and storage of urine samples that may seriously influence the results. The following factors have been recognized; (a) the time of day of urine collection (Piotrowski et al., 1975), (b) bacterial contamination, which might give rise to significant losses of mercury by volatilization (Magos et al., 1964), (c) the nature of the container (Greenwood & Clarkson, 1970), (d) contamination from mercury in workers' clothing and from the collection of urine samples under working conditions. It should be noted that urine samples do not give a reliable indication of exposure to methylmercury (Bakir et al., 1973).

Hair samples are becoming the samples of choice in determining exposure to methylmercury through diet. Depending upon the length of the hair sample, it is possible to recapitulate exposure to methylmercury for several years[b]. The concentration of mercury in hair when formed is directly proportional to the concentration of mercury in the blood, the concentration in hair being about 250 times the concentration in blood. The ratios are well established for exposure to methylmercury but only limited information is available for inorganic mercury. Attention has been drawn to the errors introduced during the collection and transportation of hair samples (Giovanoli & Berg, 1974). Usually

---

[a] Trade name of heparinized test-tube manufactured by Becton & Dickinson, USA, and used for collection of blood samples.

[b] The average rate of growth of hair is approximately 1 cm per month (Giovanoli et al., 1974; Shahristani & Shihab, 1974).

50–100 strands of hair are needed for analysis. Differential rates of growth for each strand and lateral displacement of the samples during cutting and transportation of the hair will affect the longitudinal profiles of mercury in the hair sample. Giovanoli & Berg (1974) have described a computerized procedure for the correction of these artifacts.

### 2.3.2 Analytical methods

Methods of analysis are usually classified according to the type of instrument used in the final measurement. This convenient classification will be used here. However this approach tends to belittle the role of the skill and experience of the analyst. In fact a poor method in the hands of a highly skilled analyst is more likely to yield accurate results than a good method in the hands of a poor analyst. In recent years it has become a practice to test methods by a "round robin" distribution of a standard sample. Comparison of results from the participating laboratories is more likely to give information on the competence of the analysts in the laboratory than it is to give a critical evaluation of the method itself.

Measurement of the very low levels of mercury found in the non-contaminated environment makes special demands both on the skills of the analyst and the resources of the method employed. No matter how frequently used, a method for the determination of mercury in nanogram quantities cannot be regarded as a routine procedure. Continued vigilance over the results is an absolute requirement. Furthermore, where conditions allow, it is highly desirable that the results with one method and from one laboratory be checked against those with a different method from another laboratory. One useful combination of different procedures is the analysis of total and inorganic mercury by selective atomic absorption and the selective analysis of organic mercury compounds (usually methylmercury and other short-chain mercurials) by gas chromatography (Giovanoli et al., 1974).

The literature is full of papers concerning methods of determining mercury. Several recent reviews have appeared (D'Itri, 1972; NIOSH, 1973; Burrows, 1975, Swedish Expert Group, 1971; Wallace et al., 1971; CEC Working Group of Experts, 1974). The most frequently used methods for measurements of total mercury are colorimetric (dithizone), flameless atomic absorption, and neutron activation. The flameless atomic absorption method has become the "work-horse" for measurement of environmental samples. Difficulties might arise in the measurement of mercury owing to the fact that it is strongly bound to the organic materials in most samples. Many procedures require the destruction of organic materials by wet oxidation or by high temperatures. Loss of mercury by volatilization may occur. If the wet oxidation is too mild the result will be inadequate recovery. A high reagent blank may be introduced by the chemicals used for oxidation. In certain procedures involving atomic absorption or neutron activation the digestion of the sample or heating of the sample is not necessary. These procedures have the advantage of having a low blank but problems of variable recovery or interference may arise.

The determination of mercury by colorimetric measurement of a mercury dithizonate complex has been the basis of most of the methods in the 1950s and in the 1960s. Other related methods using dithizone for measuring mercury in environmental samples have been described by Kudsk (1964) and Smart et al. (1969). The above procedures all make use of wet oxidation of the sample followed by extraction of mercury in an organic solvent as a dithizonate complex and finally the colorimetric determination of the complex itself.[a] Selectivity for mercury is obtained by adjusting the conditions of extraction. Copper is the metal most likely to interfere with mercury measurement by dithizone.

The dithizone procedure has an absolute sensitivity of about 0.5 μg of mercury. A sample size of 10 g is suitable for most digestion procedures so that mercury can be determined at the 0.05 mg/kg level in most foodstuffs and tissues.

Kudsk (1964) has described a dithizone procedure for measuring mercury in air that will measure as little as 0.05 μg of mercury. With the usual sample size of 0.1 $m^3$, the detection limit would be 0.5 $μg/m^3$. This is more than adequate sensitivity for monitoring air in the working environment with the MAC levels in force. The quoted recovery rates from foodstuffs and tissues are in the range of 85–99% and the reproducibility can yield a coefficient of variation of as low as 2%. On account of its long history of use, the dithizone procedure has been used to measure mercury in virtually all types of environmental samples including air, water, food, tissues, and soils. It suffers from the disadvantage that it is time consuming and its sensitivity is not high when compared with atomic absorption procedures.

The latest developments in atomic absorption procedures have recently been reviewed by Burrows (1975). The most commonly used method in the USA is that of Hatch & Ott (1968) as modified by Uthe et al. (1970). The procedure involves oxidative digestion ("wet ashing"), followed by reduction, aeration, and measurement of mercury vapour absorption at 253.7 nm. The detection limit is approximately 1–5 ng of mercury. The wide popularity of cold vapour atomic absorption has

---

[a] The organic material may also be destroyed by combustion in an oxygen flask (Gutenmann & Lisk, 1960; White & Lisk, 1970; and Fujita et al., 1968). This allows all biological materials to be treated alike but has the disadvantage of requiring dried material.

resulted in a large number of publications dealing with various applications of this procedure to the measurement of mercury in sediments, soils and biological samples (including foodstuffs). Of the 16 publications reviewed by Burrows (1975), 13 reported recoveries of 90% or more. The relative standard deviation was 10% or less in half of the published procedures, and was less than 20% in more than 90% of these procedures.

The measurement of very low levels of mercury in water samples requires some preconcentration. This may be achieved by dithizone extraction (Chau & Saiton, 1970; Thomson & McComas, 1973), by electro-deposition (Doherty & Dorsett, 1971) and by an amalgamation on silver wire (Hinkle & Learned, 1969; Fishman, 1970), in each case permitting detection limits of 1 ng/litre–10 ng/litre. Winter & Clements (1972) have described a procedure that will measure mercury in water in the range of 200 ng/litre and does not require preconcentration.

Magos (1971) has described a reduction technique that selectively determines total and inorganic mercury in biological samples without digestion of the material. This technique has been modified by Magos & Clarkson (1972) to permit determination of mercury in blood samples at the low levels found in unexposed populations (0.1–1.0 µg/100 ml). The technique has a sensitivity of approximately 0.5 ng of mercury. Recently it has been successfully applied to the measurement of total and inorganic mercury in hair samples (Giovanoli et al., 1974). The relative standard deviation was 2% and the recovery rates were quoted as being close to 100%. The technique has the advantage of high speed—each determination taking less than 2 minutes—high sensitivity, and the apparatus involved is light, portable, and suitable for field applications. Its widest application to date has been in the measurement of mercury in biological samples in the large Iraq outbreak (Bakir et al., 1973). Since the procedure does not require digestion of the biological sample, internal standards are used in each determination. The rates in this procedure must be checked for each new biological matrix.

The atomic absorption techniques referred to above are subject to interference. The most common interfering substances are benzene and other aromatic hydrocarbons that absorb strongly in the 253.7 nm region. Interference from a variety of organic solvents has been reported by Kopp et al. (1972).

The combustion–amalgamation method has undergone a series of developments to avoid difficulties due to interfering substances. Reference may be made to the work of Lidmus & Ulfvarson (1968), Okuno et al. (1972), and Willford (1973) who developed techniques for oxidation of the biological sample, and the trapping of mercury vapour on silver or gold followed by its release into an atomic absorption measuring device. All these methods have sensitivities down to the 1 µg/litre level and avoid the risk of interference from other substances. However, as pointed out by Burrows (1975), care must be taken in the design and operation of the combustion tube to avoid losses of volatile mercury derivatives.

In summary, a wide variety of applications of atomic absorption procedures have now been published. The technique is rapid and sensitive and the procedure is technically simple. Procedures are available for avoiding difficulties due to interfering substances. Most procedures have a detection limit in the range of 0.5–5 ng of mercury and a relative standard deviation of about 10% or less. Recovery rates are usually of the order of 95–100% depending on the technique used in the preparation of the biological sample and the rate of release of mercury from it.

Procedures for neutron activation analysis of total mercury have recently been reviewed by Wallace et al. (1971), Swedish Expert Group (1971), Westermark & Ljunggren (1972), and Burrows (1975). The method is based on the principle that when natural mercury (a mixture of stable isotopes) is exposed to a high flux of thermal (slow) neutrons, it is converted to a mixture of radioactive isotopes, principally $^{197}Hg$ and $^{203}Hg$, which have decay half-lives of 65 hours and 47 days, respectively. The Sjostrand (1964) technique has been used most in the measurement of environmental samples. After the sample has been irradiated with neutrons, a precise weight of carrier mercury is added and the sample subjected to digestion and organic destruction. On completion of digestion, mercury is isolated by electrodeposition on a gold foil and the radioactivity is determined with a gamma counter. The use of carrier mercury corrects for any losses of mercury during the digestion, extraction, and isolation procedures. The limit of detection is 0.1–0.3 ng of mercury. The sample size is 0.3 g giving a concentration limit of 0.3–1 µg/kg in most biological samples. The relative standard deviation in samples of kale, fish, minerals, oil, blood, and water is less than 10%. Samuel (unpublished data) decomposed biological material irradiated with neutrons using fuming sulfuric acid and hydrogen peroxide and after the addition of hydrogen bromide, distilled the mercury as bromide together with other trace elements. This method, which is suitable for series analysis, is characterized by high recovery (96%) and good reproducibility. Trace mercury in biological and environmental materials can also be rapidly and satisfactorily determined through isolation as mercury(II) oxide or mercury(II) sulfide after digestion and clean-up procedures following neutron activation (Pillay et al., 1971; Samuel, unpublished data).

In general, the analyst is faced with three major options in the use of neutron activation procedures; (a) destruction or non-destruction of the sample, (destruction and isolation of the mercury is usually required in samples containing less than 1 μg of mercury); (b) the choice of isotope $^{197}$Hg (if the longer-lived isotope, $^{203}$Hg, is used the sample may be allowed to stand to avoid interference from short-lived elements activated along with the mercury—however, $^{203}$Hg requires a more intense neutron flux or a longer irradiation time to achieve the same activity as the $^{197}$Hg); (c) the choice of detector (the sodium iodide (thallium) detector does not have as high a resolution as the germanium (lithium) detector, although its sensitivity is significantly higher).

Interference may come from the following elements, produced at the same time as the radioactive mercury isotopes, $^{24}$Na, $^{82}$Br, $^{32}$P, and $^{75}$Se. Interference from these isotopes may be avoided, as in the Sjostrand (1964) procedure, by chemical isolation of the radioactive isotope. However, $^{75}$Se may not be completely removed by the isolation procedures and might interfere if the sodium iodide (thallium) detector is used. The better resolution of the germanium (lithium) detector allows correction for $^{75}$Se interference through use of other lines in the $^{75}$Se spectrum. For samples containing more than 1 μg of mercury, the required selectivity can be achieved without destruction of the sample, i.e., by instrumental analysis only. One procedure is to measure the $^{203}$Hg isotope, after allowing the sample to stand for approximately one month to eliminate interference due to sodium, phosphorous, and bromine. Another procedure is to make use of the discriminating germanium (lithium) detector when the gamma irradiation from the radioactive isotope may be determined to the exclusion of most of the interfering radioactivity.

A recent non-destructive procedure for measuring mercury in coal makes use of a low-energy photon detector to estimate levels at the 100 μg/kg level with a precision of 10% (Weaver, 1973).

Burrows (1975) has recently reviewed 11 publications describing the application of neutron activation to a variety of environmental samples. Non-destructive (instrumental) determination was used in only two of these publications. In 9 of these publications the $^{197}$Hg isotope was determined. Mercury levels were reported in lake water (4 μg/litre, relative standard deviation 23%), in glacial ice (0.2 μg/kg, relative standard deviation 90%), in coal (100 μg/kg, relative standard deviation 10%), in whole blood (0.7 μg/100 ml,[a] relative standard deviation 10%), in fish (1–3 mg/kg, relative standard deviation less than 10%). Many environmental samples were measured by neutron activation, especially in Sweden, before the introduction of the atomic absorption technique (Westermark & Ljunggren, 1972).

---

[a] In this document the concentration of mercury in blood is expressed in μg/100 ml although in some original papers the values are given in μg/100 g. For practical purposes the difference of about 5% can be neglected.

Compared with other methods reviewed here, the neutron activation procedure has the following advantages; (1) high sensitivity (approximately 0.5 µg/kg); (2) no reagent blank; (3) independence from the chemical form of the element; and (4) non-destructive instrumental methods applicable to samples containing 1 µg of mercury or more. It has the disadvantages that it cannot be adapted to field use and, that if there are large numbers of samples, special radiation facilities and data processing are required. It is generally agreed that the neutron activation procedure finds its most important use as a reference method against which other procedures can be checked.

A variety of other instrumental techniques, such as X-ray fluorescence, mass spectrometry, and atomic fluorescence, for the measurement of total mercury have been reviewed by Lamm & Ruzicka (1972) and by Burrows (1975). In general, some of these methods may have a potentially higher sensitivity or selectivity for mercury. The fact is that, at the time of writing, these procedures have not yet found useful application in the measurement of mercury in environmental samples.

To summarize the present methods for the determination of total mercury in environmental samples, it would appear that the method of choice is that of flameless atomic absorption. No single procedure is appropriate, however, in all circumstances. The methods of sample handling depend upon the particular biological matrix to be analysed. Neutron activation is principally of use as a reference method against which atomic absorption methods may be checked.

### 2.3.3 Analysis of alkylmercury compounds in the presence of inorganic mercury

Techniques for the identification and measurement of alkylmercury compounds in the presence of other compounds of mercury have been reviewed recently (Swedish Export Group, 1971; Tatton, 1972; Sumino, 1975; Westöö, 1973). In general, three methods are available for the identification of alkylmercury compounds. These include (a) paper chromatography (Kanazawa & Sato, 1959; Sera et al., 1962), (b) thin layer chromatography (Johnson & Vickery, 1970; Westöö, 1966, 1967; Tatton & Wagstaffe, 1969), (c) gas–liquid chromatography (Westöö, 1966, 1967; Sumino, 1968; Tatton & Wagstaff, 1969). The paper chromatographic techniques have given way to thin-layer chromatography (TLC) for qualitative identification of the organomercurial compounds. Most quantitative work is now carried out using TLC techniques, and also gas–liquid chromatography (Westöö, 1966, 1967; Sumino, 1968; Tatton & Wagstaffe, 1969; Solomon & Uthe, 1971). However, the method of

Magos & Clarkson (1972) that selectively determines organic mercury by cold vapour atomic absorption is frequently applicable to the determination of methylmercury at levels occurring in fish and blood. Methylmercury is the only organic form of mercury present in fish. Blood samples from people exposed to methylmercury contain only inorganic mercury and methylmercury compounds. Thus the determination of organic mercury by this procedure is an accurate measure of methylmercury in these situations.

The basic procedures for samples of food, soil, and biological materials are first, homogenization of the sample, acidification by a hydrogen halide acid followed by extraction with an organic solvent, usually benzene, a clean-up step involving the conversion of the organomercurial compound to a water soluble compound usually the hydroxide or sulfate or a cysteine complex, and re-extraction with benzene. The benzene layer is now ready for analysis by thin-layer chromatography for qualitative purposes or by gas–liquid chromatography if quantitative measurements are required. A recent variant by Rivers et al. (1972) converts the organic into inorganic mercury and then makes use of cold vapour atomic absorption for final determination.

The gas–liquid chromatographic system is the one most commonly used. Problems may be encountered both in the pre-treatment of the sample and in the gas chromatographic determination itself. All these techniques involve non-destructive extraction of mercury from the sample. Thus recovery rates have to be checked for every different type of sample matrix. The efficiency of extraction of mercury is determined by both the nature of the sample matrix and the extraction procedures themselves. Von Burg et al. (1974) introduced the idea of adding a tracer amount of radioactively labelled methylmercury to the homogenate and counting the final benzene extract to check variations in the efficiency of extraction. This procedure is well worth consideration for routine use as it is most difficult to check extraction recovery rates.

Acidification of the homogenate is usually achieved by the addition of a hydrogen halide acid (usually HCl). At this point mercury(II) chloride may be added to either the homogenate or the benzene to tie up excess sulfur compounds and prevent recombination of methylmercury with sulfur. Westöö (1968) has shown that this approach may give high recovery rates but cannot be used with liver as there is a danger of methylation of the inorganic mercury. Clean-up of the first benzene extract is usually achieved by using solutions of cysteine. However, this complexing agent is subject to oxidation, particularly by substances in muds. A more suitable system in the presence of oxidizing agents is the ammonium hydroxide–sodium sulfate solution described by Westöö. No problems are usually encountered in the reextraction of methylmercury from cysteine to benzene using 3 mol/litre hydrochloric acid. However, in the extraction procedures, volumetric errors may arise especially when the concentration of hydrochloric acid is low (1 mol/litre) and when small amounts of methylmercury are extracted from large volumes (Westöö, 1973).

In gas chromatography, the main object is to produce sharp peaks and attain high sensitivity. Tatton (1972) has noted that most commercial preparations of alkylmercury salts are not pure enough to use as standards. Sumino (1973) prepares pure methylmercury from the combination of inorganic mercury with tetramethyl lead salts. The peak is identified by electron-capture detectors using tritium or nickel as the source of beta particles. These detectors are subject to overloading and not more than 100 ng of mercury should be determined at one time (Tatton, 1972). Absolute confirmation of the identity of the peak should be made by mass fragmentation methods (Sumino, 1975).

The detection limit in the Westöö procedure is approximately 1–5 µg per kilogram of sample using a 10 g sample. The precision is 3% at the 0.05 mg/kg level for fish samples. Recovery rates are generally above 90% but do vary with the sample matrix. Solomon & Uthe (1971) developed a semimicro-method for the rapid determination of methylmercury in fish tissues. Samples of about 2 g were used. A precision of 2% was reported with recovery rates of about 99%. Samples such as blood, liver, and kidney are much more difficult to extract than fish tissues.

Thin-layer chromatography usually requires, for optimum spot size, 2 µg of mercury for each type of compound.


# 3. SOURCES OF ENVIRONMENTAL POLLUTION

The sources of mercury leading to environmental pollution have been the subject of several recent reviews (Wallace et al., 1971; D'Itri et al., 1972; Joint FAO/WHO Expert Committee on Food Additives, 1972; Heindryckx et al., 1974; Korringa & Hagel, 1974). Estimates of both natural and anthropogenic sources of mercury are subject to considerable error. In the first place the levels of mercury in environmental samples such as ice from Greenland are extremely low and close to the limit of sensitivity of the analytical methods. These low values are then converted by large multiplication factors (annual total global rainfall, $5.2 \times 10^5$ km$^3$) so as to obtain values for the global sources and turnover of mercury. Enormous fluctuations may be seen in samples such as coal and oil, which are believed to be an important anthropogenic source of mercury. Values quoted by D'Itri (1972) indicate ranges of concentrations of mercury in crude oil varying by a factor of 1000 and ranges in coal even greater than this. Estimates of industrial production and consumption of mercury are subject to the vagaries of the economic market and in recent years to government regulation because of concern over mercury pollution. Nevertheless, despite all the assumptions and approximations in these procedures, the general picture that emerges from a variety of independent calculations is that the natural sources of mercury are at least as great as, and may substantially outweigh, the anthropogenic sources. However, man-made sources may be of considerable importance in terms of local contamination of the environment. For example, Korringa & Hagel (1974) have calculated that the man-made release of mercury in the Netherlands is 100 times greater than the release of mercury by natural degassing processes.

## 3.1 Natural Occurrence

A recent review by the Joint FAO/WHO Expert Committee on Food Additives (1972) quotes the major source of mercury as the natural degassing of the earth's crust and quotes figures in the range of 25 000–150 000 tonnes of mercury per year. These figures originate from a paper by Weiss et al. (1971) on concentrations of mercury in Greenland ice that was deposited prior to 1900. The most recent calculations on natural sources of mercury have been published by Korringa & Hagel (1974). These authors also made use of the figures of Weiss et al. (1971) to calculate the annual amount of mercury reaching the earth's surface due to precipitation of rainfall and arrived at a figure of approximately 30 000 tonnes. It was admitted that the sources of this atmospheric mercury are not yet clearly established but that volcanic gases and evaporation from the oceans are probably significant sources. It was also calculated by these authors that the run-off of mercury from rivers having a "natural mercury" content of less than 200 ng/litre would account for approximately 5000 tonnes of mercury per year. Measurements of the concentrations of mercury in air attached to aerosols (Heindryckx et al., 1974) indicate that soil dispersion to the atmosphere is not an important source of mercury.

Significant local contamination may result from natural sources of mercury. For example, Wershaw (1970) has shown that water sources located near mercury ore deposits may contain up to 80 µg/litre as compared with the levels of 0.1 µg/litre in non-contaminated sources.

## 3.2 Industrial Production

According to a recent review by Korringa & Hagel (1974), world production averaged about 4000 tonnes per year over the period 1900–1940. Production in 1968 was 8000 tonnes per year and, in 1973, attained 10 000 tonnes per year. Although considerable yearly fluctuations were noted, the average rate of increase since 1950 has been about 2% per year. Recent concern over environmental problems related to the use of mercury seems to have stabilized production rates and to have led to a dramatic fall in the price of mercury. For example, according to figures quoted by Korringa & Hagel (1974), the 1966 price was $452 per flask (a flask is 34.5 kg), the 1969 price had risen to $510.00 but by 1972 it had fallen dramatically to $202 per flask.

It is difficult to estimate the amount of mercury released into the environment as a result of the mining and smelting of this metal. High levels of mercury in lake and stream waters have been attributed to the dumping of materials and tailings (for review, see Wallace et al., 1971). It has been estimated that stack losses during smelting operations should not exceed 2–3%. Thus, based on a production figure for mercury of 10 000 tonnes in 1973, one might expect to find losses to the atmosphere of the order of 300 tonnes per year.

## 3.3 Uses of Mercury

Wallace et al. (1971) have attempted to give a picture of the use of mercury in the USA. They note that 26% of the mercury mined is not reusable. They point out, however, that at least from the theoretical point of view most of the remaining mercury (i.e. 74% of the mercury mined) is reusable. To what extent these theoretical possibilities are attained is debatable at the present moment.

Rauhut & Wild (1973) reported on the consumption and fate of mercury in the Federal Republic of Germany in 1971. Flewelling (1975) noted that the chloralkali industry, one of the largest users of mercury, has been able to cut losses in water effluent by at least 99% in the last two or three years; consequently losses from chloralkali plants now occur predominantly by emission into the atmosphere. Losses by volatilization into the atmosphere have been reduced (approximately 50%) by the introduction of cooling systems for effluent gases. Korringa & Hagel (1974) take a more pessimistic point of view and conclude that there is every reason to assume that by about 1975 all the 10 000 to 11 000 tonnes of mercury produced per year due to mining operations will finally find its way into the environment, predominantly via the atmosphere.

Average consumption patterns for industrialized countries have been summarized by Korringa & Hagel (1974) as follows: chloralkali plants, 25%; electrical equipment, 20%; paints, 15%; measurements and control systems, such as thermometers and blood pressure meters, 10%; agriculture, 5%; dental, 3%; laboratory, 2%; and other uses including military uses as detonators, 20%. This pattern of consumption in industrialized countries is similar to that published by D'Itri (1972) for the consumption in the USA in 1968. Included in "other uses" are mercury compounds in catalysts, preservatives in paper pulp industries, pharmaceutical and cosmetic preparations, and in amalgamation processes. The use of mercury in the paper pulp industries is dramatically declining and it was banned in Sweden in 1966 (Swedish Expert Group, 1971). Hasanen (1974) has reported that no mercury compounds have been used in the paper pulp industry in Sweden and Finland since 1968.

### 3.4 Contamination by Fossil Fuels, Waste Disposal, and Miscellaneous Industries

Industrial activities not directly related to mercury can give rise to substantial releases of this metal into the environment. The most significant source is probably the burning of fossil fuels. Heindryckx et al. (1974) calculated the following approximate figures based on reports published in 1971 and 1972 (Joensuu, 1971; Cardozo, 1972): the combustion of coal and lignite, 3000 tonnes per year; the refining and combustion of petroleum and natural gas, 400 tonnes per year; the production of steel, cement, and phosphate, 500 tonnes per year. Korringa & Hagel (1974) made similar calculations from published material (Joensuu, 1971; Filby et al., 1970; Cardozo, 1972; Weiss et al., 1971). They estimated for the year 1970, an annual release of 3000 tonnes of mercury from coal burning, 1250 tonnes from mineral oil, and 250 tonnes from the consumption of natural gas. They expected that, by 1975, a total of 5000 tonnes of mercury would be emitted from burning fossil fuels.

Smelting of metals from their sulfate ores should contribute some 2000 tonnes annually and the making of cement and phosphate and other processes involving heating should have contributed another 5000 tonnes per year by 1975.

D'Itri (1972) points out that the disposal of sewage might be an important source of environmental mercury. Calculations from data in the literature indicate that somewhere between 200 and 400 kg of mercury per million population may be released from sewage disposal units. This would amount to approximately 40–80 tonnes per year for the entire population of the USA. He further points out that sewage sludge can retain high amounts of mercury according to published studies from Sweden (6–20 mg/kg). This sludge is sometimes used as a fertilizer resulting in widespread dispersal of mercury or is sometimes heated in multiple hearth furnaces when most of the mercury would probably be released into the atmosphere. If the United States production is taken as being roughly 30% of world consumption, one might extrapolate the sewage release figure for the United States to indicate that something of the order of 1000 tonnes of mercury may be released frow sewage systems on a global scale.

The anthropogenic release of mercury has been well summarized in a recent article by Korringa & Hagel (1974) and will be briefly stated here. The total global release of mercury is taken as the sum of the global production (following their pessimistic view that all will be released into the environment) plus the release from fossil fuels and natural gas and release from non-mercury related industries.

It was calculated that by 1975 the total anthropogenic release of mercury on a global scale would be about 20 000 tonnes per year. These figures should be compared with a minimum estimated release of 25 000 to 30 000 tonnes per year from natural sources. The latter figure may, in fact, be as high as 150 000 tonnes per year, given the uncertainties in calculations on the natural global release of mercury.

# 4. ENVIRONMENTAL TRANSPORT, DISTRIBUTION, AND TRANSFORMATION

Jenson & Jernelov (1972) have suggested different types of cycle for the distribution of mercury. One cycle is global in scope and depends upon the atmospheric circulation of elemental mercury vapour. The other cycle is local and is based on an assumed circulation of volatile dimethylmercury compounds. In the global cycle most of the mercury is derived from natural sources whereas the local cycle is predominantly concerned with man-made release.

## 4.1 Distribution between Media—the Global Mercury Cycle

Recent calculations on the global circulation of mercury have been reported by Korringa & Hagel (1974). Their calculations are based principally on data giving mercury levels in ice samples collected in Greenland and in the Antarctic as reported by Weiss et al. (1971). The circulation of mercury from natural sources was calculated using a figure of 0.06 µg of mercury per kilogram of Greenland ice samples collected prior to the year 1900. Using a reported figure for the global precipitation of water as $5.2 \times 10^5$ km$^3$ per year, they estimated that minimum transport from the atmosphere to the earth should have been about 30 000 tonnes annually, prior to 1900. The contribution by dust particles was regarded as insignificant, an assumption now supported by the findings of Heindryckx et al. (1974). Based on a published figure of $4.1 \times 10^5$ km$^3$ for annual precipitation over the oceans, these authors estimated the annual delivery of mercury to the oceans as 25 000 tonnes.

Korringa & Hagel (1974) also calculated the contribution of the man-made release of mercury to the atmospheric transport cycle. They assumed that 16 000 tonnes of mercury is now released per year to the atmosphere from man-made sources and that the mercury is returned to the continental land surfaces and would soon re-evaporate to the atmosphere. The 16 000 tonnes per year would eventually find its way into the oceans and thus the annual delivery to the oceans from both natural and man-made sources would be 25 000 plus 16 000 tonnes which on a proportional basis should increase the background level from the 0.06 µg/kg observed prior to the 1900s in Greenland ice to a predicted level of 0.1 µg/kg. However, they point out that since most of the man-made release is probably in the northern hemisphere, the present level in Greenland ice should be somewhat higher than 0.1 µg/kg. They note that this estimate agrees well with the observations of Weiss et al. (1971) who found present levels in Greenland ice to range from 0.09 to 0.23 µg/kg with an average of 0.125 µg/kg. Thus, from these rough estimates, it would appear that present day "background" levels in rainwater, and presumably in the atmosphere, have a substantial component related to man-made release (approximately one-third).

Observations on "background" mercury levels in the atmosphere tend to confirm the quantitative features of this global picture (Heindryckx et al., 1974). These authors assume that 50 000 tonnes are released each year from the continental land masses, that the mercury mixes up to a height of 1 km and that, in effect, the 50 000 tonnes are located over the continental land masses that account for 30% of the earth's surface.[a] The assumption of the location of this mercury over the land masses is not in contradiction with the calculations of Korringa & Hagel (1974). It assumes only that the atmosphere above the land masses is in steady state,

---

[a] Recent studies in Sweden cast some doubt on the validity of this assumption.

and receives 50 000 tonnes of mercury a year as evaporation and loses 50 000 tonnes per year to the atmosphere over the oceans. Their figure of 50 000 tonnes per year comes from the publication of Bertini & Goldberg (1971) and agrees well with the figure of 41 000 tonnes per year as indicated above. With these assumptions, Heindryckx et al. (1974) concluded that the background continental levels of mercury vapour plus aerosols should be 10 ng/m$^3$. The assumed mixing height of 1 km is probably the maximum level and they suggest that the actual level of mercury in air would lie between 1 and 10 ng/m$^3$. These figures are in good agreement with the published air levels as indicated in section 5.1.

Korringa & Hagel (1974) estimate the amount of mercury transported by rivers to the oceans to be 5000 tonnes per year based on quoted figures of 37 000 km$^3$ of water flow via the rivers and a natural mercury content of less than 0.2 µg/litre in river water. They note that this figure does not change substantially if one takes into account the fact that most of the mercury in river water is adsorbed to suspended matter with a mercury content of 200–500 µg/kg and that some $10^{10}$–$10^{11}$ tonnes of sediment are carried each year to the oceans. In fact river transport of mercury to the oceans may be less than 5000 tonnes per year. Heindryckx et al. (1974) noted that the concentrations of mercury in the North Sea and in the coastal areas around the North Sea were far less than would be predicted if all the mercury in the rivers entering this area were, in fact, delivered into the oceans. Presumably a considerable amount of mercury observed in river water is retained in sediments in the rivers and estuaries and does not reach the ocean by normal flow of the river. Thus it would appear that the major pathway of global transport of mercury is metallic mercury transported in the atmosphere.

An important conclusion from these calculations on the global cycle of mercury is that the concentration of mercury in the oceans should not change substantially in the foreseeable future, and that the mercury concentration in the oceans has not changed significantly since the beginning of the industrial era. The amount of mercury in the oceans has been calculated as 70 million tonnes using a figure for total ocean volume of $1.37 \times 10^9$ km$^3$ and taking the average mercury content of ocean water as 50 ng/litre. Thus contrary to what has been observed for the mercury content of the atmosphere, it will be a long time before the mercury content in sea water is significantly increased. Since water is thought to remain in the surface layers of the ocean for 10–50 years, these authors concluded that the mercury resulting from man-made activities should be well distributed in the water of all the oceans and therefore should not lead to high local concentrations.

This conclusion is consistent with the findings reported in section 5.1

that mercury levels in swordfish and tuna fish caught at the beginning of the century fall within the same range as mercury levels reported in recent catches.

The origin of mercury released by natural processes is not well established. Volcanic emissions are a possible source in view of the high concentrations of mercury vapour reported in the vicinity of volcanoes (for review, see Jonasson & Boyle, 1971). The general "degassing" of the earth's surface is probably a major source (Weiss et al., 1971). Levels of metallic mercury vapour in the atmosphere over soils rich in mercury (the humus layers of topsoil) have been reported in the range of 20–200 ng/m$^3$ as compared to background levels of 5 ng/m$^3$ according to a report by Barber et al. (quoted by Vostal, 1972). Korringa & Hagel (1974) have raised the possibility that evaporation from the oceans may make a contribution to the mercury present in the atmosphere in view of the substantial quantities of water vapour that evaporate ($4.48 \times 10^5$ km$^3$). However, it seems unlikely that mercury would evaporate at the same rate as water in view of the fact that it is believed to be in a complex form in the oceans (see section 5.1). Furthermore, the observations of Williston (1968) (referred to in section 5.1) indicate that the mercury content of the atmosphere over the oceans is considerably lower than that over land (industrialized and rural areas).

The mechanisms of volatilization of mercury from the land masses are not well understood. Presumably release of mercury from volcanoes is due to the high temperatures associated with volcanic activity. Vostal (1972) has suggested two major mechanisms, firstly the reduction of mercury in soils by a chemical process depending on the local redox potential, and secondly reduction by the activity of microorganisms. The quantitative importance of these two processes is not known. Mercury-volatilizing microorganisms are known to exist and have been identified (Magos et al., 1964; Furukawa et al., 1969; Tonomura & Kanzaki, 1969).

## 4.2   Environmental Transformation—the Local Mercury Cycle

Mercury is present naturally in the environment and released from manmade sources in a variety of chemical and physical states. The principal mercury ore is cinnabar, which is mercury sulfide. Andersson (1967) has shown that mercury in soils is complexed to the organic (humus) content. Metallic mercury may be discharged into the environment from natural sources as discussed above and also from man-made sources such as chloralkali plants. A variety of organomercurial compounds are also discharged into the environment as a result of human activities (see section 3). Both the inorganic forms of mercury (such as metallic mercury vapour and cinnabar) and the organic forms of mercury are subject to conversion in the environment.

Jensen & Jernelov (1972) have summarized the major pathways of transformation. The inorganic forms of mercury ($Hg^0$ and $HgS$) undergo transformations in the environment mainly by oxidation–reduction reactions. Mercury vapour is oxidized to ionic divalent mercury ($Hg^{++}$) in water in the presence of oxygen. Concentrations as high as 40 g/litre have been attained when water saturated with oxygen was exposed to mercury vapour (Wallace et al., 1971). As pointed out by Jensen & Jernelov (1972) the oxidation of metallic mercury to inorganic divalent mercury is greatly favoured when organic substances are present in the aquatic environment.

Ionic mercury, once present in water, is capable of forming a wide variety of complexes and chelates with organic materials. Of considerable importance is its reaction with the sulfide ($S^{--}$) ion to form highly insoluble mercury(II) sulfide. This reaction is likely to occur in anaerobic aquatic environments owing to the presence of hydrogen sulfide gas. This sulfide complex of mercury is highly stable and will not normally become involved in transformation under anaerobic conditions. However, in the presence of oxygen, the insoluble mercury(II) sulfide can become oxidized to the soluble sulfite and sulfate salts of mercury which allow the metal to ionize and enter subsequent chemical reactions.

In addition to the oxidation of metallic vapour, inorganic mercury ($Hg^{++}$) can be formed by the breakdown of a variety of organic mercury compounds. The alkoxyalkylmercury compounds are very unstable in acid conditions and it has been reported (see Jensen & Jernelov, 1972) that, in humid soil (pH = 5), methoxyethylmercury has a half-life of only 3 days. Aryl- and alkylmercury compounds can all be degraded in the environment by chemical and physical processes and by biologically mediated processes.

Divalent inorganic mercury ($Hg^{++}$) can undergo two important reactions in the environment. The first is the reduction to metallic mercury vapour, a reaction that will occur in nature under appropriate reducing conditions. As mentioned above, certain bacteria, particularly of the genus *Pseudomonas*, can convert divalent mercury into metallic mercury (Magos et al., 1964; Furukawa et al., 1969). The formation of inorganic divalent mercury in nature and its reduction to metallic mercury vapour are probably key processes in the global cycle of mercury. The reduction to metallic mercury vapour must be the key step in the release of mercury because of degassing of the earth's surface. The oxidation of metallic mercury vapour to divalent ionic mercury must be the critical step in the uptake of mercury vapour in rainwater and in the oceans. Unfortunately, other than these crude generalizations, little is known of the details of the kinetics of these processes in nature.

The second important reaction that ionic divalent mercury ($Hg^{++}$) undergoes in nature is its conversion to methylmercury and dimethylmercury compounds and the interconversions between these compounds. These reactions play a critical role in the so called "local cycle" of mercury and are worth further discussion. Some countries, particularly those in Scandinavia, that used methylmercury fungicides extensively, experienced a general rise in the mercury content of their agricultural products. High levels were also noted in some species of birds. The increase corresponded with the onset of the use of methylmercury fungicides. However, it was discovered that mercury levels in fish were also high and that these fish were obtained in areas where methylmercury compounds were not used (Jensen & Jernelov, 1969). It was subsequently discovered that methylmercury was the predominant form of mercury in fish regardless of the nature of the mercury pollutant. This was the first evidence that transformations of mercury compounds must occur in the environment and that, indeed, they must be of great significance. It has now been demonstrated that biological methylation of mercury occurs in the organic sediments of aquaria and in sediments from freshwater and coastal waters of Sweden (Jensen & Jernelov, 1967, 1969; Jernelov, 1968).

Two biochemical pathways of methylation of mercury have been identified, one anaerobic the other aerobic. The anaerobic pathway involves the methylation of inorganic mercury by methylcobalamine compounds produced by methanogenic bacteria in a mildly reducing environment (Wood et al., 1968). The process is non-enzymic and is strictly anaerobic. The aerobic pathway has been described by Landner (1971) in studies of *Neurospora crassa*. His findings indicate that methylmercury bound to homocysteine becomes methylated by those processes in the cell normally responsible for the formation of methionine. In other words, the methylmercury-homocysteine complex is methylated by "mistake".

Despite the fact that an anaerobic pathway for methylmercury production is well known, it seems unlikely that significant amounts of methylmercury are formed in the aquatic environment under anaerobic conditions. The chief reason for this, as pointed out by Jensen & Jernelov (1972), is that, in natural water when oxygen is exhausted, hydrogen sulfide is formed and divalent mercury becomes bound up as mercury(II) sulfide. In this sulfide form, mercury is not available for methylation under anaerobic conditions (Jernelov, 1968; Rissanen, quoted by Jensen &

Jernelov, 1972), and methylation is slow even under aerobic conditions (Fagerstrom & Jernelov, 1971).

In an aquatic environment under aerobic conditions, it must be borne in mind that the upper sedimentary layers and sedimentary particles suspended in the water may be both aerobic and anaerobic, the exterior being well oxygenated and the interior deficient in oxygen. Thus both pathways, aerobic and anaerobic, are possible routes of methylation in water that is oxygenated.

The ability to methylate mercury is not confined to a limited number of species of microorganism. Thus, conditions that promote bacterial growth in general, will lead to enhanced methylation of mercury. The highest rates of methylation in the aquatic environment are, therefore, seen in the uppermost part of the organic sediments and on suspended organic material in water (Jernelov, 1973).

The formation of dimethylmercury from monomethylmercury compounds has been shown to occur in decomposing fish (Jensen & Jernelov, 1968), and from (originally) inorganic mercury in sediments. The anaerobic pathway using methylcobalamines is one means by which dimethylmercury can be synthesized. The reaction is greatly favoured by high pH whereas the formation of monomethylmercury is favoured by a low pH environment.

The ability to methylate mercury at a high rate correlates with the resistance of the microorganism to concentrations of inorganic mercury (for review, see Jernelov, 1973).

The observations, reviewed above, of the interconversion of the various mercury compounds in nature have led to a hypothesis for a local cycle (Jensen & Jernelov, 1972). Inorganic divalent mercury is formed either by the oxidation of metallic mercury vapour by physico-chemical processes or by the cleavage of the carbon–mercury bond in organomercurial compounds either chemically or enzymatically. The divalent ionic mercury becomes attached to sediments either suspended in the water or in the sedimentary layers. The upper sedimentary layers are biologically active but it is postulated that, with the passage of time, large quantities of inorganic mercury will penetrate down to the inorganic mineral layers of the sediments where the mercury should remain inactive. In the surface layers of the sediment, part of the inorganic mercury becomes methylated. Methylation significantly increases the ability of mercury to cross biological membranes. This is why aquatic organisms contain mainly methylmercury.

If conditions of pH are appropriate, dimethylmercury will be formed. Dimethylmercury is water insoluble, possesses a very high volatility, and is postulated to diffuse from the aquatic environment into the atmosphere.

Once in the atmosphere, it is subject to removal by rainfall. If the rain-water is acidic, the dimethylmercury is converted to monomethylmercury compounds and is thereby returned to the aquatic environment completing the cycle. In the presence of mercury(II), dimethylmercury is converted to two methylmercury molecules (Jensen & Jernelov, 1969).

Key parts of this local cycle remain conjectural. It is known that di-methylmercury compounds can be formed and that the conditions for their formation can exist in an aquatic environment. Unfortunately analytical data are sparse but Bramen & Johnson (1974) have identified both mono-and dimethyl compounds in the atmosphere both outdoors and indoors in the USA. Evidence is still lacking for methylmercury compounds in rain-water. The analytical difficulties are considerable. Nevertheless the present weight of evidence supports the existence of a local cycle for the transport of mercury involving dimethylmercury as the key intermediary for the atmospheric turnover in this cycle. The observations available today on this cycle refer to local bodies of water such as lakes and rivers and the cycle itself would represent the best available explanation for the presence of methylmercury compounds in freshwater fish.

The origin of methylmercury compounds in oceanic fish has not been well described. Inorganic mercury is available in unlimited quantities in the oceans, as has been indicated in the calculations reported in section 4.1. The site of methylation of this mercury is not known. Sediment suspended in oceanic water would seem to be a prime suspect. Methyla-tion of mercury is also known to occur in the slime covering fish but it does not occur in the fish tissues themselves (Jensen & Jernelov, 1972). It would seem an important research priority to describe the methylation pathways in ocean waters. Only then will it be possible to state whether the rate of formation of methylmercury in ocean waters and uptake in oceanic fish is related to the total deposit of mercury in the oceans (70 million tonnes) or whether it is related to a very small sub-fraction of the mercury in the oceans that may respond to man's activities more dramati-cally than the total ocean pool.

## 4.3 Interaction with Physical or Chemical Factors

The interaction of mercury with physical or chemical factors has been referred to frequently in the previous section, so that only a brief summary will be given here. In terms of the global distribution of mercury, such physicochemical factors as temperature, pH, redox potential, and chemical affinities for the organic materials in soil will interact to deter-mine the degree of volatility of mercury under specific local conditions and the rate of release of mercury from the earth's crust as elemental mercury vapour. The interplay between these factors is so complex that studies of mercury volatilization from soil and from the earth's crust, in general, do not lend themselves easily to experimental work. Once in the atmosphere, metallic mercury is liable to both physical and chemical interactions. Physically it may be adsorbed on to particulate materials in air but evidence reviewed in section 5.1 indicates that the aerosol fraction of mercury is 5% or less of the total mercury in air. Metallic mercury vapour should distribute more or less evenly between air and water providing it remains in the unoxidized metallic state (Hughes, 1957). However, the reported levels in rainwater (see section 5.1) are higher than the background level by a factor of at least 2 or 3. This is no doubt a consequence of the oxidation of metallic mercury to ionic mercury in the water in the presence of oxygen. Once deposited in the ocean from rainwater, any remaining metallic mercury should be liable to oxidation to ionic mercury whereupon it will undergo rapid chemical combination with various chemical compounds in ocean water. Sillen (1963) has estimated that the mercury may be present as negative chloride complexes (section 5.1). However, it seems probable that, because of its affinity for sulf-hydryl groups, mercury will also bind strongly to living organisms in ocean waters.

Another aspect that should be considered is the relationship between mercury and selenium. Recent data indicate that selenium compounds known to detoxify mercury, increase mercury retention in some organisms changing the tissue distribution (Parizek et al., 1971). High mercury concentrations were accompanied by high selenium concentrations in tissues of several animal species (Ganther et al., 1972; Koeman et al., 1972, 1973) and also in man (Kosta et al., 1975; Byrne & Kosta, 1974). This relationship is further discussed in section 7 of this document.

In the local cycle of mercury, the same physico–chemical factors will be operative. Oxygen tension in the aquatic environment will determine the degree of formation of insoluble mercury(II) sulfide that will limit the rate of methylation. The pH of the aquatic environment and also of the rainwater will determine the distribution of the methylated forms of mercury between dimethyl and monomethyl compounds.

## 4.4 Bioconcentration

The short-chain alkylmercurials, especially methylmercury compounds, have a strong tendency to bioaccumulation since they possess a group of properties that makes them unique among the mercury compounds. Methylmercury is very efficiently absorbed through biological membranes. In mammals, absorption of methylmercury from food is virtually complete. Methylmercury is degraded much more slowly into inorganic mercury than are the other classes of organomercurial compounds. It is excreted from living organisms much more slowly than other mercury compounds. It possesses a very high chemical affinity for the sulfhydryl group. Since this group occurs mainly in proteins in living organisms, methylmercury, once it has entered the organism, is soon converted to a non-diffusible protein-bound form. However, even though most of the methylmercury is bound to protein, a small fraction remains in a diffusible form. Methylmercury rapidly equilibrates between diffusible and non-diffusible binding sites and thus retains its mobility within animal tissues.

In view of its ability to accumulate in living organisms, one would, in general, expect to see higher concentrations of methylmercury at higher trophic levels in natural food chains. Qualitatively, this generalization appears to be true but quantitative predictions are not possible because of the complex interplay of a host of factors that influence the accumulation and movement of mercury in food chains. For example, remarkably large species differences exist in biological half-times which vary from approximately 7 days in the mouse, to 70 days in the monkey and man, 500 days in seals, and over 1000 days in some species of fish (for review, see Clarkson, 1972a).

The origin of methylmercury in terrestrial food chains is predominantly the use of mercury fungicides in the treatment of seed grain (D'Itri, 1972). The seeds are consumed by grain-eating birds or rodents and the rodents themselves become victims of the large carnivorous birds. The dramatic increase in the concentration of mercury in feathers of carnivorous birds in Sweden was associated with the introduction of methylmercury fungicides in 1940 (for review, see Swedish Expert Group, 1971). High concentrations of mercury in pheasants and other game birds are also a result of this terrestrial food chain and have led to restrictions on hunting in certain areas of North America. The replacement of methylmercury by the alkoxyalkylmercury compounds in Sweden led to a diminished level in this terrestrial food chain. Generally speaking, alkoxyalkyl- and phenylmercury compounds are either less well absorbed or more easily degraded to inorganic mercury and more rapidly excreted.

The accumulation of methylmercury compounds in aquatic food chains has been the subject of a recent review (Fagerstrom & Larsson, unpublished report). This chain or group of chains is considerably more complex than the terrestrial ones. Nevertheless, several tentative generalizations seem plausible at this time. Once methylmercury is formed in the upper sedimentary layers or in suspended sediments in water, it readily leaves the sedimentary particle (Gavis & Ferguson, 1973). The reason for this is not fully established but Fagerstrom & Larsson suggest that it may be due to the pathway of synthesis of methylmercury compounds. For example, if methylmercury is formed by the pathway proposed by Landner (1971), it will be in the form of a diffusible complex with homocysteine. In contrast, inorganic mercury in the sediment is probably bound to large macromolecules. Once methylmercury has diffused from the sedimentary particle into the water, it must be rapidly accumulated by living organisms. This accumulation is so efficient that methylmercury has never been detected in filtered water. Fagerstrom & Larsson, in reviewing recent experimental work on methylmercury accumulation, noted that this form of mercury accumulates in all species, whether plant or animal, that possess membranes for gas exchange with their aquatic environment.

The accumulation of methylmercury in food chains in freshwater systems has been proposed as a three-step process by Fagerstrom & Larsson. The first step is an accumulation by bottom fauna that are in closest proximity to the active sedimentary layers where the methylmercury is formed. Accumulation in the bottom fauna, including plankton, would be followed by accumulation in species such as the roach and finally in the large carnivorous fish such as the northern pike. The authors point out that the relative importance of uptake of methylmercury directly from water through the gill membranes, as opposed to intake from food, should depend upon the trophic level of the fish. The higher the trophic level the more important the intake from food. However, for the overall food chain, uptake through the gills is the key process. If for some reason there is a dramatic change in the environmental layers of methylmercury, the authors predict that it would take from 10–15 years for the levels in the top predators to readjust to the new environment.

These generalizations on freshwater species should be expected to apply to oceanic fish. The remarkably high levels of methylmercury seen in swordfish and tuna fish are due to a variety of factors. First these species are large carnivorous fish at the end of a food chain. They live for a relatively long time compared with other species of fish and it is well established that methylmercury levels show a positive correlation with age (and or weight) of the fish. They are highly active fish having insatiable appetites. Because of their activity, large quantities of oceanic water pass through the gill membranes each day. Thus it is possible that tuna fish, swordfish and related species have a high intake of methylmercury both from their food supply and from the surrounding water.

Accumulation of mercury in the terrestrial and aquatic food chains

(Fagerstrom & Larsson) results in risks for man mainly through the consumption of: game birds in areas where methylmercury fungicides are in use; fish from contaminated waters, especially predator species, tuna fish, swordfish and other large oceanic fish even if caught considerably off shore; other seafoods including muscles and crayfish; fish-eating birds and mammals; and eggs of fish-eating birds.

Space does not permit a full discussion of the important questions concerning the chain of mercury transport from soil to plant to domestic animals and ultimately to man. Important parameters in this transport include absorption and availability in the soil, intake and distribution in the plant, toxic effects on the plant, and intake by domestic animals and by man. The maximum amounts tolerated in the soil may be key factors in determining the possible enrichment in food chains and the ultimate hazards to man (Koronowski, 1973; Kloka, 1974).

## 5. ENVIRONMENTAL LEVELS AND EXPOSURES

The levels of mercury in the environment have been reviewed either partially or completely by: Swedish Expert Group (1971), Joint FAO/WHO Expert Committee on Food Additives (1972), Holden (1972), D'Itri (1972), Petersen et al. (1973), Bouquiaux (1974), and CEC (1974). The principal findings may be summarized as follows. The concentration of mercury vapour in the atmosphere is so low that it does not contribute significantly to human intake of mercury. A few "hot spots" may exist but these require further investigation. Concentrations of mercury in water, particularly drinking water, are also sufficiently low as not to contribute significantly to human exposure. The industrial release of methylmercury compounds into a sheltered ocean bay (Minamata Bay) and into a river (the Agano River) in Japan have led to extremely high concentrations of methylmercury in fish (up to 20 000 µg/kg wet weight) and resulted in human poisonings and fatalities. The industrial release of a variety of chemical and physical forms of mercury into inland waters has led to local pollution, to mercury levels in fish occasionally over 10 000 µg/kg but usually less than 5000 µg/kg, and to the restriction of fishing for sport and commercial fishing in these areas. The mercury level in most freshwater and oceanic fish is below 200 µg/kg. However, in large carnivorous fish such as tuna, swordfish, halibut, and shark, levels are usually above 200 µg/kg and can be as high as 5000 µg/kg wet weight. The general population face no significant hazards from the consumption of methylmercury in the diet. However, certain sub-populations, either those eating locally contaminated fish or those with an unusually high consumption of large carnivorous oceanic fish eventually develop blood levels of mercury in the range of the lowest levels associated with signs and symptoms of poisoning in the Japanese outbreak. It is estimated that the average daily intake of the general population is less than 20 µg of mercury per day in the diet. An appreciable amount of this would be methylmercury. However, individuals in certain sub-populations having unusually high exposure may ingest daily amounts of mercury of up to 200 µg, mainly as methylmercury compounds.

## 5.1  Levels in Air, Water, and Food

*Air*

The average concentration of mercury in the general atmosphere was reported by Stock & Cucuel (1934) to be 20 ng/m³. These results were confirmed by Eriksson (1967) in Sweden. Sergeev (1967) noted concentrations of 10 ng/m³ in the USSR. Fujimura (1964) reported concentrations of 0–14 ng/m³ in non-industrialized regions of Japan. The lowest reported levels are those reported by McCarthy (1968) in Denver, USA, of 2–5 ng/m³. Williston (1968) reported mercury levels in the vicinity of San Francisco, USA, of 0.5–50 ng/m³, the level depending greatly on the direction of the wind. Williston's method would have detected only mercury vapour.

Levels of particle-bound mercury have also been reported. Goldwater (1964) noted that airborne dust in New York City contained from 1 to 41 ng/m³ and that outdoors the concentration was from 0 to 14 ng/m³. Brar et al. (1969) noted that particle-bound mercury in air above Chicago ranged from 3 to 39 ng/m³. Heindryckx et al. (1974), in the most recent study, found that aerosol mercury levels corresponding to remote background levels in Norway and Switzerland were as low as 0.02 ng/m³. In a heavily industrialized area of Belgium, near Liège, the aerosol mercury levels noted were as high as 7.9 ng/m³. Other sampling stations in Belgium reported values roughly an order of magnitude below this. Unfortunately it is not known to what extent particle-bound mercury contributes to total mercury levels in the atmosphere. An indirect reference to Jervis by Heindryckx et al. (1974) indicates that aerosol mercury accounts for only 5% of total mercury in the atmosphere. All the particle-bound mercury reported by Heindryckx et al. (1974) had a particle size of less than 0.4 µm.

"Hot spots" of mercury concentration have been reported in atmospheres close to industrial emissions or above areas where mercury fungicides have been used extensively. Fujimura (1964) reported air levels up to 10 000 ng/m$^3$ near rice fields where mercury fungicides had been used and values of up to 18 000 ng/m$^3$ near a busy super highway in Japan. McCarthy et al. (1970) noted air values of up to 600 and 1500 ng/m$^3$ near mercury mines and refineries. Fernandez et al. (1966) reported maximum values of 800 000 ng/m$^3$ in a village close to a large mercury mine in Spain. The remarkably high mercury vapour levels reported by these authors indicate the need for further studies into localized high concentrations of mercury in the atmosphere.

## Water

Limited data are available for concentrations of mercury in rainwater and snow. First reported values were 50–500 ng/litre (Stock & Cucuel, 1934). Eriksson (1967) found values from 0 to 200 ng/litre. Brune (1969) noted values of approximately 300 ng/litre in rainwater in Sweden. Values for mercury in snow have been reported by Johnels et al. (1967) as 70 ng/kg and by Byrne & Kosta (quoted by Holden, 1972) as 1000–3000 ng/kg in centrifuged melted snow. It is probable that mercury levels in snow depend greatly on the collection conditions and upon how long the snow has laid on the ground. For example, Straby[a] noted values of 80 ng/kg in fresh snow but 400–500 ng/kg in snow that may have partly melted or evaporated over the winter. Analysis of ice deposited in Greenland prior to the 1900s (Weiss et al., 1971) indicates values of 60 ng/kg.

Bodies of freshwater for which there is not independent evidence for mercury contamination, contain levels of mercury of less than 200 ng/litre. Stock & Cucuel (1934) reported 10–50 ng/litre in well-water and 100 ng/litre in the River Rhine. Dall'Aglio (1968) in measurements of 300 samples from natural water in Italy found values in the range of 10–50 ng/litre. Voege (1971) reported levels up to 40 ng/litre for uncontaminated Canadian waters. Durum et al. (1971) have reported data on the concentration of mercury in surface waters of the USA. In areas where mercury mineralization was present, values of up to 200 ng/litre were seen. The results of the CEC International Symposium, reviewed by Bouquiaux (1974), indicate that the purest surface water (drinking quality) contains less than 30 ng/litre based on over 700 samples collected from drinking reservoirs in the Federal Republic of Germany. Rivers believed to have low contamination, such as the Danube, and bodies of water such as the Boden See, have values close to 150 ng/litre based on the analysis of 152 samples. The rivers in the lowland countries of Western Europe that flow

---

[a] STRABY, A. (1968) *Analysis of snow and water.* In: Westermark, T. & Ljunggeren, K., ed. *Development of analytical methods for mercury and studies of its dissemination from industrial sources.* Stockholm, Swedish Technical Research Council, mimeographed documents.

into the North Sea have mercury values in the range of 400–700 ng/litre no doubt reflecting the high industrialization of this area (Schramel et al., 1973). Reports by Hasselrot[a], Fonds (1971), and Smith et al. (1971a), indicate that mercury is predominantly particle-bound in contaminated water-ways. In the Federal Republic of Germany the mercury concentration measured was around 400 ng/litre in inland waters, between 100 ng and 1800 ng/litre in rivers, and 600 ng/litre in a sample of potable water. (Reichert, 1973; Schramel et al., 1973.)

Data for mercury concentrations in ocean waters are not as extensive as those reported for freshwater. Findings of Stock & Cucuel (1934) giving a mean value of 30 ng/litre were confirmed by Sillen (1963). Sillen, on the basis of physico-chemical arguments, suggested that most of the mercury in seawater would be present as negatively charged halide complexes. Hosohara (1961) noted the following levels in the Pacific Ocean: at the surface, 80–150 ng/litre; at a depth of 500 metres, 60–240 ng/litre and at a depth of 3000 metres, 150–270 ng/litre. Levels reported at the CEC International Symposium (reviewed by Bouquiaux, 1974) were 20 ng/litre in 14 samples from the English Channel but were as high as 150 ng/litre in samples taken from the Belgian shoreline and the Waddenzee in the Netherlands. Other references such as Burton & Leatherland (1971) and Leatherland et al. (1971) also support the general rule that oceanic levels are below 300 ng/litre. Higher concentrations have been produced as a result of local contamination such as in Minamata Bay where Hosohara et al. (1961) have reported values up to 600 ng/litre.

In view of questions, discussed earlier, on the total mercury content of the ocean, the stability of mercury levels in the ocean over the past 50 years, and on the high mercury levels in species of oceanic fish, the paucity of data on oceanic levels of mercury is remarkable. This would seem to be one area for future studies of environmental levels of mercury. These efforts should include attempts to analyse the different physical (particulate, or soluble) and chemical (inorganic, or methyl) forms of mercury.

*Food* (except fish)

Smart (1968) has reviewed data concerning mercury concentrations in foods and the most recent data from Europe have been summarized by Bouquiaux (1974). Mercury levels in milk products (81 samples from the Federal Republic of Germany and the United Kingdom) ranged from 0 to 40 µg/kg with a median value of 6 µg/kg. Levels in eggs (440 samples, taken from Denmark, the Federal Republic of Germany and the United

---

[a] HASSELROT, T. (1971) Mercury in fish, water, and bottomless sediments. Investigations at the research laboratories of the National Swedish Environment Protection Board (mimeographed document).

Kingdom, ranged from 0 to 100 µg/kg with most of the values between 10 and 20 µg/kg. Levels in meat, meat products, and prepared meat products (318 samples from the United Kingdom) ranged from 0 to 50 µg/kg with most values lying between 10 and 20 µg/kg. Various kinds of cereal and flour (2133 samples, taken from the Federal Republic of Germany and the United Kingdom) ranged from 0 to 20 µg/kg with most values being close to 3 µg/kg. Mercury levels in cereal products from the same countries (52 samples) ranged up to 50 µg/kg with most values close to 20 µg/kg. Vegetables and fruits (288 samples) from Belgium, the Federal Republic of Germany, and the United Kingdom had mercury levels up to 50 µg/kg with most values close to 7 µg/kg. The analysis of nearly 1400 foods, excluding fish, in Canada during 1970 showed mercury residues to be less than 60 µg/kg in bread, flour, grains, and eggs and less than 40 µg/kg in meats and vegetables (Somers, 1971).

A Swedish Expert Group (1971) has reviewed Swedish experience on the effects of widespread use of methylmercury fungicides on food levels of mercury. As a result of a ban on the use of methylmercury fungicides, food levels fell by a factor of three. For example, the mercury levels in Swedish hen eggs (whole) averaged 29 µg/kg prior to April 1966. Between October 1967 and September 1969, following the ban on methylmercury fungicides instituted in 1966, the level in Swedish hen eggs fell to 9 µg/kg.

The chemical form of mercury in foodstuffs other than fish has not been well identified. The reason is that the levels are, in general, so low as to preclude gas chromatographic identification. However, Westöö (quoted by a Swedish Expert Group, 1971) has noted that methylmercury accounts for over half the total mercury in samples of pork chop and liver, filet of beef, and egg white. Inorganic mercury can account for more than half the total mercury in pig kidney, pig brain, ox liver, and egg yolk.

*Fish*

The earliest reported mercury levels for freshwater fish are those of Stock & Cucuel (1934) and Raeder & Snekvik (1949) and range from 30 to 180 µg/kg wet weight. Upper limits for mercury levels have been quoted as, 200 µg/kg wet weight (Lofroth, 1969) for Sweden, 150 µg/kg (Sprague & Carson, 1970) for Canada, and 100 µg/kg (Ui, 1967) for Japan. These are probably to be regarded as normal levels, i.e. for fish in uncontaminated water. The WHO Regional Office for Europe (1973) has summarized references indicating that fish caught in contaminated freshwater areas may have values of 200–5000 µg/kg and, where the water is heavily polluted, values may be as high as 20 000 µg/kg.

The CEC International Symposium (Bouquiaux, 1974) quote levels in freshwater fish caught in Western Europe as ranging from 0 to 1000 µg/kg with most values being between 200 and 400 µg/kg wet weight. Canned fish, excluding tuna taken from several Western European countries (597 samples), had values up to 500 µg/kg with an average close to 50 µg/kg wet weight. Canned tuna from the same areas (1798 samples) had values ranging up to 4000 µg/kg with most values falling into the range of 200–500 µg/kg. Salmon appears to have remarkably low levels of mercury. Measurements of some 260 samples of Atlantic Ocean, Canadian, and Baltic Sea salmon had mercury levels ranging up to 150 µg/kg with most values being close to 50 µg/kg. On the other hand, pike caught in contaminated rivers in Denmark had average mercury values of 5000 µg/kg, results which are in agreement with experiences summarized by a Swedish Expert Group (1971) in contaminated freshwater areas in Sweden and Finland. The concentration of mercury in marine fish showed marked variations. Not all the factors responsible for these variations are understood but it is generally realized that the species of fish, the geographical location, and the age and/or weight of the fish are important. The highest values of mercury are usually seen in those fish at the end of a long foodchain such as the large carnivorous species.

The concentration of mercury in marine fish has been the subject of intense study in recent years. The first measurements reported by Stock & Cucuel (1934) and Raeder & Snekvik (1941) are in agreement indicating levels from 44 to 150 µg/kg wet weight. The most recent reports (Peterson et al., 1973; Bouquiaux, 1974) indicate that mercury levels in most species of oceanic fish fall in the range of 0–500 µg/kg wet weight with most values close to 150 µg/kg wet weight (more than 1600 samples). The most important exceptions to this rule are swordfish, tuna fish, and halibut, whose values usually range from 200 to 1500 µg/kg (reviewed by the Joint FAO/WHO Expert Committee on Food Additives, 1972). Skipjack, white tuna, and yellowfin tuna (911 samples) ranged from 0 to 1000 µg/kg with most values ranging from 200 to 300 µg/kg. These samples were caught in the Atlantic, Pacific, and Indian Oceans. Bluefin tuna from the Bay of Biscay (285 samples) ranged from 200 to 800 µg/kg with most values close to 500 µg/kg. The same species caught in the Mediterranean Sea (136 samples) ranged from 500 to 2500 µg/kg with most values close to 1100 µg/kg. Big-eye tuna (20 samples from various origins) had mercury values ranging from 400 to 1000 µg/kg. Over 5200 samples of tuna, variety not specified but originating from Italy, had levels in the range of 0–1750 µg/kg with most values ranging from 300 to 500 µg/kg wet weight.

Swordfish caught in the western Atlantic (210 samples) had mercury values ranging from 50 to 4900 µg/kg with a mean value of 1150 µg/kg. 40 samples of swordfish, originating near Italy, had values ranging from 650 to 1750 µg/kg with most values close to 1100 µg/kg wet weight.

The geographical location appears to be important. This is illustrated by mercury analysis of cod (Dalgaard-Mikkelsen, 1969). Samples recovered from the strait between Denmark and Sweden, which is heavily contaminated, had values up to 1290 μg/kg; cod caught in the area of Greenland had values of 12–36 μg/kg whereas North Sea cod had values in the range of 150–195 μg/kg wet weight. Peterson et al. (1973) quote evidence that halibut caught in the southern areas of the Northern Pacific had higher mercury levels than those caught in the North. Beckett & Freeman (quoted by Peterson et al., 1973) in a study of 210 swordfish from six areas extending from the Caribbean Sea to the Grand Banks noted significant variations from one area to another in average mercury levels.

Metabolic differences may also affect mercury levels. For example, Barber et al. (1972) noted differences in mercury content in different species of benthopelagic fish despite the fact that they had identical feeding habits and ecological requirements and were exposed to mercury in the same area for the same length of time.

The age (or weight) of the fish appears to be an important determinant of mercury levels. A positive correlation between mercury concentrations and the weight of the fish has been demonstrated by Beckett & Freeman (quoted by Peterson et al., 1973) for swordfish, halibut, benthopelagic morid (Barber et al., 1972), spiny dogfish (Forrester et al., 1972), blue marlin (Rivers et al., 1972), and tuna (quoted by Peterson et al., 1973). In the last study, mercury levels were measured in 88 yellowfin tuna whose sizes ranged up to 100 kg. Tuna having weights below 25 kg had mercury levels not exceeding 250 μg/kg; tuna having body weights below 50 kg had mercury levels not exceeding 500 μg/kg. Tuna with body weights above 60 kg had values ranging up to 1000 μg/kg. However, large variations in mercury content were noted in tuna with body weights in the range of 60–100 kg. A relationship between mercury content and body weight has previously been noted for freshwater fish (Johnels, 1967; Kleinart, 1972; Bache et al., 1971).

Mercury content may also differ with the sex of the fish. For example, Forrester et al. (1972), in studies of spiny dogfish on the coast of British Columbia, noted that males had a higher mercury content than females for a given body weight. These authors suggested that this difference may be due to the fact that the males grow more slowly than the females.

Mercury in fish appears to be predominantly in the form of methylmercury. Swedish measurements of freshwater fish, summarized by a Swedish Expert Group (1971), indicated that virtually all of the mercury is present in the form of methylmercury compounds. Smith et al. (1971b) confirmed these findings for fish on the North American continent and for swordfish and tuna fish. Exceptions to this rule are Pacific marlin caught off the coast of Hawaii where methylmercury accounts for only a small fraction of the total mercury (Rivers et al., 1972) and also lake trout where methylmercury seems to account for only 21–35% of total mercury (Bache et al., 1971).

Interpretation of the results of observations on museum specimens of tuna fish and swordfish caught at the turn of the century (Miller et al., 1972) indicates that mercury levels in these species of fish have not changed significantly throughout the twentieth century. Specimens from preserved fish of this age are necessarily limited. In seven samples of tuna reported by Miller et al. (1972), the mercury concentrations ranged from 180 to 640 µg/kg. These compare with present values in tuna ranging roughly from 200 µg/kg to over 1000 µg/kg wet weight. Given this variation, it is true to say that there is no statistically significant difference between samples caught in 1900 and those caught in 1970. However, because of the wide range of values, the data at present available do not preclude the possibility that some change may have taken place and that the change might be quite substantial.

## 5.2  Occupational Exposures (See also section 8.1.1)

Occupational exposure to elemental mercury vapour is still the principal hazard to human health when mercury is considered. More than 50 specific occupations or trades involving frequent exposure to mercury have been described by Gafafer (1966). Diseases caused by mercury or its toxic compounds are classical occupational diseases and in most countries are notifiable and qualify for compensation. Reporting of occupational poisoning by mercury has been inadequate, as is the case with all other occupational diseases, particularly in developing countries where there is evidence that large numbers of workers are exposed to high concentrations of mercury leading to poisoning. Occurrence of occupational mercury poisoning in a wide variety of industries in different parts of the world has been reported. In accordance with the information available, most people exposed to elemental mercury vapour appear to be employed in the mining industry, or in chloralkali plants (McGill et al., 1964; Ladd et al., 1966; West & Lim, 1968; Smith et al., 1970) and in the manufacturing of instruments where mercury finds application. These publications, all appearing within the last ten years, indicate that mercury levels in air may attain values as high as 5 mg/m$^3$. The highest mercury concentrations in air are reported in papers on exposure in mining operations. The concentration of mercury in urine may attain levels as high as 2175 µg/litre.

In mining for metals other than mercury (e.g. copper), mercury ore may be present in the mine and give rise to occupational exposure. Donovan (1974) has reported levels of mercury in urine samples (91 samples, number of workers not stated) ranging from 30 to 700 µg/litre in a non-mercury related mining operation. In the two years (1972–73), seven urine samples were found with mercury levels in excess of 250 µg/litre and some of the miners were admitted to hospital.

Ladd et al. (1964) have reported on occupational exposure to phenyl-mercury compounds. Air mercury concentrations ranged up to 0.1 mg/m³ and urinary mercury levels ranged from 1 to 788 µg/litre. A total of 67 workers were involved in these studies. Phenylmercury compounds continue to be used as fungicides in the paint industry (for review, see Goldwater, 1973) so that occupational exposure to phenylmercury compounds is still significant.

The Swedish Expert Group (1971) have summarized reports on occupational exposures to methyl- and ethylmercury compounds. All these reports were published within the period 1940–60 except for the reports on laboratory personnel published by Edwards in 1865 and 1866. Restrictions on the agricultural application of ethyl- and methylmercury compounds by various industrialized countries probably accounts for the lack of recent reports on occupational exposure.

## 5.3 Estimate of Effective Human Exposure

The daily intake of elemental mercury vapour by the general population may be calculated from the published data on ambient air levels discussed above and on the assumption that 80% of the inhaled mercury vapour is retained and that the daily ventilation in the average person is 20 m³ of air. The ambient air level, except in polluted areas, appears to be of the order of 20 ng/m³ and appears not to exceed 50 ng/m³ (see section 5.1). Assuming an ambient air level of 50 ng/m³, the average daily intake of metallic mercury vapour would amount to 1 µg/day due to inhalation. The average daily intake of those sub-groups of the general population living in specially polluted areas is difficult to estimate with any accuracy. If we use the figures of McCarthy (1970), it is possible to find mercury levels as high as 0.0015 mg/m³ close to points of emission. Individuals living continuously in these areas would have intakes of 30 µg/day. Daily intake from occupational exposure is almost impossible to estimate because of the wide variation in exposure conditions in industry (see section 5.2). Assuming that, generally, the time-weighted average threshold limit value of 0.05 mg/m³ (ACGOH, 1976) is being followed, average occupational exposure would lead to an average daily intake of 300 µg of mercury or less, assuming a ventilation of 10 m³/day at work and 225 working days per year. The published reports are insufficient to estimate occupational daily intake from other forms of mercury. The proposed guideline of 0.1 mg/m³ for phenylmercury (MAC Committee, 1969) should lead to an intake in workers exposed to phenylmercury compounds of 500 µg/day or less.

The intake of mercury from drinking water by the general population is more difficult to estimate but it is probably very low in comparison with intake from diet. The major problem is that the chemical form of mercury in water has not always been identified and the efficiency of absorption from the gastrointestinal tract depends greatly on the form of mercury. Methyl-mercury compounds are absorbed almost completely whereas absorption of inorganic mercury may be 15% or less. In making the following calculations the worst case will be assumed, namely that all mercury in drinking water is methylmercury. It will also be assumed that the daily intake of water in adults is 2 litres/day (Joint FAO/WHO Expert Committee on Food Additives, 1972). Published reports indicate that pure well-water and drinking water from reservoirs have mercury levels not exceeding 50 ng/litre (see section 5.1). Thus the daily intake of mercury from drinking water would not normally exceed 0.1 µg/day. However, drinking water in certain areas may derive either from natural waters such as those reported in Italy that have levels as high as 300 ng/litre because of exposure to mineralized mercury deposits, or from rivers in heavily industrialized areas reported to have values up to 700 ng/litre (see section 5.1). Taking the highest reported figure and assuming that mercury is not removed during purification of the water, the highest daily intake would be close to 1.4 µg/day. The advised upper limit for mercury in drinking water is 1 µg/litre (World Health Organization, 1971) which would allow intakes of up to 2 µg/day from this source.

The intake of mercury from food is the most difficult of all to estimate because of the different levels of mercury in different classes of foodstuffs and different dietary habits of individuals in the general population. The one important generalization that emerges is that the intake of mercury as methylmercury is related to fish intake. Thus normal levels for intake of mercury cannot be stated in general without some reference to the fish intake of the population in question.

Over the past forty years, various estimates have been made on the intake of mercury by the general population assuming that fish intake is close to the average values for that population. These reports have been reviewed by a Joint FAO/WHO Expert Committee on Food Additives (1972) and indicate that the range of daily intake of mercury in the general population is from 1 to 20 µg/day. The most complete reviews of dietary intake published to date are those of a Swedish Expert Group (1971) and Jonsson et al. (1972). The reports refer specifically to the Swedish population. It was noted that the intake of mercury in the diet from sources other than fish in Sweden is about 5 µg/day and that the methylmercury content is not known precisely. The median supply of methylmercury from fish is stated to be 5 µg/day or less. As fish consumption exceeds the median value for Sweden, the daily intake of methylmercury will increase in proportion. It was noted that the average daily intake of fish flesh was 30 g, that 10% of the adult men might consume between 80 and 100 g and that a few individuals may consume as much as 500 g/day.

Epidemiological studies summarized by a Swedish Expert Group (1971) indicate that in fishermen and their families, daily intakes of methylmercury can rise to values of 200 µg/day and that one individual had an unusually high intake of 800 µg/day. Another example of a Swedish fish eater with very heavy methylmercury exposure has now been published (Skerfving, 1974b).

Dietary intake of mercury in other countries is not as well documented as that in Sweden. Recent studies reported in a CEC Symposium (Bouquiaux, 1974) indicate that average dietary intake in the United Kingdom, based on total diet samples, is less than 20 µg/day. Observations on fish eating groups, such as fishermen based in American Samoa, indicate that blood mercury levels of up to 20 µg/100 ml can be obtained through fish intake (Clarkson et al., 1975). Such blood levels would be equivalent to a daily intake of between 200 and 300 µg/day of methylmercury in fish. McDuffie (1973) has reported on intakes of mercury in dieters in the United States who consume substantial amounts of tuna and swordfish. He estimated that in the 40 dieters, who had the highest daily intake of fish, 25% consumed 9–16 µg/day, that the second quartile consumed 17–26 µg/day, the third quartile consumed 27–38 µg/day and that the highest quartile consumed 40–75 µg/day. On the basis of radio-chemical measurements, Diehl & Schellenz (1974) estimate the total intake of mercury with food in the Federal Republic of Germany to be between 57 and 192 µg per person per week.

Some industrial countries appear to have an average daily intake of less than 20 µg/day but sub-groups in these countries with unusually high fish intakes (dieters, fishermen's families) may have intakes rising to 75 µg/day (dieters) and even to 800 µg/day (an extremely heavy fish-eater in Sweden).

In countries depending greatly on fish as the major source of dietary protein, there is a great need for dietary studies including the measurement of mercury in the diet of these populations. Initial studies from a South

American country indicate that coastal villages have populations that are comparable to the Swedish fishermens' families in terms of daily intake of methylmercury (Turner et al., 1974).

## 6. METABOLISM OF MERCURY

### 6.1 Uptake

#### 6.1.1 Uptake by Inhalation

Inhalation is the most important route of uptake for elemental mercury vapour. From what is known of the general principles governing pulmonary retention of vapours, the high diffusibility and appreciable lipid solubility of metallic mercury vapour should ensure a high rate of absorption in the alveolar regions of the lung (Task Group on Metal Accumulation, 1973). Calculations made by Nordberg & Skerfving (1972) indicate that mercury vapour should be distributed between air and body tissues in the proportion of 20 to 1 in favour of tissue deposition. Experiments on animals confirm that the major site of absorption is alveolar tissue where virtually complete absorption of the vapour takes place (Magos, 1967; Berlin et al., 1969; Hayes & Rothstein, 1962). If mercury vapour is completely absorbed across the alveolar membranes, one would expect that, owing to the physiological dead space, 80% of the inhaled vapour would be retained. This has been confirmed by observations in man where retention of the inhaled vapour was in the range of 75–85%, at mercury concentrations between 50 and 350 $\mu g/m^3$. (Teisinger & Fierova-Bergerova, 1965; Kudsk, 1965a). The retention of mercury vapour in man can be reduced by moderate amounts of alcohol (Kudsk, 1965b). Magos et al. (1973) have shown that the action of alcohol is due to the inhibition of oxidation of the vapour in the red blood cells and other tissues. More recently Magos et al. (1974) have shown that the herbicide, aminotriazole, has a similar action to that of alcohol.

No specific data are available on the monoalkylmercury compounds. However, it is generally believed that absorption is high, of the order of 80% of the inhaled amount (Task Group on Metal Accumulation, 1973). Ostlund (1969a, 1969b) reported a high retention of inhaled dimethylmercury in mice. The inorganic and organic compounds of mercury may also exist in the atmosphere in particulate form (see section 5). No detailed studies have been reported on pulmonary retention and clearance of mercury aerosols. In general, one would expect that aerosols of mercury should follow the general physical laws governing deposition in the respiratory system.

Particulates with a high probability of deposition in the upper respiratory tract should be cleared quickly. For particulates deposited in the lower respiratory tree, longer retention will be expected, the length of which will depend on solubility, among other factors (Task Group on Lung Dynamics, 1966). Approximately 45% of a mercury(II) oxide aerosol having a mean diameter of 0.16 μm was cleared in less than 24 hours and the remainder cleared with a half-time of 33 days according to experiments on dogs by Morrow et al. (1964). Information on pulmonary retention of aerosols of the organomercurials is lacking. Pulmonary absorption of monoalkylmercury must be significant to judge from the incidents of poisoning resulting from occupational exposures to dusts and vapours of the alkylmercury fungicides. It should be noted that the gastrointestinal route may include those particulates of mercury compounds that have been cleared from the lung in the bronchociliary tract.

### 6.1.2  Uptake by ingestion

The general principles underlying the gastrointestinal absorption of mercury and its compounds are not clearly understood. Probably the formation of soluble salts and complexes is a prerequisite for absorption of metals ingested from food.

Liquid metallic mercury has long been considered to be poorly absorbed from the gastrointestinal tract. Based on the data of Bornmann et al. (1970), in animals given gram quantities by mouth, Friberg & Nordberg (1973) have calculated that less than 0.01% of an administered dose of metallic mercury was absorbed. Persons who had accidently ingested several grams of metallic mercury showed increased blood levels of mercury (Suzuki & Tanaka, 1971).

The efficiency of absorption from food depends greatly upon the type of mercury compound (Clarkson, 1972a). Studies on mice revealed that the absorption of inorganic salts of mercury from food was 15% or less in contrast with 80% or more in the case of phenyl- or methylmercury compounds. Observations on volunteers given tracer doses of inorganic mercury revealed that the efficiency of absorption was the same with both free and protein-bound mercury. The absorption from food in these volunteers was an average of about 7% (Rahola et al., 1973).

Aberg et al. (1969) and Miettinen (1973) have reported on the absorption of radioactive methylmercury compounds in volunteers given oral doses. The absorption of the administered dose was 95% irrespective of whether the methylmercury was administered as a salt dissolved in water or in a protein-bound form. Information on the absorption in humans of other organic compounds of mercury including the other short-chain alkylmercurials is not available. As episodes of accidental poisoning due to ingestion of food contaminated with ethylmercury compounds have occurred, absorption must be significant.

The Task Group on Metal Accumulation (1973) considered the possibility that the gastrointestinal absorption of one metal may be influenced by the presence of another. Studies on animals and animal tissues (Sahagain et al., 1966, 1967) suggest the possibility that some interaction may occur between zinc, manganese, cadmium, and inorganic mercury.

### 6.1.3 Absorption through skin

Debate has persisted throughout most of the present century about the importance of skin as a route for entry of metallic mercury into the body. Early studies on man (Juliusberg, 1901) and animals (Schamberg et al., 1918), where inhalation of mercury vapour was prevented, indicated that appreciable skin absorption of metallic mercury took place. It would appear that metallic mercury can cross the skin barrier but to what extent is not known.

Studies on experimental animals reveal that inorganic salts of mercury, principally mercury(II) chloride, may be absorbed in significant amounts through skin. For example, Friberg et al. (1961) and Skog & Wahlberg (1964) indicate that 5% of mercury in a 2% water solution of mercury(II) chloride was absorbed through intact skin of guinea-pig over a 5-hour period. Such a penetration rate, if applicable to man, could result in absorption of substantial amounts of mercury under conditions of high exposure.

Friberg et al. (1961) and Wahlberg (1965) have demonstrated in guinea-pigs that methylmercury dicyandiamide was absorbed from a water solution through intact skin, the rate was more or less the same as that for mercury(II) chloride reported above. No information is available on animals with respect to ethyl- or other alkylmercury compounds.

No quantitative data are available for skin absorption of the short-chain alkylmercurials in man. People have been poisoned by administration of methylmercury compounds locally to the skin such as methylmercury thioacetamide (Tsuda et al., 1963; Ukita et al., 1963; Okinaka et al., 1964; Suzuki & Yoshino, 1969; Suzuki et al., 1970). The methylmercury compound was absorbed in sufficient amounts to cause severe poisoning although the possibility of some inhalation exposure cannot be excluded.

## 6.2 Distribution in the Organism

Details on the organ distribution of mercury have been recently reviewed (Clarkson, 1972a; Nordberg & Skerfving, 1972). New publications since that time have not substantially changed the general picture. Methylmercury and its homologous short-chain alkylmercurials, which are much more uniformly distributed throughout the body than are the other organomercurials, and inhaled elemental mercury vapour are distinguished from other types of mercury compound in their ability to cross the blood–brain barrier and placenta rapidly.

Organ distribution is not only affected by the type of mercury compound ingested or inhaled but also changes with time after exposure. For example, the phenylmercurials are subject to rapid conversion in the body to inorganic mercury so that the distribution of mercury following administration of these compounds and related organomercurials approaches that of inorganic mercury with increasing time after exposure (for details see Clarkson, 1972b).

The distribution between cells and plasma (the red cell/plasma ratio) depends upon the form of mercury to which the subject is exposed. Studies on fish-eating populations reported by Birke et al. (1972) and on a heavily exposed population in Iraq (Bakir et al. 1973) indicate that the cell to plasma ratio for methylmercury is approximately 10, as was found in human volunteers given tracer doses of radioactive methylmercury (Aberg et al., 1969; Miettinen, 1973). The red cell to plasma ratio in human volunteers given radioactive inorganic mercury salts was 0.4 (Miettinen, 1973).

The distribution of mercury between hair and blood tends to follow a constant ratio in people exposed to methylmercury (Table 1). In various populations having a broad range of dietary methylmercury intake from fish, the concentration of total mercury in hair is proportional to the concentration in whole blood. The ratio of hair to blood concentration is about 250 as determined by linear regression analysis. The data in Table 1 are from populations of individuals, most of whom probably have a steady concentration of methylmercury in hair and blood. In the Iraq epidemic, hair and blood concentrations underwent rapid changes. Two cases have been reported in Iraq in which blood and hair concentrations were measured when both were declining following cessation of heavy exposure. (Amin-Zaki et al., in press.) The ratios of hair to blood concentrations were constant and the value of the ratio was close to 250. However it should be noted that, when hair and blood concentrations are changing, it is important to choose the segment of hair for analysis that corresponds to the blood sample. Depending on the length of hair segment used for analysis and the rate of growth of hair, there is a delay about 2–4 weeks between the time of sampling the blood, and the emergence of the appropriate segment of hair above the scalp (Amin-Zaki et al., in press).

Table 1. Relationship between concentrations of mercury in samples of blood and hair in people having long-term exposure to methylmercury from fish

| No. of subjects | Whole blood $(x)$ (mg/kg) | Hair $(y)$ (mg/kg) | Linear regression | References |
|---|---|---|---|---|
| 12 | 0.004–0.65 | 1–180 | $y = 280x - 1.3$ | Birke et al. (1972) |
| 51 | 0.004–0.11 | 1–30 | $y = 230x + 0.6$ | Swedish Expert Group (1971) |
| 50 | 0.005–0.27 | 1–56 | $y = 140x + 1.5$ | Swedish Expert Group (1971) |
| 45 | 0.002–0.80 | 20–325 | $y = 260x + 0$ | Tsubaki (1971) |
| 60 | 0.044–5.5 | 1–142 | $y = 230x - 3.6$ | Skerfving (1974b) |

In people occupationally exposed to metallic mercury vapour, the red cell to plasma ratio may be as high as 2 (Lundgren et al., 1967; Suzuki et al., 1970; Einarsson et al., 1974). Work on experimental animals has shown that the ratio was higher in animals given radioactive vapour compared with those given salts of inorganic mercury.

Studies on a variety of experimental animals indicate that the kidney is the chief depository of mercury after the administration of inorganic salts and exposure to elemental mercury vapour. Over 50% of the body burden of mercury can be found in the kidneys of rats exposed to mercuric salts and metallic mercury vapour a few days after receiving the dose. This percentage may rise to 90% or more as the length of time after exposure increases (Rothstein & Hayes, 1960; Hayes & Rothstein, 1962; Trojanovska, 1966). However, it should be noted that in experimental animals, the brain levels of mercury following exposure to elemental mercury vapour were ten times higher than brain levels after equal doses of inorganic salts (Berlin et al., 1966; Magos, 1967; Nordberg & Serenius, 1969). A more uniform distribution of methylmercury throughout the body also results in much higher brain levels for a given body burden of mercury as compared with inorganic salts.

Little information is available on the distribution of mercury in human organs following exposure to elemental mercury vapour. Takahata et al. (1970) and Watanabe (1971) have reported mercury levels in the brain several times higher than those in the liver and other organs (except the kidney) of miners with long-term exposure to high concentrations of mercury vapour. These concentration ratios were maintained even several years after cessation of exposure. High mercury concentrations in the thyroid and pituitary glands in persons connected with mercury mining have been reported (Kosta et al., 1975). It should be noted that organ distribution of mercury after inhalation of elemental mercury vapour can be dramatically affected by moderate intakes of alcohol and small doses of the herbicide aminotriazole, as shown in animals (Magos et al., 1973, 1974). These agents reduce levels in the lung and increase levels in the liver several-fold.

The percentage of the body burden of methylmercury found in the brain is much higher in primates than in other animal species (Swedish Expert Group, 1971). Observations on human volunteers given tracer doses of radioactive methylmercury (Aberg et al., 1969) indicate that 10% of the radioactivity in the whole body is located in the posterior part of the head. Probably not all of this represents methylmercury in the brain but would include methylmercury attached to the hair. Studies by Miettinen's group (quoted by a Swedish Expert Group, 1971), on volunteers given tracer doses of radioactive methylmercury, indicate that an initial rapid distribution throughout the body is followed by a further slow redistribution of methylmercury to the brain.

## 6.3 Elimination in Urine and Faeces

Urine and faeces are the principal routes of elimination of mercury from the body. The contribution of each pathway to total elimination depends upon the type of mercury compound and the time that elapses after exposure. Experiments in animals indicate that elimination of inorganic mercury by the gastrointestinal tract depends on the size of the dose and the time after exposure. The faecal route is dominant soon after exposure. The urinary route is favoured when high doses are given (Prickett et al., 1950; Friberg, 1956; Rothstein & Hayes, 1960; Ulfvarson, 1962; Cember, 1962; Phillips & Cember, 1969; Nordberg & Skerfving, 1972).

Data obtained on rats subjected to a single exposure of labelled $^{203}$Hg vapour indicated that about 4 times more mercury was eliminated in the faeces than in the urine (Hayes & Rothstein, 1962). In prolonged exposure of rats, the proportion changed in favour of urinary excretion (Gage, 1961). In workers exposed to mercury vapour, the output of mercury in urine slightly exceeded that in the faeces (Tejning & Ohman, 1966). High individual variation and great fluctuation from day to day were the principal features of urinary excretion in workers under similar exposure conditions (Goldwater et al., 1963; Jacobs et al., 1964). There is evidence that, on a group basis, urinary excretion is roughly proportional to exposure (air concentration) to elemental vapour (MAC Committee, 1969). Occupational exposure of at least 6 months, 5 days per week at average air concentrations of mercury of 0.05 mg/m$^3$, should lead to average urinary concentrations of mercury of about 150 µg/litre.

Piotrowski et al. (1975) have reported changes in urinary rates of excretion in workmen following exposure to elemental mercury vapour. They noted that urinary excretion could be described by a two-term exponential equation with rate constants equivalent to half-times of 2 and 70 days. The short half-time compartment accounted for about 20–30% of the excretion rate under conditions of steady-state excretion. Piotrowski et al. (1975) suggested that there is variation in urinary mercury excretion in individuals and that this can be greatly reduced by collecting the urine samples at the same time in the morning.

Mercury exhalation found in animals after exposure to the elemental vapour (Clarkson & Rothstein, 1964) has also been confirmed in man (Hursh et al., 1975). This pathway of excretion accounted for about 7% of the total excretion of mercury in volunteers following inhalation of a tracer dose. Recent observations indicate that the concentration of mercury in sweat may be sufficiently high to be taken into account in the overall mercury balance in workers exposed to elemental mercury vapour (Lovejoy et al., 1974).

The faecal route is most important in the elimination of mercury after acute or chronic dosing with methylmercury. Studies on human volunteers (Aberg et al., 1969; Miettinen, 1973) indicate that approximately 90% of the elimination takes place via the faeces. This proportion does not change with time after exposure. Concentrations of total mercury in urine showed no correlation with blood mercury in people heavily exposed to methylmercury (Bakir et al., 1973).

## 6.4 Transplacental Transfer and Secretion in Milk

The transplacental movement of mercury in women exposed to elemental mercury vapour has not been studied thoroughly.

Experiments on animals reveal that after brief (approximately 20 minutes) exposure to radioactive elemental mercury vapour, the radioactive mercury easily penetrates the placental barrier (Clarkson et al., 1972). These authors report that, after equal exposure of pregnant rats, the fetal uptake was 10–40 times higher after exposure to elemental mercury vapour than to inorganic salts. In contrast, the placental content of mercury after exposure to elemental mercury vapour was only about 40% of that after exposure to inorganic salts of mercury.

The alkylmercuric compounds have been known for some time to penetrate the placenta readily as indicated from studies on experimental animals (for review, see Swedish Expert Group, 1971). In a recent study,

Childs (1973) noted that the level of methylmercury in the fetus may be twice that in the maternal tissues when low levels of methylmercury are fed to rats in a tuna fish matrix. At higher dose levels, the ratio between fetal and maternal tissues becomes close to unity. Transplacental movement of methylmercury in women has been sufficient to cause several cases of prenatal poisonings in various countries (Engleson & Herner, 1952; Harada, 1968; Bakulina, 1968; Snyder, 1971; Bakir et al., 1973; Amin-Zaki et al., 1974a). Tejning (1970) has reported methylmercury levels in fetal blood cells to be 30% higher than in maternal cells in studies on women having normal pregnancies and a low to moderate fish intake. The relatively higher concentrations in fetal blood have been confirmed in a study by Suzuki et al. (1971). It was noted that the plasma levels in both types of blood were similar and that differences arose only in terms of concentrations in the red blood cells.

Information on the transplacental movement of other compounds of mercury in women is lacking. Animal experiments indicate that those compounds rapidly converted to inorganic mercury in the body, such as phenylmercury compounds, behave in this respect like inorganic mercury (for review, see Clarkson, 1972b).

Mercury has been reported in breast milk in women exposed to methylmercury from fish (Harada, 1968; Skerfving, 1974a) and from bread contaminated with methylmercury fungicides in the 1971–72 outbreak in Iraq (Bakir et al., 1973). In Iraq, it was noted that levels of total mercury in milk correlated closely with levels in whole blood and averaged 5% of simultaneous concentrations in maternal blood. The total mercury in milk consisted of two fractions identified as inorganic mercury (40%) and methylmercury (60%). Skerfving's (1973) observations on 15 lactating mothers exposed to methylmercury in fish are in general agreement with the findings in Iraq except that methylmercury accounted for only 20% of the total mercury in milk.

Despite the relatively low concentration in milk as compared with maternal blood, the suckling infants accumulated high concentrations of mercury in their blood if their mothers were heavily exposed (Amin-Zaki et al., 1974b). Some Iraqi infants, exposed only through maternal milk, had blood levels in excess of 100 µg/100 ml. In prenatally exposed infants, intake of methylmercury by suckling is one factor responsible for the slower decline in blood levels as compared with the mother (Amin-Zaki et al., 1974a).

## 6.5 Metabolic Transformation and Rate of Elimination

The most dramatic example of metabolic transformation is the conversion of metallic mercury to divalent ionic mercury in the body. This oxidation reaction has been shown to take place *in vitro* in the red cells (Clarkson et al., 1961). More recent studies indicate that it probably takes place in most other tissues (for details, see Kudsk, 1973). The process is enzyme mediated and the catalase complex is the most likely site of biochemical oxidation (Kudsk, 1973; Magos et al., 1974).

Studies on the biotransformation of elemental mercury make it possible to develop a picture of the role of the oxidation process in the accumulation of mercury vapour in the body and its transport to the site of action (for details, see Clarkson, 1972a). Elemental mercury vapour, after inhalation, is absorbed into the blood stream. Despite the rapid oxidation that has been shown to take place in the red blood cells, some elemental mercury remains dissolved in the blood long enough for it to be carried to the blood–brain barrier and to the placenta. Its lipid solubility and high diffusibility allow rapid transit across these barriers. Tissue oxidation of the mercury vapour in brain and fetal tissues converts it to the ionic form which is much less likely to cross the blood–brain and placental barriers. Thus oxidation in these tissues serves as a trap to hold the mercury and leads to accumulation in brain and fetal tissues.

Most studies on the metabolic transformation of organomercury compounds have concentrated on measurements of the rate of cleavage of the carbon–mercury bond. There is no evidence in the literature supporting the possibility of the synthesis of organomercury compounds in human or mammalian tissues.

The absolute rates of cleavage of the carbon–mercury bond in man or experimental animals is not known. The relative rates of cleavage of different mercury compounds have been estimated by measurements of the amounts of inorganic mercury deposited in tissues following single doses of organomercury compounds. In general, these studies reveal that the phenyl-(aryl) and the methoxyethyl compounds are converted rapidly to inorganic mercury in the body (for reviews, see Gage, 1974; Clarkson, 1972a). The short-chain alkylmercurials are converted more slowly to inorganic mercury with the methylmercury compounds being converted the most slowly of all. The phenyl- and methoxyethylmercurials are probably converted to inorganic mercury more or less completely within a few days whereas methylmercury can be detected in human tissue months after exposure has stopped (Bakir et al., 1973). Suzuki et al. (1973) have reported the only case in which the metabolic conversion of ethylmercury has been studied in man. Proportional values of inorganic mercury to total mercury ranging from 12 to 69% were detected in red cells, plasma, brain, spleen, liver, and kidney in a patient exposed for about 3 months to ethylmercurythiosalicylate.

The role of biotransformation in determining the toxicity of organomercurials is not well understood (for discussion, see Clarkson, 1972b). The rapid conversion of phenylmercury to inorganic mercury probably accounts for the fact that, in chronic studies on animals, the effects of this organomercury compound on kidneys were similar to those of inorganic mercury (Fitzhugh et al., 1950).

The conversion of organic to inorganic mercury may increase or decrease the total rate of excretion of mercury from the body. If the intact molecule of an organomercurial is more rapidly excreted than inorganic mercury, biotransformation will decrease the overall excretion rate. This has been demonstrated in the case of the diuretic, chlormerodrin, where the intact molecule is almost completely excreted within 24 hours, but inorganic mercury remains in the animal for much longer periods (Clarkson et al., 1965). The phenyl- and methoxyethylmercury compounds are excreted at a rate similar to that of inorganic mercury according to studies on experimental animals. In the case of methylmercury, biotransformation may play an important part in determining the rate of excretion of total mercury from the body (Swensson & Ulvarson, 1968, 1969). Inorganic mercury accounts for approximately 50% of the total mercury in faeces, the principal pathway of excretion following single or chronic doses of methylmercury compounds. Methylmercury undergoes extensive enterohepatic recirculation in rats but inorganic mercury does not (Norseth & Clarkson, 1971). Thus a small rate of metabolic transformation in the liver leading to biliary excretion of inorganic mercury could make an important contribution to the faecal elimination of mercury.

### 6.6 Accumulation of Mercury and Biological Half-time ("Metabolic Model")

The body accumulates a metal when uptake exceeds elimination. At a certain stage a steady state may be reached when uptake and elimination are equal. A common way to express the accumulation is in terms of biological half-time. The biological half-time for mercury would be the time taken for the amount of mercury in the body to fall by one-half. The concept of biological half-time is meaningful, however, only if the elimination can be approximated to a single exponential first-order function. This will be true if the distribution and turnover of a metal in different tissues of the body are faster than the elimination from the body as a whole. If elimination from one organ is slow compared with that from other organs then the calculation of a biological half-time for the whole body may be completely misleading from the toxicological point of view (Task Group on Metal Accumulation, 1973; Nordberg, 1976).

Studies on experimental animals and volunteers indicate that, for methylmercury compounds, the elimination can be approximated to a single exponential first-order function (Miettinen, 1973; Aberg et al., 1969; for reviews, see Clarkson, 1972b; Swedish Expert Group, 1971; Task Group on Metal Accumulation, 1973). Observations on experimental animals indicate that the elimination of mercury after exposure to mercury vapour, inorganic mercury salts, and the phenyl and methoxyethyl compounds does not follow such a pattern and thus the accumulation and elimination of mercury ("the metabolic model") is much more complex. The pattern of elimination of these mercury compounds, when administered to animals, is dose- and time-dependent (Rothstein & Hayes, 1960; Ulfvarson, 1962; Piotrowski et al., 1969).

In cases where the elimination of a metal such as methylmercury follows a single exponential first order function, the concentration in an organ at any time can be expressed by the following equation;

$$C = C_o \cdot e^{-b \cdot t} \tag{1}$$

$C$ = concentration in the organ at time $t$
$C_o$ = concentration in the organ at time $o$
$b$ = elimination constant
$t$ = time

The relation between the elimination constant and the biological half-time is the following:

$$T = \ln 2 / b$$

$T$ = biological half-time
$\ln 2$ (natural logarithm of 2) = 0.693

If data on exposure and absorption of the metal are known, then it will be possible to predict the body burden of the metal at constant exposure over different time periods. If a constant fraction of the intake is taken up by a certain organ, the accumulated amount in that organ can also be calculated. The following expression gives the accumulated amount of metal in the total body (or organ):

$$A = (a/b)(1 - \exp(-b \cdot t)) \tag{2}$$

$A$ = accumulated amount
$a$ = amount taken up by the body (or organ) daily

At steady state the following applies:

$$A = a/b \tag{3}$$

In other words, the steady state amount in the body (or organ) $A$ is proportional to the average daily intake and inversely proportional to the elimination rate. The latter point will be taken up later (section 9) in discussing hazards to man, as large individual variations in elimination rates imply large individual variation in steady state body burden, even in people having the same average daily intake.

Equations (1), (2), and (3) are illustrated graphically in Fig. 1. During

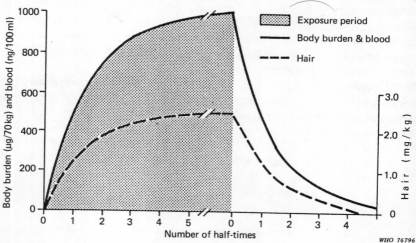

Fig. 1. The changes in the body burden and hair and blood concentrations of mercury during constant daily exposure (shaded area) and after exposure. This calculation was based on a daily intake of 10 µg of methylmercury during the exposure period, an elimination half-time of 69 days, and a hair to blood concentration ratio of 250.

the period of steady daily intake (assumed to be 10 µg/70 kg body weight), the amount in the body rises rapidly at first, reaching half its maximum (steady state) value in a time equivalent to one elimination half-time (assumed to be 69 days for methylmercury in man). After an exposure period equivalent to 5 elimination half-times (approximately one year for methylmercury), the body burden is within 3% of its final steady state value. The steady state value is one hundred times the average daily intake assuming an elimination half-time of 69 days. On cessation of exposure, the body burden will immediately begin to fall following an exponential curve that is an inverse image of the accumulation curve. Thus the body burden will have returned to within 3% of pre-exposure values in 5 half-times.

In this example, it is assumed that the hair to blood ratio is constant and equal to 250 and that 1% of the body burden is found in 1 litre of blood in a 70-kg man.

That this model provides a reasonable approximation to the accumulation of methylmercury in man over a wide range of daily intakes is indicated by the data in Tables 2 and 3. Data on elimination rates for the whole body reported by Aberg et al. (1969) on 5, and by Miettinen (1973) on 15 volunteers were in good agreement indicating average values close to the value of 69 days used in Fig. 1. An average value of $50 \pm 7$ days for clearance half-time from blood was reported by Miettinen (1973) in volunteers receiving a single tracer dose. Blood clearance values are difficult to measure accurately with tracer doses owing to the low counting rates in the blood samples. Skerfving (1974b) reported clearance from whole blood ranging from 58 to 87 days in 4 people having high intake (up to 5 μg/kg body weight) of methylmercury from fish, one individual had a clearance half-time of 164 days. Bakir et al. (1973) reported that patients having very high blood levels (over 100 μg/100 ml) in Iraq, had clearance half-times in the same range (45–105 days, mean 65 days). "Clearance" from hair is estimated by analysis of consecutive short (0.2–1 cm) segments of hair samples and plotting the mercury concentrations against the distance from the scalp on semilogarithmic paper (for details, see Birke et al., 1972). A straight line is usually obtained, the slope of which is equivalent to a biological half-time if the growth rate of the hair is known. "Clearance" half-times from hair are assumed to reflect clearance half-times for blood. Data from a fish eating population (daily intake up to 5 μg/kg) and on a highly exposed population in Iraq (daily intake up to 50 μg/kg) are compatible with this assumption given the wide range of individual variations (Table 2).

The relationship between steady state body burden ($A$) and average daily intake is given by equation (3); using data derived from tracer observations on volunteers (Aberg et al., 1969; Miettinen, 1973), one

Table 2. Mercury intake and clearance

| No. of subjects | Hg intake (μg/kg/day) | Clearance half-times (days) | | | References |
| --- | --- | --- | --- | --- | --- |
| | | Body | Blood | Hair | |
| 5 | tracer | 70 | — | — | Aberg et al. (1969) |
| 15 | tracer | 76 (52–93) | 50 | — | Miettinen (1973) |
| 5 | up to 5 | — | — | (33–120) | Birke et al. (1972) |
| 5 | up to 5 | — | see [a] (58–164) | — | Skerfving (1974) |
| 16 | up to 50 | — | 65 (45–105) | — | Bakir et al. (1973) |
| 48 | up to 50 | — | — | 72[b] (35–189) | Shahristani & Shihab (1974) |

[a] One person had a biological half-time of 164 days. The other four were in the range of 58–87 days.

[b] The data were distributed bimodally. One group accounting for 89% of the samples had a mean value of 65 days and the other group had a mean value of 119 days.

would predict that the steady state blood level ($y$ ng/ml) is numerically equal to the average daily intake ($x$ µg/day/70 kg body weight) as indicated in Table 3. This calculation assumes a 69-day elimination half-time from the whole body, that 1% of the body burden is found in 1 litre of blood in a 70-kg "standard man". Observed steady state relationships between blood

Table 3. Relationship of steady state blood concentrations to daily intake of methyl mercury[a]

| No. of subjects | Time of exposure | Ave. Hg intake (µg/day/70 kg B.W.) | Steady blood concentration (ng/ml) | References |
|---|---|---|---|---|
| | | ($x$) | ($y$) | |
| 6+26[b] | years | 0–800 | $y = 0.7x + 1$ | Birke et al. (1967) |
| 139+26[b] | years | 0–400 | $y = 0.3x + 5$ | Tejning (1967, 1969a, 1969b, 1969c) |
| 6+14[b] | years | 0–800 | $y = 0.8x + 1$ | Birke et al. (1972) |
| 725[c] | years | 0–800 | $y = 0.5x + 4$ | Estimated from Kojima & Araki (unpublished data) |
| 22 | years | 0–800 | $y = 0.5x + 10$ | Skerfving (1974b) |
| 30[d] | 1–2 months | 0–2340 | $y = 0.8x$ | Estimated from Shahristani & Shihab (1974) and Shahristani et al. (1976) |
| 15 | single tracer dose | | $y = 1.0x$ | Estimated from Miettinen (1973) |

[a] For details of these calculations, see text.

[b] None or low fish consumers.

[c] Estimated from data on hair concentrations and daily intake. The hair to blood concentration ratio was assumed to be 250 and the average body weight of the population under study to be 60 kg.

[d] Estimated from data on hair concentrations and daily intake. The hair to blood concentration ratio was assumed to be 250.

concentration ($y$) and daily intake ($x$) are given in Table 3 for several populations. These populations consist of fishermen and their families who had had a high dietary intake of fish for many years. The range of intake between different individuals is high—up to 800 µg/day. The relationship between blood concentrations and average daily intake was found to be linear for each population studied. Linear regression analysis reveals that the observed relationship between $y$ and $x$ is less than that predicted by tracer studies. The coefficient of $x$ lies between 0.5 and 0.8 in the fish eating populations as compared with the predicted value of unity from tracer studies.

Given the difficulties in the accurate measurement of dietary intake and the uncertainty in tracer studies based on counting blood samples, it is likely that differences between the observed and the tracer values are not real. This conclusion is supported by the fact that the Iraqi populations (Table 3, Shahristani et al., 1974), having an extremely high dietary intake yielded a factor of 0.8 suggesting that the relationship between $y$ and $x$ is not substantially changed at high doses.

In summary, a considerable body of evidence exists to support the linearity of the metabolic model for methylmercury in man. No definitive evidence is yet available that refutes this conclusion. However, we cannot exclude the possibility that the mean values of the parameters of the metabolic model could change by about a factor of two over a wide dose range. We have, however, taken the predicted value from the tracer data, since this approach would offer a greater margin of safety in estimates of hazards to human health.

Biological half-times for other mercury compounds are not well established and this is particularly true for those organs that are of toxicological importance. It seems, however, that the biological half-time for the greatest part of retained salts of inorganic mercury has an average value of about 40 days (Miettinen, 1973). In five female volunteers the average half-time was 37 days and in a similar number of males the average half-time was 48 days.

The biological half-time of both methyl and inorganic salts of mercury does not appear to be affected whether the compound is administered in an ionic form or bound to protein (Miettinen, 1973).

Limited information is available on biological half-times of mercury in the body following exposure to elemental vapour. Five volunteers inhaled radioactive mercury vapour for 10–15 minutes and were subjected to whole body counting for up to 43 days after exposure (Hursh et al., 1975). Elimination from the body followed a single exponential process having a biological half-time of 58 days with a range of individual values of 35–90 days.

As noted above, whole body elimination half-times may not be a reliable guide to accumulation in specific organs. For example, the fact that Takahata et al. (1970) and Watanabe (1971) found high mercury concentrations in the brain in an individual 10 years after cessation of exposure indicates that accumulation in the brain does not follow the same kinetics as seen by whole-body counting. Observations on experimental animals also indicate that the half-time in brain is longer than in other organs (Task Group on Metal Accumulation, 1973).

### 6.7 Individual Variations—Strain and Species Comparisons

As a general rule, the processes of absorption by inhalation or ingestion do not appear to be subject to large species and strain differences (for detailed review, see Clarkson, 1972b). Elemental mercury vapour and methylmercury compounds are well absorbed across the pulmonary epithelium and the gastrointestinal tract, respectively, in a variety of animal species. Distribution in the body tissues is subject to species

81

differences. A very pronounced example is the case of the red cell to plasma ratios of methylmercury where the ratio can be as high as 300 in the rat and as low as 10 in primates. Differences in red cell to plasma ratio may account for species differences in blood to brain ratios (Vostal, 1972). The blood to brain ratio has been reported to be approximately 10–20 for the rat, approximately unity for the cat, 0.5 for the dog and pig, and 0.1 for primates (from data reviewed by a Swedish Expert Group, 1971). Careful and accurate quantitative comparison of species differences and distribution is not possible because of the different experimental conditions in these studies.

The oxidation of elemental mercury vapour to ionic mercury and the cleavage of the carbon–mercury bond in a variety of mercurials has been described for several different species of animal. However the observations in this field are not adequate to allow quantitative comparison of metabolic rates of breakdown in different species.

The rate of elimination of mercury from the body is subject to wide species variation (for detailed review, see Clarkson, 1972a). In general, animals of small body weight tend to excrete mercury more rapidly than larger animals and the cold blooded species, particularly fish, appear to retain mercury for an extremely long time.

Species differences in the elimination of methylmercury have been reported. The mouse and the rat have half-times between 8 and 16 days as compared with 70 days in primates. The seal is reported to have a half-time of 500 days and fish and crustaceans appear to have half-times ranging from 400 to over 1000 days. These species differences indicate that we cannot extrapolate parameters describing the metabolic fate of mercury in animals to that in man. Furthermore, since these parameters determine the amount of mercury accumulated in the body, it would appear that quantitative information on toxicities cannot be directly extrapolated from animals to man.

The half-time of clearance of mercury from blood and hair varies considerably between individuals exposed to methylmercury (Birke et al., 1972; Skerfving, 1974). In the outbreak in Iraq the half-time of clearance from blood ranged from 40–105 days in 16 subjects (Bakir et al., 1973) and from hair the range was from 35–189 days (Shahristani & Shihab, 1974). The Iraqi data on blood samples (Bakir et al., 1973) were obtained about 1–2 months after the termination of exposure.

Average biological half-times for groups of at least 15 individuals seem to be remarkably constant. The average half-time in blood was 65 days (Bakir et al., 1973), in hair 72 days (Shahristani & Shihab, 1974), and in the whole body in 15 subjects given a tracer dose, 76 days (Miettinen, 1973).

# 7. EXPERIMENTAL STUDIES ON THE EFFECTS OF MERCURY

## 7.1  Experimental Animal Studies

### 7.1.1  Acute studies

Little information is available on the acute toxicity of elemental mercury vapour to animals. Ashe et al. (1953) reported evidence of damage to brain, kidney, heart, and lungs in rabbits exposed to mercury vapour at a mercury concentration of 29 mg/m$^3$ of air. This concentration under the circumstances of their experiments would represent an atmosphere saturated with mercury vapour. The first effects were seen within 1 hour of exposure and subsequent severe changes resulted after longer exposures.

Information on the $LD_{50}$ (the dose of mercury that kills half the test population) has been reviewed by a Swedish Expert Group (1971). The results reported for different mercury compounds are not easily comparable since different animal species and different routes of administration have been used for the test. Nevertheless, despite all these differences, the $LD_{50}$ lies between approximately 10 and 40 mg/kg body weight for all compounds tested to date including inorganic mercury, arylmercury, alkoxyalkyl- and alkylmercury compounds. The remarkable similarity in $LD_{50}$s of these various types of mercury compound is probably due to the fact that when given in acute massive doses, mercury in whatever chemical form will denature proteins, inactivate enzymes, and cause severe disruption of any tissue with which it comes into contact in sufficient concentration. The symptoms of acute toxicity usually consist of shock, cardiovascular collapse, acute renal failure and severe gastrointestinal damage. A variety of complexing and chelating agents, all of which contain sulfhydryl groups can modify the $LD_{50}$s of mercury and its compounds (for review, see Clarkson, 1972a). These agents are most effective when given either prior to the mercury dose or in the few hours following a single dose of mercury. The importance of time of administration is to be expected since the effects of these agents are to reduce the reactivity of mercury in the body and to do so before irreversible damage has been inflicted on the tissue (see also, section 9.2).

Irreversible damage is used in this document to define cellular or organ damage that is not repaired even after the cessation of exposure; some improvement in function or in the condition of the poisoned tissue may occur but recovery is never complete. In the case of reversible damage, regeneration of cells and restoration of function takes place after the cessation of exposure.

### 7.1.2 Subacute and chronic studies

#### 7.1.2.1 *Reversible damage*

This section deals with toxic effects known to be reversible, at least up to a certain dose and/or duration of exposure. It should be noted, however, that at higher doses or longer duration of exposure, the damage can surpass the stage of reversibility.

Studies by Trahtenberg (1969) (reviewed by Friberg & Nordberg, 1973) indicate that exposure of rats to concentrations of elemental mercury vapour in the range of 0.1–0.3 mg/m$^3$ for over 100 days increased uptake of radioactive iodine by the thyroid. Friberg & Nordberg (1973) also refer to unpublished observations by Aveckaja which, under different conditions including prolonged exposure, indicate the opposite effect. Kournossov reported in 1962 (reviewed by Friberg & Nordberg, 1973) effects on the behaviour of rats at mercury concentrations in air as low as 0.005 mg/m$^3$. Studies by Armstrong et al. (1963) on pigeons showed irreversible behavioural changes only at vapour levels well in excess of recommended maximum allowable concentrations. It is clear that many more studies need to be carried out on behavioural and other subtle changes resulting from exposure to mercury vapour at these low concentrations.

Fitzhugh et al. (1950) studied the toxicity of mercury(II) chloride and phenylmercury acetate when added to the diet of rats for periods of up to 2 years. Morphological changes were induced in kidney tissue at approximately the same mercury levels for both compounds. The similarity in the effects of the two compounds on the kidney is probably due to the fact that phenylmercury compounds are rapidly converted to inorganic mercury in animal tissues.

The main effect of alkylmercurials is the irreversible action on the central nervous system. However, Lucier et al. (1971, 1972, 1973) have reported that subacute doses (with no neurological signs) caused a marked decrease in rat liver mixed-function oxidase activity. This effect was shown to be due to an increased degradation rate of cytochrome P-450 *in vivo*. Methylmercury also depressed the activity of enzymes dependent upon cytochrome P-450. Ultrastructural changes involving the endoplasmic reticulum in liver have been reported by Chang & Yamaguchi (1974) and these effects were reversible.

In rats, administration of methylmercury can produce kidney damage manifested by tubular degeneration in the distal convoluted tubules after daily doses of 10 mg/kg for 7 days (Klein et al., 1972). With lower doses of methylmercury, morphological and functional damage is produced in kidney tissue in the absence of any signs of neurological dysfunction

Fowler, 1972a, 1972b; Magos & Butler, 1972; Klein et al., 1973). Ultrastructural studies by Fowler (1972a) showed that female rats given 2 mg/g methylmercury in their diet were more sensitive to methylmercury than males. The primary lesion was characterized by extrusion of cytoplasmic masses from proximal tubular cells. It has been suggested that the nephrotoxic effect is due to inorganic mercury split from methylmercury *in vivo* (Klein et al., 1973).

#### .1.2.2 *Irreversible damage*

With the exception of massive doses of inorganic compounds or prolonged exposure to extremely high concentrations of elemental mercury vapour, as in the experiments of Ashe et al. referred to below, the effects of inorganic mercury on tissues are generally reversible.

Ashe et al. (1953) reported microscopically detectable changes in the organs of dogs, rabbits, and rats exposed to concentrations of elemental mercury vapour ranging from 0.1 to 30 mg/m³ for different periods of time. Severe damage was noted in kidneys and brains at mercury levels in air of 0.9 mg/m³ after an exposure period of about 12 weeks. After longer periods of exposure to 0.1 mg/m³, no microscopically detectable effects could be seen.

The short-chain alkylmercurials are primarily neurotoxic in man. After a single dose there is a latency period of days or weeks before signs of poisoning occur. Many of the signs of methylmercury poisoning observed in man can be reproduced in animals under appropriate conditions. For example, Berlin et al. (1973) noted that a sudden visual disturbance occurred in monkeys given subacute doses of methylmercury. Prolonged exposure to methylmercury resulted in a gradual constriction of the visual field and impaired motor coordination and possibly sensory disturbances. Neurological signs of damage have also been produced in the mouse, rat, ferret, cat, and dog by feeding them methylmercury compounds (Chang et al., 1974; for review, see Swedish Expert Group, 1971). The toxicity of methylmercury to animals does not appear to be affected whether it is given to them as a pure chemical, e.g. methylmercury chloride, or whether it has accumulated naturally as in fish such as the Northern Pike (Swedish Expert Group, 1971; Albanus et al., 1972).

Grant (1973), in his studies of primates experimentally poisoned with methylmercury, reported findings confirming those of Hunter & Russell 1954). Neuronal damage and destruction was observed in the visual cortex, and the granular layer of the cerebellum. The dose-rate of methylmercury, the period of dosing, and the animal species all influence the pattern of pathological damage. For example, in monkeys after short periods of exposure to high doses, there is an abrupt visual change over two days leading to blindness. This is accompanied by damage to the neurons in the visual cortex. Longer exposure to lower daily doses of methylmercury leads to more generalized damage to the cortex and is accompanied by gradual onset of visual changes and other signs of central nervous involvement such as ataxia.

Recent studies on rats reviewed by Somjen et al. (1973a) have confirmed the findings of Hunter et al. (1940) that the earliest neurological effects in these animals is damage to the peripheral sensory nerves. Later the disease affects other parts of the central and peripheral nervous systems. There is now evidence that the primary site of the disease is the cell bodies in the dorsal root ganglia with secondary deterioration in their fibres (Chang & Hartman, 1972a, 1972b). Consistent with this interpretation, Somjen et al. (1973b) found that the spinal dorsal root ganglia contained the highest concentrations of mercury. Electrophysiological investigations confirmed the findings drawn from morphological evidence that the cell bodies in the spinal ganglia are the primary sites of action (Somjen et al., 1973a).

Morphological, electrophysiological and biochemical changes have been demonstrated in animals prior to the onset of overt signs of poisoning. These phenomena, especially with respect to morphological changes, have been referred to as "silent damage". For example, Nordberg et al. (1971) and Grant (1973) noted that morphological damage was present in certain of the test monkeys before signs of visual impairment could be detected. Somjen et al. (1973a) reported electrophysiological manifestations of methylmercury intoxication in rats preceding clinical signs. Yoshino et al. (1966) noted a decreased uptake of amino acids in brain slices taken from animals given high doses of methylmercury at a time before signs of poisoning had appeared. Cavanagh & Chen (1971) reported that incorporation of amino acids into protein was impeded in spinal root ganglia of rats, treated with methylmercury before signs of poisoning were present. Chang & Hartman (1972c) noted damage to the blood–brain barrier as early as 12 hours after a dose of 1 mg/kg of methylmercury chloride to rats. If these observations on experimental animals may be extrapolated to man, the possibility must be considered that significant damage to the central and peripheral nervous systems may take place prior to the onset of clinical signs and symptoms.

Methylmercury compounds have been demonstrated to disturb mitosis in the plant cell, in human leucocytes treated *in vivo*, and in human cells in tissue culture. The short-chain alkylmercurials cause chromosome breakage in plant cells and point mutations in *Drosophila* (for detailed reviews, see Swedish Expert Group, 1971; Ramel, 1972).

Clegg (1971) has given a detailed review of the embryotoxicity of the ort-chain alkylmercurials. In general, animal experiments confirm the idea derived from epidemiological observations in the Minamata epidemic that much more damage was inflicted on the fetus than on the mother. Spyker et al. (1972) have recently reported on performance deficits in mice treated prenatally with methylmercury. The alkylmercury compounds may also damage the gametes prior to fertilization (Khera, 1973). Virtually no information is available on the morphological and biochemical factors related to prenatal damage in experimental animals. The results, however, do point to incipient hazards to human fetuses exposed before birth.

In discussing the biological effects of methylmercury compounds, species differences should be considered. Although man, monkeys, and pigs may become blind at high exposures, similar visual disturbances in rats were not detected (Albanus et al., 1972; Charbonneau et al., 1974). Pronounced morphological changes are seen in the peripheral nervous system of rats (Somjen et al., 1973a), but again, were not detected in cats (Albanus et al., 1972; Charbonneau, 1974). However, one cannot exclude the possibility that qualitative differences, reported in studies of different species including man, may reflect differences in degree and intensity of exposure in man and in experimental conditions in animal studies.

### .1.2.3 *Interactions with physical and chemical factors*

Parizek & Ostadalova (1967) reported that selenite salts could protect experimental animals against the toxic effects of inorganic mercury. Selenium also depressed the passage of inorganic mercury into fetuses and its secretion into milk (for review, see Parizek et al., 1969, 1971, 1974). Parallelism between tissue concentrations of mercury and selenium has been reported in human subjects exposed to elemental mercury vapour (Kosta et al., 1975). Several recent publications have claimed that selenite added to the diet protects experimental animals against methylmercury compounds (Ganther et al., 1972; El Bergerami et al., 1973; Potter & Metrone, 1973). Ganther & Sunde (1974) have reviewed evidence indicating that the content of selenium in tuna fish is sufficiently high to provide substantial protection against methylmercury. However, the protective factor in tuna fish, whether selenium or some other substance, has yet to be isolated.

Toxicological interactions have been reported between methylmercury and the chlorinated hydrocarbon pesticide, dieldrin. Rats dosed with both dieldrin and methylmercury, showed less morphological damage of the pars recta tubule than animals given only methylmercury. However, there was degeneration in proximal tubular cells (Fowler, 1972b). It has been demonstrated recently that phenobarbital administration increases the biliary excretion of methylmercury compounds (Magos & Clarkson, 1973).

Estrogenic hormones (Lehotzky, 1972) and spironolactone (Selye, 1970) protect the kidney from methoxyethylmercury salts and mercury(II) chloride, respectively. The mechanism of these actions is unknown.

The question of the interaction of the physical and chemical factors on the toxicity of methylmercury is important and should be a major priority in future research studies. Extrapolation of epidemiological and toxicological data from populations in Japan and in Iraq suffering from methylmercury poisoning is fraught with difficulties when the possible interactions of local factors are not taken into account.

### 7.1.3 Biochemical and physiological mechanisms of toxicity

A physiological basis for the action of mercury and other heavy metals had already been propounded prior to 1967 (for reviews, see Hughes, 1957; Passow et al., 1961; Peters, 1963; Webb, 1966). Two general concepts on the mechanisms of action of mercury and other heavy metals are discussed in these reviews. The first dates back to the 1940s and is attributed to Peters, 1963. The toxic sequelae of heavy metal action on tissues result from a primary "biochemical lesion" whereby a critical enzyme or metabolic process is inhibited. Unfortunately, despite a considerable amount of research work (see Webb, 1966), it has not been possible to locate the biochemical lesion associated with the toxic actions of mercury.

An alternative general concept mainly proposed by Passow et al. (1961) is that the cell membrane is the first site of attack by heavy metals. Topographically, this would seem reasonable. Furthermore, the membrane is known to contain sulfhydryl groups that are essential to the normal permeability and transport properties of the cell membrane. These same sulfhydryl groups are known to have a very high affinity for mercury and its compounds. Passow et al. (1961) summarized a great many experimental studies that support this general idea. However, it must be admitted that most experimental work testing this idea is based on *in vitro* studies on isolated cells and tissues so that the role of membrane damage in the pathogenesis of heavy metal poisoning remains to be established. The effect of mercury in intercellular membranes is also of some interest.

The affinity of mercury for thiol groups in proteins and other biological molecules is far in excess of its affinity for other biologically occurring ligands (Clarkson, 1972b). As pointed out by Rothstein (1973), the affinity of mercury(II) cations for the sulfhydryl groups of proteins creates a "severe logistics problem" for those interested in elucidating the mechanisms of action of mercurials. "Although mercurials are highly specific for sulfhydryl groups, they are highly unspecific in terms of proteins. Almost all proteins contain sulfhydryl groups that are metal-reactive. Furthermore, because most sulfhydryl groups are important in most protein functions, mercurials can disturb almost all functions in which proteins are involved. Thus almost every protein in the body is a potential target". In other words, the mercurials are potent but non-specific enzyme poisons. Mercury will inflict cellular damage wherever it accumulates in sufficient concentrations. This reasoning has given rise to the idea that the selective toxicity of mercury is related to its selective distribution. In general, there seems to be some truth in this. Inorganic mercury compounds are avidly accumulated by the kidney which is the target organ for this compound.

Studies by Somjen et al. (1973b) on the microdistribution of mercury in the nervous system also lend credence to the importance of distribution of methylmercury in that the spinal root ganglia, the site of peripheral nerve damage, are also the area of highest accumulation of methylmercury in rats.

However, it seems that distribution factors alone cannot give a complete explanation for the toxicity of methylmercury. The kidney is always the site of the highest accumulation of mercury irrespective of the form of the mercury compound involved. For example, kidney levels of methylmercury are much higher than brain levels and yet kidney damage, except in the rat, is much less than that seen in the central nervous system.

In recent years, interest has arisen in the biocomplexes of mercury in the body. The toxicity of any mercury compound will be determined by its chemical activity close to its site of action. For example the chloride salts of mercury compounds when added *in vitro* to tissue preparations are highly toxic, whereas when mercury is added in the presence of sulfhydryl compounds the toxicity is very much less (for review of this subject, see Clarkson & Vostal, 1973). Thus the chemical state of combination of mercury in plasma and other body fluids may be of primary importance in determining the particular site of action of the mercurial.

Mercury accumulated in the kidney is contained there partly in form of a metallothioneine-like complex (Jakubowski et al., 1970; Wisniewska, et al., 1970). In the rat, binding by this protein is especially effective in repeated exposure to mercury(II) chloride, owing to the induction of higher levels of the metallothioneine-like protein by mercury (Piotrowski et al., 1974a, 1974b). This form of binding probably also occurs in the case of exposure to elemental mercury vapour since this exposure results in enhancement of the metallothioneine level in the kidneys of rats (Sapota et al., 1974). Binding of inorganic mercury in other organs may also involve storage of a similar form, as found in the liver (Wisniewska et al., 1972) and brain of the rat (Sapota et al., 1974).

The binding of mercury by metallothioneine-like protein of the kidneys is enhanced by the presence of cadmium (Shaikh et al., 1973) and therefore may play an important role in man whose kidneys accumulate, in normal conditions, considerable amounts of cadmium (Piscator & Lind, 1972). The above applies also to organic mercurials, which are rapidly converted into inorganic mercury, as in the case of phenylmercury acetate (Piotrowski & Bolanowska, 1970).

The primary biochemical lesions associated with mercury poisoning have not yet been established. Virtually nothing is known of the biochemical disturbances associated with exposure to metallic mercury vapour. Studies referred to in section 7.1.2.2 by Cavanagh & Chen (1971) and Yoshino et al. (1966) suggest that protein synthesis may undergo an early biochemical change preceding clinical signs and symptoms of methylmercury poisoning. This may give an explanation for the latent period associated with this form of mercury poisoning.

# 8. EFFECTS OF MERCURY ON MAN—
# EPIDEMIOLOGICAL AND CLINICAL STUDIES

## 8.1 Epidemiological Studies

### 8.1.1 Occupational exposure to mercury vapour, alkylmercury vapour and other compounds

Occupational exposures to elemental mercury vapour have been the subject of recent reviews by Friberg & Nordberg (1972, 1973) and NIOSH (1973). Many studies dating back to the 1930s have related the frequency of signs and symptoms of mercury poisoning to exposure. These studies, involving observations of more than one thousand individuals, indicate that the classical signs and symptoms of elemental mercury vapour poisoning (objective tremors, mental disturbances, and gingivitis) may be expected to appear after chronic exposure of workers to air concentrations of mercury above 0.1 mg/m$^3$ (Neal et al., 1937; Smith & Moskowitz, 1948; Smith et al., 1949; Friberg, 1951; Bidstrup et al., 1951; Vouk et al., 1950; Kesić & Heusler, 1951; Baldi et al., 1953; Seifert & Neudert, 1954; McGill et al., 1964; Ladd et al., 1966; Copplestone & McArthur, 1967; Smith et al., 1970). The industries involved included the chloralkali industry, the manufacture of thermometers and graduated scientific glassware, the repair of DC electrical meters, the mining and milling of mercury, the manufacture of artificial jewellery, the felt hat industry and others (NIOSH, 1973). Most of the publications referred to above do not report time-weighted average exposures and few give information as to the physical and chemical forms of mercury in the atmosphere. Different methods of measurement of mercury in air were employed some of which measured only mercury vapour, while others attempted to include particulate forms of mercury. Most of the studies, if not all, assumed that exposure occurred only during the working day. However, evidence has now come to light that, in certain industries, metallic mercury may be entrapped in the clothing and contaminate the home, particularly in those industries actually handling liquid metallic mercury (West & Lim, 1968; Danzinger & Possick, 1973).

Effects of elemental mercury vapour, other than those designated as classical mercurialism, have been reported (Smith et al., 1970; Trahtenberg, reviewed by Friberg & Nordberg, 1973). The study of Smith and co-workers involved observations on 567 workers exposed to mercury in chloralkali plants. The air concentrations of mercury (measured by a mercury vapour meter) ranged from less than 0.01 to 0.27 mg/m$^3$ and time-weighted averages were calculated for each worker. A significant increase in the frequency of objective tremors was noted at mercury levels in air above 0.1 mg/m$^3$ in agreement with previous reports on occupational exposure. However, a significant increase was observed at mercury concentrations in air of 0.06–0.1 mg/m$^3$ in such non-specific signs and symptoms as loss of appetite, weight loss, and shyness.

Studies related to assessment of the occurrence of a so-called "asthenic–vegetative syndrome" or "micromercurialism" have been reported by Trahtenberg (1969). This syndrome may occur in persons with or without mercury exposure. For a diagnosis of mercury-induced asthenic vegetative syndrome Trahtenberg (1969) (reviewed by Friberg & Nordberg, 1972) required that not only neurasthenic symptoms should be present but as supporting evidence three or more of the following clinical findings; tremor, enlargement of the thyroid, increased uptake of radio-iodine in the thyroid, labile pulse, tachycardia, dermographism, gingivitis, haemotological changes, and excretion of mercury in the urine which was above normal or increased 8-fold after medication with unithiol.

Trahtenberg, in her monograph (1969) considered that this syndrome would be more frequently found in persons exposed to mercury concentrations between 0.004–12 mg/m$^3$ but, upon detailed scrutiny of her data (see review by Friberg & Nordberg, 1972), there does not seem to be any difference between exposed and control groups that can be related to mercury exposure.

Studies on the prevalence of a similar syndrome defined as a combination of insomnia, sweating, and emotional lability have been reported by Trahtenberg, et al., quoted by Trahtenberg (1969), in workers exposed to mercury levels of 0.006–0.01 mg/m$^3$ and temperatures of 40–42°C in the summer and 28–38°C in the winter. They found 28–50% prevalence of this syndrome in this exposed group and only 13% of the same syndrome in a control group exposed to only 38–42°C temperatures. Details about the selection and evaluation of these workers are not known.

The studies of Bidstrup et al. (1951) and Turrian et al. (1956) also indicate that psychological disturbances may be seen at air concentrations of mercury below 0.10 mg/m$^3$. Thus it is impossible, at this time, to establish a lower exposure limit at which no effects occur. There is a continuous need for research studies on the effects of exposure of people to mercury vapour concentrations below 0.1 mg/m$^3$.

Short-chain alkyl compounds have been the subject of recent reviews by a Swedish Expert Group (1971) and Kurland (1973), but no new information has appeared in the literature on occupational exposures to methylmercury or other short-chain alkylmercury compounds. Following the description of the first two cases of occupational exposure to diethylmercury compounds in 1865 by Edwards (1866), occupational exposures have been infrequent and usually limited to a few individuals. For example, exposure has occurred in laboratory personnel, workers, and farmers involved in either the production of alkylmercury fungicides or their application to cereal seeds, in people in seed testing institutes, and in workers in pulp mills and saw mills. Exposure is presumed to be mainly by inhalation of the vapour or dust but it is possible that, in some cases, absorption of the liquid preparation of the fungicide may have occurred through the skin.

Occupational exposures to alkylmercury compounds, although not important numerically or epidemiologically, have been the occasion for the description of the signs and symptoms of poisoning. For example, four workmen exposed to methylmercury fungicide were the basis of the now classic reports of Hunter et al. (1940) and Hunter & Russell (1954) which gave the detailed pathology of methylmercury poisoning in man.

Occupational exposures to the phenyl- and methoxyethylmercury compounds have been the subject of a recent review (Goldwater, 1973). The hazards from industrial exposure to these compounds appear to be very low. Ladd et al. (1964), in a study of 67 workers occupationally exposed to phenylmercury compounds, found no evidence of adverse health effects. Mercury levels in air in this study were mainly below

0.1 mg/m³. Comparison of readings with a mercury vapour detector and estimates of total mercury in the atmosphere revealed that elemental mercury vapour was the principal form of mercury. Goldwater (1964, 1973) makes reference to seven workers who had spent approximately six weeks preparing and packaging a batch of material containing methoxyethylmercury chloride. Four weeks after they had completed this task their whole-blood mercury levels were in the range of 34–109 µg/100 ml with an average of 65 µg/100 ml. At no time did any toxic sign or symptom appear. These workers may also have had a limited exposure to phenylmercury compounds.

At this point it is worth while to quote from the Task Group on Metal Accumulation (1973) indicating the need for carefully controlled studies of occupationally exposed groups. "There is a need in industrially exposed populations for standardized, wherever possible collaborative, epidemiological studies, where cohorts can be followed in time and where groups can be related to each other. With some occupational exposures to the less common metals, only small groups may be available for study in any one country, so that international collaboration in epidemiological studies would again be of value".

### 8.1.2  General population

Epidemics of poisoning in the general population due to exposure to phenyl- and methoxyethylmercury compounds have not been reported. Two outbreaks of poisoning due to elemental mercury vapour occurred in the 19th century, one due to a fire in the mercury mines in Idria, the other being caused by spillage of metallic mercury in a British warship in the early 1800s (Bidstrup, 1964). Fernandez et al. (1966) have reported that, in the village of Almaden, the site of the large mercury mines in Spain, air levels exceeded 0.1 mg/m³. However, there are no reports as yet about the health status of the population in the village.

Methyl- and ethylmercury compounds have been the cause of several major epidemics of poisoning in the general population due either to the consumption of contaminated fish or to eating bread prepared from cereals treated with alkylmercury fungicide. The two major epidemics of methylmercury poisoning in Japan in Minamata Bay (Katsuna, 1968) and in Niigata (Niigata Report, 1967) were caused by the industrial release of methyl- and other mercury compounds into Minamata Bay and into the Agano River followed by accumulation of the mercury by edible fish. The median level of total mercury in fish caught in Minamata Bay at the time of the epidemic has been estimated as 11 mg/kg fresh weight and in the Agano river in Niigata as less than 10 mg/kg fresh weight (Swedish Expert Group, 1971).

A recent report by Tsubaki (1971) indicates that follow-up observations on exposed people in Niigata revealed a much larger number having mild signs and symptoms than the original 46 that had been reported. These milder cases may only have had paraesthesia. By 1971 a total of 269 cases of methylmercury poisoning had been reported in Minamata and Niigata, of which 55 proved fatal. By 1974, more than 700 cases of methylmercury poisoning had been identified in Minamata and more than 500 cases had been identified in Niigata (personal communication, Tsubaki, 1975).[a] The two Japanese epidemics have been the subject of intensive studies on the effects of methylmercury on man and have resulted in important conclusions concerning dose–response relationships (Swedish Expert Group, 1971).

Epidemics resulting in the largest number of cases of poisoning and of fatalities have been caused by the ingestion of contaminated bread prepared from wheat and other cereals treated with alkyl- (methyl- or ethyl-) mercury fungicides. The largest recorded epidemic took place in the winter of 1971–72 in Iraq resulting in the admission of over 6000 patients to hospital and over 500 deaths in hospital (Bakir et al., 1973). Previous epidemics have occurred in Iraq (Jalili & Abbasi, 1961), in Pakistan (Haq, 1963), in Guatemala (Ordonez et al., 1966), and on a limited scale in other countries (Snyder, 1971). Reports on these epidemics have resulted in interesting clinical findings but quantitative studies relating exposure to effects have been reported only on the recent epidemic in Iraq (Bakir et al., 1973; Kazantzis et al., 1976; Mufti et al., 1976; Shahristani et al., 1976).

In the Iraqi outbreak, the mean methylmercury content of the wheat was 7.9 mg/kg with most samples falling between 3.7 and 14.9 mg/kg. The mean methylmercury content of wheat flour samples was 9.1 mg/kg with a range of 4.8–14.6 mg/kg in 19 samples (Bakir et al., 1973). The average weight of the home-made loaves was about 200 g with a moisture content of about 30% of the fresh weight (Damluji, 1962). The range of daily intake of bread varied widely. In an epidemiological survey of a heavily affected village, Mufti et al. (1976) reported that the average total ingested dose of a group of 426 people was about 150 mg of mercury but some people may have consumed as much as 600 mg. The average daily intake of contaminated loaves was 3.2 loaves although individual variation was large, some people eating up to 10 loaves per day. The daily intake of

---

[a] It has not been possible for this group to review data on the new cases reported since the publication of the Niigata and Minamata Reports.

methylmercury would vary greatly. The average daily intake of mercury in this village would be 80 µg/kg assuming a body weight 50 kg for the population, with extremes of daily intake attaining 250 µg/kg. In the most severely affected group, reported by Bakir et al. (1974), the highest daily intake of mercury was about 130 µg/kg body weight. The average period of consumption for groups of patients reported by Bakir et al. (1973) ranged from 43–68 days. Mufti et al. (1976) reported mean consumption periods in villages to be about 32 days but some people continued for up to 3 months. Birke et al. (1972) and Skerfving (1974b) have reported on

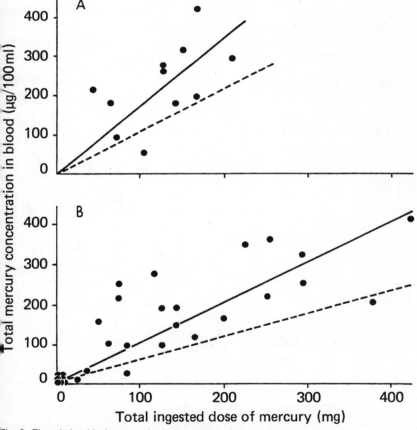

Fig. 2. The relationship between the concentration of the total amount of mercury in the blood and the estimated total amount of mercury ingested from contaminated bread. The solid line was drawn from linear regression analysis. The dotted line is the predicted concentration in the blood estimated from published data by Miettinen (1973) on 15 volunteers given a single oral dose of labelled methylmercury. (A) Patients aged 10–15 years. (B) Patients over 18 years of age. From *Mercury, Mercurials and Mercaptans, 1973.* Courtesy of Charles C. Thomas, Publisher, Springfield, Illinois.

families in Sweden consuming fish containing mercury levels of 0.3–7 mg/kg. Daily intake ranged up to approximately 5 µg/kg body weight. In two cases, intake was as high as 10–20 µg/kg. The highest recorded blood level of mercury was 1.2 µg/g of red cells or approximately 60 µg/100 ml of whole blood. A total of 188 people were referred to in these studies. No signs or symptoms of poisoning attributable to methylmercury were noted.

Clarkson et al. (1975) and Marsh et al. (1974) have offered preliminary information on 163 fishermen based in American Samoa who ingested unusually high amounts of fish containing methylmercury. Data on daily intake of mercury were not reported but the mercury levels in blood in this population ranged as high as 28 µg/100 ml. No signs or symptoms of poisoning could be ascribed to methylmercury.

Turner et al. (1974) have reported on neurological examinations of 186 persons living in two fishing villages in northern Peru. Concentrations

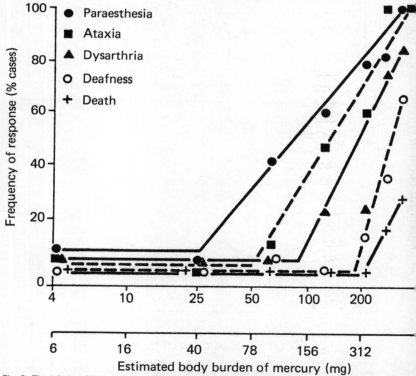

Fig. 3. The relationship between frequency of signs and symptoms and the estimated body burden of methylmercury as reported by Bakir et al. (1973). Both scales of the abscissa refer to body burdens of methylmercury (mg) at the cessation of exposure. The two scales represented different methods of calculating the body burden as discussed in the text (section 8). Copyright 1973 by the American Association for the Advancement of Science.

96

of total mercury, methyl- and inorganic mercury were measured in blood samples from 141 of these villagers. The concentration range of methyl-mercury in blood was from 1.1 to 27.5 µg/100 ml with a mean of 8.9 µg/100 ml. The mean intake of fish was approximately 10 kg per family per week and the average family size was six.

Fifty-one persons from a "control" village were also examined. The mean intake of fish was 1.0 kg per family per week and the mean family size was 6.4. Methylmercury concentrations in blood averaged 0.99 µg/100 ml with a range of 0.33–2.5 µg/100 ml. No correlation was observed between blood levels of methylmercury and the frequency of signs and symptoms usually associated with methylmercury poisoning (paraesthesia, ataxia of gait and limbs, impaired vision, and deafness).

Paccagnella et al. (1974) have reported blood concentrations of mercury in a community in the Mediterranean island of S. Peitro (Cagliari). The average dietary weekly intake of fish was 300 g. Tuna fish, with an average total content of mercury of 1.23 mg/kg and a methylmercury content of 0.92 mg/kg was an important dietary item. Other fish in their diet

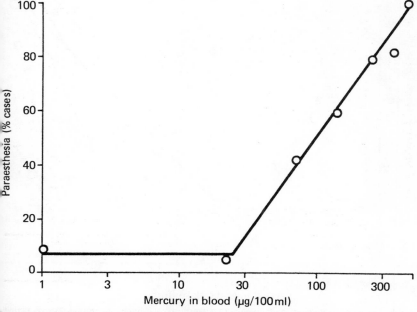

Fig. 4. The frequency of paraesthesia as a function of the concentration of mercury in blood, 35 days after cessation of exposure. The graph uses data from Table 4 of Bakir et al. (1973). The mean blood concentrations are computed as the logarithmic means for each cohort in their table. The line connecting the first two points was assumed to be horizontal. The line connecting the other points was computed by least squares linear regression analysis. Copyright 1973 by the American Association for the Advancement of Science.

had an average total mercury concentration of 0.33 mg/kg. The concentration of total mercury in blood was measured in 115 people aged from 10 years upwards. The mean mercury concentration in blood in the males was approximately 8 µg/100 ml and the maximum recorded concentration was 23 µg/100 ml. The females had average mercury levels in blood of 6 µg/100 ml with a maximum level of 24 µg/100 ml. Three individuals in the sample population of 115 had blood levels higher than 20 µg/100 ml.

Epidemiological studies show that these exposed populations may be classified into three categories distinguished by intensity and duration of exposure to the short-chain alkylmercurials. Populations consuming contaminated grain had a high daily intake of mercury (reaching over 200 µg/kg for brief periods averaging 1–2 months). The outbreaks in Japan fall into the second category where daily mercury intakes ranged from 5 to 100 µg/kg with a median of 30 µg/kg/day with exposure times lasting from several months to years, in Niigata, although doubts on the accuracy

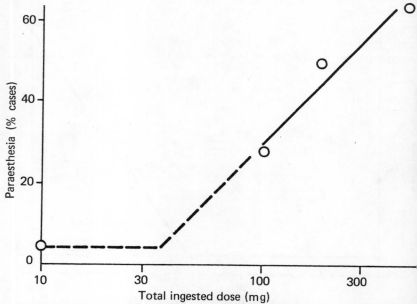

Fig. 5. The relationship between frequency of paraesthesia and the total ingested dose of methyl-mercury-data from Mufti et al. (1976). Each loaf contained 1.4 mg of mercury as methylmercury according to Bakir et al. (1973). The broken lines are extrapolations. The horizontal line is taken to represent background frequency of paraesthesia—using the figure of 4% as given in Mufti et al. (1976)—for the group receiving a total of 1–49 loaves. The line connecting points at higher ingested doses was drawn by least squares linear regression analysis and was extrapolated to intersect the horizontal line.

of those figures have been expressed by a Swedish Expert Group (1971). The third category includes populations having unusually high fish intakes for years if not for most of their lives, such as in parts of Sweden (Skerfving, 1974) the Samoan fishermen (Marsh et al., 1974), and in Peru (Turner et al., 1974). A small proportion of those in this category may attain mercury levels up to 5 µg/kg/day or even higher. Quantitative studies relating frequency of signs and symptoms to various indices of exposure have been reported for all three categories so that it is now possible to compare the effects of differences in intensity and duration.

Before discussing these studies, it would be well to point out some of the attendant difficulties. Methylmercury and the other short-chain alkylmercurials produce effects unique among the mercury compounds. However, some, if not all, of these effects can be caused by agents other than mercury or by certain disease states. Thus, in examining the population for neurological changes, it must be borne in mind, that there could be, at least in theory, many causes of the observed neurological effects other than methylmercury itself. Dose–response relationships derived from these epidemiological studies imply a cause–effect relationship. In fact, the only proof we have that methylmercury caused certain effects in these populations is that (1) these effects coincide in time with exposure to methylmercury, (2) the frequency of these effects in a given population increases with increasing exposure to methylmercury, and (3) the major

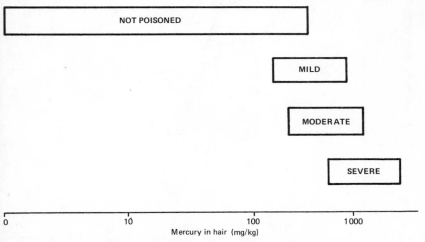

Fig. 6. This diagram is based on data reported by Shahristani et al. (1976) on an Iraqi population that consumed contaminated bread. Hair samples were divided into 1 cm segments and the concentration of mercury measured in each segment by neutron activation analysis. The mercury concentration in the segment with the highest concentration is compared to the severity of signs and symptoms of methylmercury poisoning. The length of each box is the range of maximum mercury concentrations in hair.

signs have been reproduced in some animal models. One of the key problems in these studies is to distinguish between the background frequency of a sign or symptom and the increase in that frequency due to increased exposure to methylmercury.

The studies in the Iraqi population (typical of category one) were made 2–3 months after cessation of exposure and in most cases, sometime after the onset of signs and symptoms in the patients. Thus the investigators were faced with the problem of recapitulation of exposure and of determining the dose received by the individual.

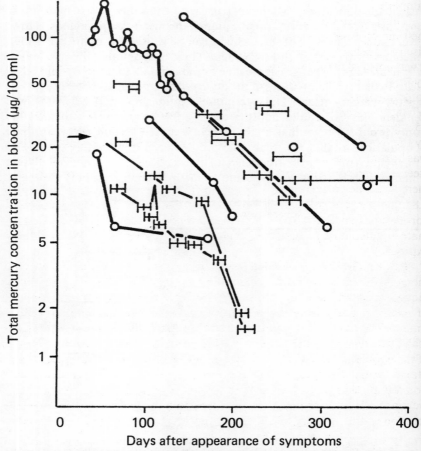

Fig. 7. Concentration of mercury in samples of blood collected from patients suffering from methylmercury poisoning in the Niigata outbreak. Samples from the same patients are connected by a straight line. The arrow indicates the estimated time of onset of symptoms. Data is taken from a report of a Swedish Expert Group (1971).

Bakir et al. (1973) classified the population into cohorts according to blood levels of mercury. The first blood samples were collected at various times after cessation of exposure. They were corrected to an average point in time corresponding to 65 days after cessation of exposure using a clearance half-time of 65 days. Back extrapolation to times earlier than days was not attempted because the pattern of mercury clearance from blood was not established for this period. In fact it was noted that in 11 cases, blood mercury concentrations did not exhibit statistically significant decline during the first 20 days of March 1972. These individuals had stopped consumption of bread 45 days before collection of the first blood sample.

Thus, instead of attempting to back extrapolate the blood samples to the time of onset of symptoms (for example in the Niigata studies to be discussed below), a different approach was made to recapitulate exposure. It was noted that 58 individuals (approximately half the population studied by Bakir et al., 1973) gave sufficient information on their consumption of contaminated bread to allow estimation of the ingested dose of mercury. When the blood levels in these individuals, corrected to the time point of 65 days after exposure, were plotted against the ingested dose as reported by the patients, the relationship between blood levels and estimated dose was linear for both adults and children, the results for the children giving a steeper slope consistent with a smaller volume of distribution of methylmercury. The correlation coefficient for people over 18 years of age was 0.85 and for people of 10–15 years it was 0.89 (Fig. 2). This empirical correlation between blood mercury and ingested dose was used to estimate ingested dose from observed blood levels (corrected to 65 days after exposure).

Bakir et al. (1973) proceeded to estimate the average amount of methylmercury for each group of the people by use of the exponential equation (equation 1) discussed in section 6. They assumed complete absorption of methylmercury from the diet and that the average elimination half-time was 70 days. This estimated body burden, plotted on a logarithmic scale, was related to the frequency of signs and symptoms in each group of the population (see Fig. 3). The signs and symptoms were paraesthesia, ataxia, visual changes, dysarthria, hearing defects, and death. It was noted that there was a background frequency of signs or symptoms that were not related to the mercury level, and that the frequency increased in relation to the mercury levels, at high doses of mercury. The threshold[a]

---

[a] The phrase "threshold body burden" is meant to indicate the value of the body burden at which effects due to methylmercury become detectable above the background frequency. It is not intended to mean that methylmercury does not produce effects in some individuals below this level.

body burden for this mercury-related increase in frequency was estimat
for each sign of symptom. Paraesthesia had the lowest threshold bo
burden. From the dose–response curve, the onset of symptoms w
estimated to occur at approximately 25 mg of mercury or 0.5 mg/kg bo
weight.

Bakir et al. (1973) noted that the empirical relationship between blo
level and ingested dose did not conform with the relationship expect
from Miettinen's tracer studies on volunteers. Bakir et al. (1973) point
out that this difference may be due "either to differences in conditions
exposure between the Iraqi patients and the volunteers given labell
methylmercury, or to underestimations of dose in Iraq, or to both causes
They noted that, in 14 patients, the average exposure period report
by the patients was 48 days as compared with 66 days calculated from ha
analysis of the same patients. Thus, an alternative dose–response r
lationship was plotted by increasing the estimates of ingested dose by
factor of about 1.6, which made their empirically observed bloo
ingested dose relationship identical to that calculated from Miettine
This yielded a threshold body burden for paraesthesia of 40 mg of mercu
or 0.8 mg/kg body weight.

An alternative calculation may be made directly from the blood leve
to try to estimate the minimum concentration of mercury in blood
which paraesthesia became detectable. If the mean values (estimated
geometric means) of the blood mercury for each cohort, as reported b
Bakir et al., are plotted on a logarithmic scale against the frequency
paraesthesia, the relationship has basically the same pattern as observe
when body burden was used (Fig. 4). The horizontal portion of the lir
relates to the background frequency at a mean blood mercury level
between 1.0 and 22 µg/100 ml. The next points lie significantly above th
background frequency level. A least-squares linear regression line throug
these points intercepts the background level at a mercury level of approx
mately 24 µg/100 ml. However, the estimated threshold value is for mercur
concentrations in the blood 65 days after cessation of exposure. If correc
tion is made using a 65 day clearance half-time, the actual threshold leve
could be twice as high, i.e. 48 µg/100 ml. Evidence noted above suggeste
that blood mercury may have been cleared at longer half-times than 6
days. Thus the actual threshold mercury value for blood probably lie
between 24 and 48 µg/100 ml.

Mufti et al. (1976) and Kazantzis et al. (1976) have reported the result
of a survey of 956 persons in a heavily affected village in Iraq, on a
additional 207 persons living nearby, and on 1014 persons in a contro
village that did not receive the treated grain. Mufti et al. (1976) reporte
on 427 persons who had eaten contaminated bread. They were divide

to groups according to the total consumption of contaminated bread loaves per day × period of consumption) and the frequency of parasthesia as reported for each group. The total quantity of mercury consumed can be estimated from the mercury content of each load.

Figure 5 is a plot of the frequency of paraesthesia in this population against the log of the mean total ingested dose for each group plotted on a logarithmic scale. A horizontal line is drawn through the point corresponding to 4% paraesthesia assuming this to be the background frequency in this population. The points at higher ingested doses lie significantly above this line. A linear regression line drawn through these points interrupts the background frequency at a total infested dose of 37 mg of mercury. Assuming 50 kg to be the average body weight for this population (Mufti et al., 1976), this threshold dose would be 0.7 mg/kg. However, during the period of consumption (average 32 days), some excretion of methylmercury took place, so that the maximum body burden must have been less than the total ingested. The difference would be small in view of the short period of consumption.

Thus the study of this large population, carried out approximately 6 months after the study by Bakir et al. (1973), would also be compatible with a body burden of mercury in the range of 0.5–0.8 mg/kg body weight.

Shaharistani et al. (1976) have reported on 184 persons in rural Iraq, 43 of whom consumed the contaminated bread.[a] The signs and symptoms were classified as mild, moderate, and severe. People classified as mild cases complained of numbness of the extremities and had slight tremors and mild ataxia. Moderate cases had difficulty of hearing, tunnel vision, and partial paralysis. The severe cases generally suffered from a combination of the following; complete paralysis, loss of vision, loss of hearing, loss of speech, and coma. The dose was expressed as the peak hair concentration of mercury determined by neutron activation of consecutive 1-cm segments of the hair sample, and graphical determination of the peak concentration as described by Giovanoli & Berg (1974).

Shahristani et al. (1976) did not formulate the usual dose–response relationships. Instead the population was classified according to the signs and symptoms of poisoning into four groups (no symptoms, mild, moderate, and severe symptoms). Figure 6 is redrawn from a similar figure presented by Shahristani et al. to indicate the range of peak mercury concentrations in hair for each group. The group having no signs and symptoms attributable to mercury poisoning had hair values in the range 0–300 mg/kg, the mild group in the range of 120–600 mg/kg, the moderate

---

[a] The clinical observations were made by a local physician in the rural district and by a resident physician of the hospital where the hair samples were collected (Shahristani, personal communication).

in the range 200–600 mg/kg and the severe in the range 400–1600 mg/k
Unfortunately, insufficient information is given in the paper to allo
formulation of the usual dose–response relationship in which the popul
tion is classified according to dose, and so these results cannot be cor
pared quantitatively with the results of Bakir et al. (1973) and Mufti et a
(1976). However, they do indicate that mild cases have been reporte
with peak hair levels as low as 120 mg/kg.

Shahristani et al. gave sufficient information on the 30 cases in th
group they studied to allow estimation of their daily intake of methy
mercury from contaminated bread. It was possible to compare th
mercury concentrations in hair, as they increased during the ingestic
period, with the daily intake of mercury. It was found that the concentra
tion of mercury in hair was related to daily intake by an exponenti
equation similar to that described in section 6 relating body burden t
daily intake. Thus they were able to calculate the ratio of mercury cor
centration in hair (mg/kg) to average concentration of mercury in th
body (mg/kg). The average ratio was found to be 137 with a range c
82–268 in the 30 individuals. Using this average ratio, the hair level c
120 mg/kg at which mild symptoms were first observed would be equiva
lent to a body burden of mercury of 0.8 mg/kg body weight.

Observations on the Niigata outbreak of methylmercury poisonin
included figures on concentrations of mercury in samples of blood an
hair as well as detailed clinical reports. A Swedish Expert Group (1971
estimated the blood levels in patients at the time of onset of symptom
This involved a graphical procedure in which the concentration of mercur
in samples of blood, collected from the patient at various times afte
admission to hospital, were plotted against the time of collection o
semilogarithmic paper (Fig. 7). The decline in blood levels of mercur
corresponded to a clearance half-time of 70 days, although there wa
considerable scatter of the data. Back extrapolation of time of onset c
symptoms revealed that the lowest group (about 3 patients) had bloo
mercury levels in the range of 20–40 µg/100 ml.

Data on hair concentrations in the Niigata outbreak were analyse
and discussed by the Swedish Expert Group (1971). One hair sample
collected close to the time of onset of symptoms, contained mercury a
52 mg/kg. Unfortunately, no corresponding blood sample was available
Accurate comparison of hair to blood concentrations in the Niigat
samples was not possible because the hair specimens were not representa
tive of the current blood levels. Tsubaki (1971) has presented data relatin
fish consumption to hair values in the Niigata outbreak which als
identified cases of poisoning (also published by the Swedish Expert Group
1971). The hair concentrations cannot be related to the signs and symp

ms because they were not extrapolated back to the time of onset of
mptoms.

Several fish-eating populations (category 3) have been examined for
gns and symptoms of poisoning, and determinations made of concentra-
on of mercury in samples of blood and hair. A Swedish population of
sh eaters had blood levels of mercury up to 56 µg/100 ml. Skerfving
972) reported a dose–response curve calculated from data for this
opulation, which had no cases of mercury poisoning, and for the Niigata
ses reviewed above. The frequency of signs and symptoms was related
concentrations of mercury in the hair. The apparent threshold effect
orresponded to 50–90 mg/kg in the hair and thus to a blood mercury level
f 20–36 µg/100 ml.

The studies by Turner et al. (1974) and Marsh et al. (1974) on popula-
ons of fish eaters in Peru and Samoa are consistent with the relationship
ublished by Skerfving (1972). Thus Turner et al. noted that 8 people
ad mercury blood levels between 20 and 30 µg/100 ml and Marsh et al.
lso found 2 people having blood levels in the same range.

Paccagnella et al. (1974) could not find any connexion between the
revalence of neurological defects and concentrations of mercury in
amples of blood and hair in the high fish consumers on S. Pietro Island.
No neurological deficits were reported in the three individuals having
lood mercury levels between 20 and 24 µg/100 ml. However the dose–
esponse data reviewed above indicate that the risk of poisoning at these
evels is small.

Table 4 records the results obtained in studies on the populations
iscussed above. The studies on the Niigata population and the report of
hahristani et al. on the Iraqi population identify cases of methylmercury
oisoning. The quoted levels in hair and blood are those seen in the most
ensitive individuals in that population. The results (Table 4) indicate
at such "sensitive" individuals may exhibit symptoms of methylmercury

able 4. Summary of concentrations of mercury in samples of blood and hair and the body burden
f mercury associated with effects (usually paraesthesia) in the most sensitive group in the
opulation[a]

| Population | Total No. studied | Mercury concentration | | Mercury in the body (mg/kg) | References |
| | | Blood (µg/100 ml) | Hair (mg/kg) | | |
| --- | --- | --- | --- | --- | --- |
| Niigata | 17 | 20–40 | 52 | — | Swedish Expert Group (1971) |
| Iraq | 184 | — | 120 | 0.8 | Shahristani et al. (1976) |
| Iraq | 125 | 24–48 | — | 0.5–0.8 | Bakir et al. (1973) |
| Iraq | 427 | — | — | 0.7 | Mufti et al. (1976) |

[a] The numbers quoted in this table should not be considered independently of the accom-
anying test.

poisoning at blood mercury levels in the range of 20–40 μg/100 ml and hair levels of 50–60 mg/kg. Unfortunately these studies do not indica the percentage of the general population that is sensitive. Studies on hi. fish consumers in Sweden, Peru, Samoa, and Italy, suggest a low prob bility of symptoms at mercury levels of 20–40 μg/100 ml blood. At lea 15 persons having blood levels in this range did not exhibit symptoms poisoning.

The studies by Bakir et al. (1973) and Mufti et al. (1976) on the Ira population followed a different approach. Individual cases of methy mercury poisoning were not reported. Instead, the population w divided into groups according to observed levels of mercury in the blo and according to estimated dose, and the frequency of specific signs ar symptoms was reported for each group. Using this approach, it is nece sary to distinguish between the background frequency of signs ar symptoms and the increase of frequency due to methylmercury. Th figures quoted in Table 4 represent a graphical estimate (see Fig. 3, 4, ar 5) of the blood level, a body burden or dose of methylmercury where th frequency of paraesthesia emerges above the background level. Variatic in observed background frequencies ranged from 4% (Fig. 5) to 9.5 (Fig. 3), thus setting a practical limit to the accuracy of such graphic estimates. Thus the numbers quoted in Table 4 are compatible wit frequencies of paraesthesia due to methylmercury of 5% or less.

The data in Table 4 apply only to neurological signs and symptoms i adults. They do not apply to infants exposed either prenatally or in th early postnatal period.

Skerfving et al. (1974) have reported on the cytogenetic effects c methylmercury in 23 people exposed through intakes of various amoun of fish containing methylmercury (0.5–7 mg/kg) and in 16 people with low or moderate intake of mainly oceanic fish. The mercury levels in bloo in the "exposed" subjects ranged from 1.4 to 11.6 μg/100 ml and in th "non-exposed" from 0.3 to 1.8 μg/100 ml. A statistical relationship wa found between frequency of chromosome breaks and blood mercur levels. In a study carried out in Iraq (Firman, unpublished report), n statistically significant difference was noted in an exposed group con pared with a control population with regard to chromosome damage.

## 8.1.3 Children and infants with *in utero* exposure

Of the cases of poisoning reported from Minamata (Harada, 1968), 2 were due to prenatal exposure to methylmercury. These infants had sever cerebral involvement (palsy and retardation) whereas their mothers ha mild or no manifestations of poisoning. However, there is a possibilit at slight symptoms in the mothers might have been overlooked (Harada, 71). It should also be noted that the infants were examined some years ter birth and that no mercury levels are available either for the mothers for the affected infants. A case of prenatal methylmercury poisoning s been reported for a family that consumed meat from pigs that ate ain treated with methylmercury fungicide (Pierce et al., 1972). The other was exposed in early pregnancy and had a hair mercury level of 6 mg/kg. It was stated that the mother had no symptoms other than a ight slur of speech which occurred during two weeks in early pregnancy. he child, which was delivered normally after an uneventful pregnancy, hibited tremulous movement of the extremities in the first few days of e and subsequently developed myclonic convulsions (Snyder, 1971). t one year of age the infant exhibited normal physical growth but could ot sit up and was blind.

Cases of prenatal poisoning have been referred to in a preliminary port on the Iraqi epidemic where it was noted that blood levels in the fants at birth and in the first few months after birth could be consider-bly higher than those of the mother (Bakir et al., 1973). Quantitative ata on the prenatal exposure of the infants is not yet available. These bservations, however, have led to the belief that prenatal life in man is ore sensitive than adult life but the difference in the degree of sensitivity as not yet been quantitatively established.

Amin-Zaki et al. (1974a) have reported clinical examinations of 15 fant–mother pairs in which the infant was exposed prenatally to methyl-ercury. Clinical manifestations were evident in 6 of the mothers and in t least 6 of the infants. Five of the infants were severely affected having ross impairment of motor and mental development. However, in only infant–mother pair was the infant affected and the mother free of signs nd symptoms.

These observations from the 1971–72 Iraq outbreak must be regarded s preliminary. It may take time for some of the consequences of prenatal xposure to manifest themselves. Harada (1971) scrutinized the chromo-omes of 7 victims with congenital Minamata disease and 1 infant victim vith severe noncongenital Minamata disease, and reported that the hromosome patterns were within normal range.

## 8.2 Clinical Studies of Effects of Mercury-binding Compounds

Attempts to treat mercury poisoning have generally involved the use of antidotes that reduce the amount of mercury in the target tissue, either by forming an inactive complex with mercury, or by enhancing its removal from the tissue. Such antidotes are of course used in conjunction with general supportive therapy. Ideally the antidote should have a sufficient high affinity for mercury so that nontoxic doses are able to remove mercury from tissue binding sites. The mercury chelate so formed should be less toxic than mercury and preferably should be rapidly excreted. The agent should be metabolically stable so that dosing should not be too frequent and preferably the agent should be given by mouth. These antidotes are most effective when given early after exposure to mercury. Clearly the removal of mercury is without much advantage if irreversible damage has already occurred.

The first effective antidote, 2,3-dimercaptopropanol (British Anti-lewisite—BAL), is a sulfur-containing compound (a dithiol molecule possessing a remarkably high affinity for divalent ionic mercury) and was developed on the basis that mercury and other heavy metals combine with sulfur groups in the body (for detailed review, see Levine, 1970 This compound is life saving in cases of acute mercury(II) chloride poison-ing, alleviates symptoms from overdoses of mercury diuretics, and dramatically relieves certain symptoms of acrodynia. For alkylmercury poisoning BAL is contraindicated since it increases brain mercury levels (Berlin & Rylander, 1964; Magos, 1968). It also does not alleviate neuro-logical disorders caused by mercury vapour exposure (Hay et al., 1963; Glomme & Gustavson, 1959). Unithiol (2,3, dimercaptopropansulfonate is a water soluble derivative of BAL that is apparently more effective in mobilizing mercury (Trojanowska & Azendzikowski, in press; Dutkiewicz & Oginski, 1967). Furthermore unithiol does not produce redistribu-tion to the brain, as has been observed after BAL treatment. Unithiol effective in the treatment of occupational mercurialism (Fesenko, 1969 but there are no reports on its effects on alkylmercury poisoning.

The penicillamines (D-penicillamine and N-acetyl-DL-penicillamine are effective in increasing the excretion of mercury after exposure to mercury vapour and in relieving the symptoms of chronic mercury vapour poisoning (Smith & Miller, 1961; Parameshvera, 1967). Fatal brain damage can be prevented in the offspring of rats treated with methyl-mercury, by D-penicillamine according to the experiments of Matsumoto et al. (1967). Recent literature reports (Bakir et al., 1973; Suzuki & Yoshino, 1969) indicate that the penicillamines are capable of mobilizing mercury from tissues and increasing the excretion of mercury in cases of methylmercury poisoning in man. Thus it appears that the penicillamines offer advantages over BAL in that they are orally effective, less toxic, and effective in treating mercury vapour poisoning and probably alkylmercury poisoning, when administered immediately following exposure.

A slight increase in the urinary excretion of mercury has been noted in methylmercury poisoned patients with "Minamata disease" (Katsuna, 1968), after administration of EDTA.

Thioacetamide increases urinary excretion of mercury in animals dosed with mercury(II) chloride but probably one of the major causes of this effect is kidney damage caused by the combination of the toxic effects of thioacetamide and mercury leading to an increased exfoliation of renal tubular cells (Trojanowska et al., 1971).

Some interesting new ideas in the realm of antidotes to mercury are worth noting. Aaseth (1973) described the application of large molecular weight mercaptodextran in the successful treatment of mercury(II) chloride poisoning in animals. This agent does not enter the intracellular spaces and achieves removal of mercury from the body without redistribution. However, its effectiveness is limited by the time of administration. For example, if given more than 2 hours after exposure to mercury, this compound is totally ineffective whereas BAL is still useful.

A second approach is to give a nonabsorbable mercury-binding compound in the diet, in order to trap the mercury that is secreted in the bile, to prevent its reabsorption, and to greatly increase the faecal elimination (Takahashi & Hirayama, 1971; Clarkson et al., 1973a). A polystyrene resin containing fixed sulfhydryl groups has been shown to increase mercury excretion in experimental animals given methylmercury, and to reduce blood mercury levels in the victims of methylmercury poisoning in Iraq (Clarkson et al., 1973a; Bakir et al., 1973). There is, however, variation in response among patients. More recently it has been demonstrated that phenobarbital can increase the biliary excretion of methylmercury compounds (Magos et al., 1974).

A new technique for the removal of methylmercury directly from blood has been proposed for use in methylmercury poisoning (Kostyniak et al., 1975). The simultaneous combination of extracorporeal regional complexation of methylmercury with haemodialysis has been effective in producing a rapid removal of mercury in both experimental animals and man (Kostyniak et al., 1974; Al-Abbasi et al., 1974, unpublished reports).


## 8.3 Pathological Findings and Progression of Disease

The signs and symptoms of acute toxicity, severe gastrointestinal damage, shock, cardiovascular collapse, and acute renal failure, after large doses of divalent mercury and pulmonary irritation after inhalation of massive doses of the vapour, reflect the fact that all mercury compounds are chemically reactive, can denature proteins, inactivate enzymes, and disrupt cell membranes, leading to cellular death and the destruction of any tissue with which they come into contact in sufficient concentration. The pathological effects following long-term exposure to lower doses of mercury compounds are more subtle and depend upon the type of mercury compound to which the subjects are exposed. The remainder of this section will be concerned with long-term exposure, as this type of exposure is most important in helping to assess the hazards to man of the presence of mercury in food and air.

Pathological findings on human subjects exposed to mercury and its compounds have been reviewed by a Swedish Expert Group (1971) and by Friberg & Vostal (1972).

Pathological findings demonstrate that methyl- and ethylmercury compounds are primarily neurotoxic and produce similar types of lesion in man. The main pathological features consist of the destruction of neurological cells in the cortex particularly in the visual areas of the occipital cortex and various degrees of damage to the granular layer in the cerebellum. Damage to the peripheral nerves may occur as indicated by clinical signs but no definitive pathological observations are available for man. Takeuchi (1970) has reported on changes in the diameter of the peripheral nerves in patients suffering from heavy exposure to methylmercury in the Minamata Bay epidemic. However, Von Burg & Rustam (quoted by Bakir et al., 1973) could not find any changes in conduction velocities in the patients in Iraq who received very high exposure to methylmercury.

Brain concentrations of mercury associated with the onset of pathological changes following exposure to the vapour or to doses of the alkylmercury compounds are not fully known. Takahata et al. (1970) demonstrated mean mercury levels of 11 mg/kg wet weight in the brains of two workers, poisoned by exposure in a mercury mine, who had had no known exposure to mercury for 10 years prior to death. Studies on occupationally poisoned individuals (Swedish Expert Group, 1971), and the patients who died in the Minamata epidemic indicate that the onset point of signs and symptoms corresponds to an average brain level of approximately 5 mg/kg wet weight. These findings are in general agreement with threshold methylmercury concentrations in studies on experimentally poisoned animals.

### 8.3.1 Psychiatric and neurological disturbances

Trahtenberg (as reviewed by Friberg & Nordberg, 1972) reported minor psychiatric disturbances such as insomnia, shyness, nervousness, and dizziness in workers exposed to elemental mercury vapour concentrations of the order of 0.1 mg/m$^3$. The classical literature (for review, see Friberg & Nordberg, 1973) contains detailed accounts of the consequences of long exposure to higher concentrations of elemental mercury vapour where the full syndrome of erethism is seen. Individual variation in exposed people is the rule but the most commonly reported syndrome includes loss of memory, insomnia, lack of self-control, irritability and excitability, anxiety, loss of self-confidence, drowsiness, and depression. In the most severe cases delirium with hallucinations, suicidal melancholia, or even manic-depressive psychoses have been described.

The presence of tremor is one of the most characteristic features of mercurialism and usually follows the minor psychological disturbances referred to above. With continuing exposure to elemental mercury vapour, the tremor develops gradually in the form of fine trembling of the muscles interrupted by coarse shaking movements every few minutes. It may be seen in the fingers, but also on the closed eyelids, lips, and on the protruding tongue. The frequency is of the order of 5–8 cycles per second. It is intentional and stops during sleep. On cessation of exposure, the tremor gradually disappears. Dramatic alterations in the steadiness of the handwriting may be seen in persons suffering from mercurial tremor.

The most common signs and symptoms in cases of poisoning due to methyl- or ethylmercury compounds are paraesthesia, loss of sensation in the extremities and around the mouth, ataxia, constriction of the visual fields, and impairment of hearing. In the Japanese experience the effects of alkylmercury poisoning are usually irreversible but the coordination may improve after rehabilitation (Kitagawa, 1968). On the other hand, in Iraq, improvement in motor disturbances was often spontaneous. In addition, paraesthesia was often reversible in Iraq but a persistent symptom in Japan (Damluji, 1974; Tsubaki, 1971).

### 8.3.2  Eye and visual effects

Occupational exposure to elemental mercury vapour causes the appearance of a greyish-brown or yellow haze on the anterior surface of the lens of the eye (Atkinson, 1943). It appears usually after long-term exposure and the depth of colour depends upon the length of time and the air concentration of mercury to which the worker has been exposed. The presence of this coloured reflex may or may not be associated with signs and symptoms of poisoning.

The narrowing of the visual fields is a classic sign of poisoning due to short-chain alkylmercurial compounds (Hunter et al., 1940). In cases of severe exposure, the constriction may proceed to complete blindness.

### 8.3.3 Kidney damage

Kazantzis et al. (1962) reported four cases of proteinuria in two groups of workmen exposed to elemental mercury vapour. Exposure conditions were not specified, but all four cases were excreting mercury in urine in excess of 1000 µg/litre at the time of the first examination. The proteinuria disappeared after the workers were removed from exposure. Joselow & Goldwater (1967) noted that the mean urinary protein concentration (90 mg/litre) in a group of workers exposed to elemental mercury vapour was significantly higher than the mean protein concentrations (53 mg/litre) in a nonexposed group. The urinary protein correlated with urinary mercury levels.

Renal involvement after exposure to methylmercury compounds is very rare (Bakir et al., 1973). Cases of renal damage have been reported in an outbreak of ethylmercury poisoning (Jalili & Al-Abbasi, 1961).

### 8.3.4 Skin and mucous membrane changes

Dermatitis has been reported after occupational exposure to phenylmercurials (for review, see Goldwater, 1973). However dermatitis may result from exposure to inorganic mercury (Hunter, 1969). Sensitivity to metallic mercury in tooth fillings has resulted in facial and intra-oral rashes. Skin reactions due to sensitivity to phenylmercury have also been reported (for review, see Clarkson, 1972b).

Dermatitis has been reported after skin contact with the alkylmercurials (for review, see Swedish Expert Group, 1971). Oral ingestion of methyl- and ethylmercury compounds may also result in this condition, as observed in the Iraqi epidemics (Jalili & Al-Abbasi, 1961; Damluji et al., 1976).

# 9. EVALUATION OF HEALTH RISKS TO MAN FROM EXPOSURE TO MERCURY AND ITS COMPOUNDS

## 9.1 General Considerations

In order to control risks to human health from the presence of mercury in the environment, it is necessary to attempt to define the degree of risk in any given environmental situation. The health risk evaluation depends in part on a knowledge of dose–effect, dose–response relationships. It also depends on a knowledge of the variation in exposure (or intake) in any given situation. A general review of sampling and analytical techniques (section 2), and of environmental sources and exposure levels (sections 3 and 5) has been presented in this document. However, specific assessment of exposure or intake must be made by the public health authorities responsible for any given population or group. This section of the report is concerned, therefore, primarily with a summary of those dose–effect, dose–response relationships that are relevant to human populations in both occupational and general environmental exposure to mercury and its compounds.

Species differences in the metabolism and toxicity of mercury and its compounds are so great that this health risk evaluation is based primarily on data for man. Animal data have been included or considered only when data for man are lacking, and have generally been used in a qualitative way. Thus animal data may in certain cases indicate the potential for genetic effects, or that a certain stage of the life cycle, for example, the fetus appears to be the most sensitive stage. General patterns of deposition of mercury in the body, as seen from animal experiments, have been considered to be broadly applicable to man in a qualitative sense.

Quantitative estimates of the dose–response relationship in man are fraught with many difficulties. Observations of occupationally-exposed persons should offer the best possibilities for well controlled studies. However, the exposure range and population sizes are limited and difficulties are usually encountered in obtaining accurate estimates of time-weighted average exposure to airborne mercury. The most difficult situation for controlled study arises in the case of methylmercury where our knowledge derives primarily from observations on the unexpected outbreaks of poisoning from contaminated food in Japan and Iraq. Generally the studies commenced some time after the start of the outbreak or after the end of the exposure. Attempts had to be made either to recapitulate exposure by the back-extrapolation of blood levels, or to estimate the ingested dose from the patient's ability to remember. Doubts have also been expressed on the reliability of analytical methods in the earlier Japanese outbreaks.

Estimates will be made of minimum effect exposures or concentrations in indicator media since these numbers may be of value to authorities in setting safety standards. These minimum effect figures (e.g. Tables 5 and 6) should be viewed in terms of the overall dose–response relationship, namely, that as the dose is decreased, so also is the probability of poisoning, and that the minimum effect level is that dose (expressed as exposure level, daily intake, concentration in indicator media) that is associated with the first detectable effect in the population under study. The effect will be present at some specified frequency in the population. The value of the minimum effect dose is the dose derived from the observed dose–response relationship. (It will be subject to more statistical uncertainty than, for example, the dose giving 50% frequency of the effect.) Its value will depend on the size of the population under study; the larger the population, the more likely it will be that effects will become evident in the more sensitive individuals.

### 9.1.1  Elemental mercury vapour

The central nervous system is the critical organ for the toxic effects of inhaled elemental mercury. Effects on the kidney such as proteinuria have been reported but only at doses higher than those associated with the onset of signs and symptoms from the central nervous system. No information is available with regard to mercury levels in the central nervous system at the time of onset of signs and symptoms or at death in man. Furthermore, we do not have indicator media such as urine or blood the mercury levels of which would reflect those in the brain. Observations on animals exposed to elemental mercury vapour reveal large regional differences in distribution within the brain so that average brain concentrations, even if they were available for man, might not be of much value.

Our evaluation of the health risks from exposure to elemental mercury vapour in man has therefore to be based on an empirical relationship between air levels (exposure) and the frequency of signs and symptoms in exposed populations. Concentrations of mercury in urine and blood are related, on a group basis, to average air concentrations.

A complete metabolic model relating air levels to absorption, accumulation, and excretion of mercury in man following exposure to elemental mercury vapour is not available. However, some useful generalizations are beginning to emerge from recent observations. Most of the inhaled vapour (approximately 80%) is retained in the lung. Once absorbed into the blood stream, it is rapidly oxidized to ionic mercury. The limited information on biological half-times in man suggests that a workman exposed to a constant average concentration of mercury vapour in his working environment would not reach a state of balance (steady state) until after one year of exposure. Consequently one would expect the concentrations of mercury in blood or urine to exhibit a consistent relationship to air levels after the worker had been exposed for at least one year. Unfortunately, most of the publications in the literature do not indicate the period of employment of the worker. However, general experience in occupational health studies (for review, see MAC Committee, 1969) reveals that exposure of workers to an average air concentration of mercury of 0.05 mg/m³ are associated, on a group basis, with blood levels of approximately 3.5 µg/100 ml, and with urinary concentrations of 150 µg/litre. Linear relationship has been observed between urinary and blood concentrations of mercury in people occupationally exposed to elemental mercury vapour. Thus, measurements of mercury concentrations in the working atmosphere and in samples of blood and urine from the workmen may be used as an index of exposure. These values in turn may be compared with the frequency of clinical signs and symptoms in workers experiencing different degrees of exposure. Such studies, reported in detail in section 8 and briefly reviewed below, form the basis for establishing "threshold limit values" or maximal allowable air concentrations of mercury in occupational exposure.

Early studies dating back to the 1930s indicate that cases of poisoning occurred at atmospheric mercury levels above 0.1 mg/m³ (for details, see section 8.1.1). Recent data also demonstrate that there was an increase in complaints of appetite-loss and insomnia in a group of workers exposed to time-weighted average air concentrations of mercury between 0.06 and 0.1 mg/m³ as compared with two lower exposure groups (0.01 and 0.05 mg/m³). These findings are in agreement with data published in the 1940s and 1950s indicating that mercury intoxication occurred in workers exposed to mercury in air concentration less than 0.2 mg/m³, but no data were given on the lower exposure limits.

At this time it is difficult, if not impossible, to establish a lower limit at which no effects occur. Studies reviewed in section 8.1.1 indicate that mental disturbances may be seen at extremely low mercury concentrations in air. However, the problem in the interpretation of these reports is that, as the air concentration of mercury decreases, it becomes more difficult to correlate effects with exposure with any degree of confidence. For example, it would appear that effects of mercury levels below 0.05 mg/m³ have not been unequivocally established.

Concentrations of mercury in blood and urine equivalent to average mercury concentrations in air of 0.05 and 0.1 mg/m³ are given in Table 5. In general these relationships are not seen in indivduals but only when averaged over a substantial number of workers.

An occupational limit for mercury in air concentration of 0.05 mg/m³ would be equivalent to an ambient air level for the general population of approximately 0.015 mg/m³. This calculation is based on the assumptions of a daily ventilation of 10 m³ during working hours and 20 m³ for a 24-hour day and that there are 225 working days in the year. The data of

Smith et al. (1970) would indicate that the probability of seeing adverse effects at air levels of, 0.05 mg/m³ for occupational exposure, and 0.015 mg/m³ for continuous environmental exposure, is low for such symptoms as loss of appetite, weight loss, and shyness. However, this calculation does not take into account the sensitive groups in the general population. It should be noted that concentrations found in ambient air are far below these levels (see section 5).

Table 5. The time-weighted average air concentrations associated with the earliest effects in the most sensitive adults following long-term exposure to elemental mercury vapour. The table also lists the equivalent blood and urine concentrations[a]

| Air (mg/m³) | Blood (μg/100 ml) | Urine (μg/litre) | Earliest effects |
|---|---|---|---|
| 0.05 | 3.5 | 150 | non-specific symptoms |
| 0.1–0.2 | 30 | 600 | tremor |

[a] Blood and urine values may be used only on a group basis owing to gross individual variations. Furthermore, these average values reflect exposure only after exposure for a year or more. After shorter periods of exposure, air concentrations would be associated with lower concentrations in blood and urine.

### 9.1.2 Methylmercury compounds

The estimate of risks to human health from methylmercury compounds is important for several reasons. First, many thousands of people have been poisoned following accidental consumption of food contaminated with methylmercury fungicides or the consumption of fish contaminated by industrial release of methylmercury. Second, methylmercury probably accounts for a significant part of mercury in the human diet and is especially important in fish and fish products. Third, the risk–benefit calculations with regard to methylmercury in fish are of critical importance in those countries and areas of the world where fish is an important dietary source of protein, or where the fish industry is of economic importance.

This evaluation of risk from exposure to methylmercury compounds is based primarily upon data in man. The data consist of observations on the frequency of signs and symptoms in populations exposed to a wide range of mercury intake, observations on the concentrations of mercury in hair and blood samples, and on estimates of dietary intake (section 8).

In estimates of risks to human health, the custom has been followed of attempting to determine the lowest concentration in indicator media or the lowest daily intake associated with the onset of toxic signs and symptoms in man, along with information on maximal intakes which produce no effects. This procedure has been followed in estimating the data presented in Table 6. The levels in blood and hair, and the amount

# WORLD HEALTH ORGANIZATION

## ENVIRONMENTAL HEALTH CRITERIA 1
## MERCURY

### CORRIGENDUM

Page 116 — Table 5 should be corrected to read as follows :

Table 5. The time-weighted average air concentrations associated with the earliest effects in the most sensitive adults following long-term exposure to elemental mercury vapour. The table also lists the equivalent blood and urine concentrations [a]

| Air (μg/m³) | Blood (μg/100 ml) | Urine (μg/litre) | Earliest effects |
|---|---|---|---|
| 05 | 3.5 | 150 | non-specific symptoms |
| 1–0.2 | 7–14 | 300–600 | tremor |

[a] Blood and urine values may be used only on a group basis owing to gross individual variations. Furthermore, these average values reflect exposure only after exposure for a year or more. After shorter periods of exposure, air concentrations would be associated with lower concentrations in blood and urine.

the body burden of mercury associated with the onset of signs and symptoms are taken from Table 4.

The reports on the Minamata outbreak to the effect that infants were born having cerebral palsy due to methylmercury whereas their mothers lacked or had only slight symptoms led to the belief that the fetus was the stage of the life cycle that was most sensitive to methylmercury. Studies in animals, exposed during the gestation period to methylmercury led to qualitative confirmation of this conclusion.

Table 6 lists our conclusions on concentrations of total mercury in indicator media associated with the earliest effects of methylmercury in the most sensitive group in the adult population. As discussed below, the prevalence of the earliest effects would be expected to be approximately %. The equivalent long-term daily intake quoted in Table 6 was calculated on the most conservative relationship (quoted in Table 3) i.e. that the steady state blood concentration (ng/ml) is numerically equal to the average daily intake ($\mu$g/day/70 kg body weight).

However, Nordberg & Strangert (1976) have proposed an alternative approach to estimating risks of poisoning on long-term exposure. The relationship between long-term daily intake and steady state blood levels (and hair levels and body burden) depends, *inter alia*, on the biological half-time of methylmercury in man. Biological half-time times are subject to individual variation as discussed in section 6. Thus an individual having a long biological half-time (a slow excretor of methylmercury) would accumulate higher steady state levels than one having a short biological half-time. Thus the statistical distribution of biological half-times should be taken into account in estimates of risk of poisoning.

Variations also occur in threshold concentrations for the appearance of signs and symptoms in individuals in the population. These may be estimated from empirical relationships relating frequency of signs and symptoms to concentrations of methylmercury in indicator media (blood and hair) or body burdens (e.g. Fig. 3, 4, and 5). On the assumptions that the distribution of biological half-times was normal, that the distribution of individual threshold values for paraesthesia was log-normal, and that these distributions were independent of each other, Nordberg & Strangert (1974) estimated the overall probability of an individual developing symptoms of paraesthesia. Their calculations were based on data of Shahristani & Shihab (1974) and gave the distribution of biological half-times estimated from analysis of hair samples after the Iraqi outbreak, and the statistical distribution of individual threshold values estimated from the data of Bakir et al. (1973) on the relationship of frequency of paraesthesia to body burdens. Their results indicated that with a long-term daily intake of 4 $\mu$g/kg would yield the risk of paraesthesia of about

8%. Subsequent calculations based on the data by Mufti et al. (19?
would indicate a risk of between 3% and 4% for the same daily intake

The figures estimated by Nordberg & Strangert may somewhat ove
estimate the risk of poisoning. The errors to be expected in both fie
and laboratory observations would tend to decrease the slope in dos
response relationships. This would lead to an overestimate of the varian
of threshold values in the general population. Nevertheless, the estima
of risk by Nordberg & Strangert is in reasonable agreement with the mo
empirical estimates discussed in section 8 and indicates that, with a lon
term daily intake as listed in Table 6, the prevalence of the earliest effec
could be expected to be 5% or less.

Table 6. The concentrations of total mercury in indicator media and the equivalent long-term daily intake of mercury as methylmercury associated with the earliest effects in the most sensitive group in the adult population[a, b]

| Concentrations in indicator media | | |
|---|---|---|
| Blood ($\mu$g/100 ml) | Hair ($\mu$g/g) | Equivalent long-term daily intake[c] ($\mu$g/kg body weight) |
| 20–50 | 50–125 | 3–7 |

[a] The prevalence of the earliest effects could be expected to be approximately 5%.

[b] The WHO Task Group specifically urged that this table should not be considered independently of the text in section 8.

[c] A Japanese group has recently concluded that a daily intake of mercury of 5 $\mu$g/kg is the "minimal toxic dose", following a ten-year follow-up study of the Minamata outbreak (Research Committee on Minamata Disease, 1975).

Occupational hazards have arisen mainly from airborne concentra
tions of alkylmercurials. Skin contact has also been noted, but the quanti
tative importance of skin contact and percutaneous absorption cannot
be estimated. Hazards from occupational exposures can be estimate
only by reference to data already discussed above with respect to dieta
intake of methylmercury compounds. The data summarized in Table
indicate that the first effect (paraesthesia) associated with long-term inta
of methylmercury arises at an intake level of approximately 5 $\mu$g/kg bo
weight per day. Assuming a daily ventilation at work of 10 m$^3$ of air, 80
retention of the inhaled mercurial, 225 working days to the year, t
average time-weighted air concentrations that would give rise to th
intake would be 0.07 mg/m$^3$. Consideration of occupational risks shou
also take into account the possibility of a high but brief exposure 
methylmercury compounds.

The estimates in Table 6 apply only to adults. As discussed previousl

---

[a] Nordberg & Strangert, personal communication.

prenatal life may be the stage of the life-cycle most sensitive to methyl-mercury. It would be prudent therefore to follow the advice of the MAC Committee (MAC, 1969), not to expose females of child bearing age occupationally to methylmercury compounds.

Studies by Skerfving et al. (section 8) have indicated that chromosome breaks may be associated with exposure to methylmercury. There are, however, other studies in which no such relationship was found. Further-more the health significance of chromosome breakage is not known. However, as reviewed elsewhere in this criteria document, experiments on animals and other forms of life do indicate the potential for genetic damage by methylmercury.

### 9.1.3 Ethylmercury compounds and other short-chain alkylmercurials

Insufficient information is available to allow risk calculations based on data from human exposure to ethyl- or higher short-chain alkyl-mercurial compounds. Suzuki et al. (1973) have reported observations on five patients poisoned with ethylmercury compounds. The picture of distribution between plasma and red cells and observations on autopsy tissue indicate that the metabolism and patterns of distribution of ethyl-mercury are generally similar to those of methylmercury compounds. However, in one individual the clearance-time from blood was only 10 days. Evidence reviewed in section 6.6 indicates that ethylmercury com-pounds may be more rapidly converted to inorganic mercury in the body than methylmercury compounds. Thus, these limited observations suggest that ethylmercury compounds are probably less hazardous than the methylmercury compounds in so much as they remain in the body for a shorter time because of transformation to inorganic mercury and more rapid excretion. Thus, standards set for methylmercury compounds will probably be sufficient to control the hazards from other short-chain alkyl-mercurial compounds.

### 9.1.4 Inorganic mercury, aryl- and alkoxyalkylmercurials

The risk to human health from long-term ingestion of inorganic, aryl-, and alkoxyalkyl- compounds of mercury in the diet is difficult to estimate because there are not any recorded cases of human poisoning under these circumstances. Data from animals cannot be used for exact quantitative extrapolation because of species differences in the metabolism and toxicity of these compounds. However, certain qualitative conclusions based on animals can probably be extrapolated to man. For example, the aryl-mercurials are rapidly converted to inorganic mercury in mammals. The penetration of mercury across the blood–brain and placental barriers is less after doses of inorganic and aryl compounds than after equivalent doses of elemental mercury vapour and short-chain alkylmercurials. Animal studies indicate that the kidney is the critical target organ for exposure to inorganic, alkyl-, and alkoxy-alkylmercurials. Kidney involvement appears to be minimal in workers exposed to concentrations of mercury vapour ($0.05–0.1$ mg/m$^3$) that elicit the first signs and symptoms of damage to the central nervous system. Thus guidelines for health protection, set for long-term exposure to elemental mercury vapour, should offer an even greater safety margin for equivalent exposures to inorganic, alkyl, and alkoxyalkyl compounds. In fact, recognizing the lower toxic potential of these forms of mercury, the MAC Committee (1969) advised a maximum allowable concentration for occupational exposure of $0.1$ mg/m$^3$—twice as high as that for elemental mercury vapour.

In the discussion of guidelines for exposure to elemental mercury vapour, it was concluded that long-term exposure of the general population to $0.015$ mg/m$^3$ was equivalent, in terms of average daily mercury intake, to the occupational limit of $0.05$ mg/m$^3$. Assuming an average pulmonary retention of 80%, the average daily amount entering the blood stream is 280 µg based on a daily ventilation of 20 m$^3$.

The equivalent daily intake in diet of phenylmercury compounds would also be about 240 µg, as animal studies indicate virtually complete absorption in the gastrointestinal tract. The alkoxyalkyl and aryl compounds are probably absorbed equally well from food. Tracer studies on volunteers indicate that approximately 10% of inorganic mercury compounds are absorbed from the diet, so that dietary intakes approximately ten times greater than those of phenylmercury compounds would offer no greater risk of poisoning.

A daily intake of mercury of 240 µg is in the same range as the daily intake listed for methylmercury compounds in Table 6. The biological half-time for inorganic mercury, based on tracer studies in man, appears to be less than that for tracer doses of methylmercury. Animal data suggest that aryl and alkoxyaryl compounds have biological half-times lower than that for methylmercury and similar to that for inorganic mercury. Thus the long-term ingestion of inorganic, aryl, and alkoxyaryl compounds should offer no greater hazards and probably substantially less than the hazards from ingestion of methylmercury compounds.

## 9.2 Summary and Guidelines

In the case of elemental mercury and alkylmercury the Task Group was able to construct tables (Tables 5 and 6) that related exposure to symptoms as well as to concentrations in indicator media in the human body. In the case of inorganic mercury, arylmercurials, and alkoxyalkyl mercurials, this could not be done because of the inconsistency in the animal data available as well as a paucity of data in man.

The tables for elemental mercury and alkylmercury are given above. In constructing these tables, the Task Group evaluated the results of studies summarized in this document, and drawing on their experience and judgement, identified the concentration and amounts of mercury associated with certain observed effects. There is insufficient information available to permit precise quantification of this risk. Usually, the proportion that may be expected to be affected is small.

The expected health effects for elemental mercury at a level in air of 0.05 mg/m$^3$ have been quoted only for occupational exposure assuming 8 hours per day and 225 working days a year. The equivalent environmental mercury levels in air for continuous exposure would be approximately 0.015 mg/m$^3$ to give the same degree of risk. The urine and blood values, of course, would be the same as those quoted in Table 5. Even though the figures do not only take into account specific sensitive groups, it is highly unlikely that the concentration in the general environment approaches levels of toxicological significance.

The ranges of minimum effect values quoted in Table 6 for methylmercury reflect the uncertainty in estimations.

Although it was not possible to identify even approximate minimum effect values for inorganic, aryl-, and alkoxyalkylmercurials, the Task Group concluded that the limited experience for occupational exposure suggested that these forms of mercury were probably less hazardous than either elemental mercury vapour or methylmercury compounds. Thus the figures for occupational exposure to elemental mercury vapour given in Table 5 would serve as conservative figures for occupational exposure to these forms of mercury and those in Table 6 would offer conservative figures for dietary intake.

A Joint FAO/WHO Expert Committee on Food Additives (1972) established a provisional tolerable weekly intake of 0.3 mg of total mercury per person of which no more than 0.2 mg should be present as methylmercury (expressed as mercury); these amounts are equivalent to 5 µg and 3.3 µg, respectively, per kg of body weight. Where the total mercury intake in the diet is found to exceed 0.3 mg per week, the level of methylmercury compounds should also be investigated. If the excessive intake is attributable entirely to inorganic mercury, the above provisional limit for total mercury no longer applies and will need to be reassessed in the light of all prevailing circumstances.

---

AASETH, J. (1973) *Acta Pharm. Toxicol.*, **32**: 430.

ABERG, B. ET AL. (1969) *Arch. environ. Health*, **19**: 478.

ACGIH (1976) *Threshold limit values for chemical substances and physical agents in the workroom environment with intended changes for 1976.* Cincinnati, American Conference of governmental industrial hygienists, pp. 94.

ALBANUS, L. ET AL. (1972) *Environ. Res.*, **5**: 425.

AMIN-ZAKI, L. ET AL. (1974a) *Pediatrics*, **54**: 587.

AMIN-ZAKI, L. ET AL. (1974b) *J. Pediatr.*, **85**: 81.

AMIN-ZAKI, L. ET AL. (in press) *Am. J. Dis. Children.*

ANDERSSON, A. (1967) *Grundforbattring*, **20**: 95.

ARMSTRONG, R. D. ET AL. (1963) *J. Am. ind. Hyg. Assoc.*, **24**: 366.

ASHE, W. F. ET AL. (1953) *Arch. ind. Hyg. occup. Med.*, **7**: 19.

ATKINSON, W. (1943) *Am. J. Ophthalmol.*, **26**: 685.

BACHE, C. A. ET AL. (1971) *Science*, **172**: 951.

BAKIR, F. ET AL. (1973) *Science*, **181**: 230.

BAKULINA, A. V. (1968) *Sov. Med.*, **31**: 60.

BALDI, G. ET AL. (1953) *Med. Lav.*, **44**: 161.

BARBER, R. T. ET AL. (1972) *Science*, **178**, 636.

BERLIN, M. & RYLANDER, R. (1964) *J. Pharm. exp. Ther.*, **146**: 236.

BERLIN, M. ET AL. (1966) *Arch. environ. Health*, **12**: 33.

BERLIN, M. ET AL. (1969) *Arch. environ. Health*, **18**: 42.

BERLIN, M. ET AL. (1973) *The uptake and distribution of methyl mercury in the brain of Saimiri sciureus in relation to behavioral and morphological changes.* In: Miller, M. W. & Clarkson, T. W. ed. *Mercury, mercurials and mercaptans.* Springfield, C. C. Thomas, p. 187.

BERTINI, K. K. & GOLDBERG, F. D. (1971) *Norw. Hyg. Tidschr.*, Suppl. 4, p. 65.

BIDSTRUP, P. L. (1964) *Toxicity of mercury and its compounds*, Amsterdam, Elsevier Publishing Co., p. 34.

BIDSTRUP, P. L. ET AL. (1951) *Lancet*, **2**: 856.

BIRKE, G. ET AL. (1972) *Arch. environ. Health*, **25**: 77.

BORNMANN, G. ET AL. (1970) *Arch. Toxicol.*, **26**: 203.

BOUQUIAUX, J. (1974) In: *Proceedings of the International Symposium on the problems of contamination of man and his environment by mercury and cadmium, Luxembourg, 3–5 July, 1973*, CEC Luxembourg, p. 23.

BRAMEN, R. S. & JOHNSON, D. L. (1974) In: *XIth Annual NSF-RANN Trace Contaminants Conference, California, 29–31 August, 1974*, National Technical Information Service, US Department of Commerce, Springfield, Virginia, p. 75.

BRAR, S. S. ET AL. (1969) *Thermal neutron activation analysis of airborne particulate matter in Chicago metropolitan area*, Washington DC, National Bureau of Standards, p. 43.

BRUNE, D. (1969) *An. Chem. Acta.*, **44**: 15.

BURROWS, W. D. (1975) In: Krenkel, D. A., ed. *Heavy metals in the aquatic environment*, Oxford, Pergamon.

BURTON, J. D. & LEATHERLAND, T. M. (1971) *Nature (Lond.)*, **231**: 440.

BYRNE, A. R. & KOSTA, L. (1974) *Talanta*, **21**: 1083.

CARDOZO, R. L. (1972) *Chem. Weekbl.*, **68**: 9.

CAVANAGH, J. B. & CHEN, F. C. K. (1971) *Acta Neuropathol.*, **19**: 216.

CEC, Working Group of Experts (1974) *Non-organic Micropollutants of the Environment. Methods of Analysis. Report.* Luxembourg, Commission of the European Communities, VF/1966/74é.

CEMBER, H. (1962) *Am. ind. Hyg. Assoc. J.*, **23**: 304.

CHANG, L. W. & HARTMANN, H. A. (1972a) *Acta Neuropathol.*, **20**: 122.

CHANG, L. W. & HARTMANN, H. A. (1972b) *Acta Neuropathol.*, **20**: 316.

CHANG, L. W. & HARTMANN, H. A. (1972c) *Acta Neuropathol.*, **21**: 179.

CHANG, L. W. & YAMAGUCHI, S. (1974) *Environ. Res.*, **7**: 133.

CHANG, L. W. ET AL. (1974) *Acta Neuropathol.*, **27**: 171.

CHARBONNEAU, S. M. ET AL. (1974) *Toxicol. appl. Pharmacol.*, **27**: 569.

CHAU, Y. K. & SAITOH, H. (1970) *Environ. Sci. Technol.*, **4**: 839.

CHILDS, E. A. (1973) *Arch. Environ. Health*, **27**: 77.

CLARKSON, T. W. (1972a) *Recent advances in toxicology of mercury with emphasis on the alkyl mercurials.* In: Goldberg, L. ed. *Critical reviews in toxicology*, Vol. 1, Cleveland, The Chemical Rubber Co., issue 2, p. 203.

CLARKSON, T. W. (1972b) *Ann. Rev. Pharmacol.*, **12**: 375.

CLARKSON, T. W. ET AL. (1961) *Studies on the equilibration of mercury vapor with blood*, Rochester, University of Rochester Atomic Energy Project, U.R. 582, p. 64.

CLARKSON, T. W. & ROTHSTEIN, A. (1964) *Health Phys.*, **10**: 1115.

CLARKSON, T. W. & VOSTAL, J. (1973) *Mercurials, mercuric ion and sodium transport*, In: Lant, A. F. & Wilson, G. M., ed. *Modern diuretic therapy in the treatment of cardiovascular disease*, Amsterdam, Excerpta Medica, p. 229.

CLARKSON, T. W. ET AL. (1965) *Br. J. Pharmacol.*, **24**: 1.

CLARKSON, T. W. ET AL. (1972) *Biol. Neonat.*, *(Basel)*, **21**: 239.

CLARKSON, T. W. ET AL. (1973a) *Arch. environ. Health*, **26**: 173.

CLARKSON, T. W. ET AL. (1975) In: Krenkel, D. A., ed. *Heavy metals in the environment*, Oxford, Pergamon.

CLEGG, D. J. (1971) In: *Proceedings of the Symposium on Mercury in Man's Environment, Ottawa, 15–16 February 1971*, Ottawa, Royal Society of Canada, p. 141.

COPPLESTONE, J. F. & MCARTHUR, D. A. (1967) *Br. J. ind. Med.*, **24**: 77.

COTTON, F. A. & WILKINSON, L. (1972) *Advanced inorganic chemistry—a comprehensive text*, New York, Wiley Inter-Science Publications. Third edition.

DALGAARD-MIKKELSEN, S. (1969) *Norw. Hyg. Tidschr.*, **50**: 34.

DALL'AGLIO, M. (1968) *The abundance of mercury in 300 natural water samples from Tuscany and Altium.* In: Ahrens, L. H., ed. *Origin and distribution of the elements*, New York. Pergamon Press, p. 1065.

DAMLUJI, S. (1962) *J. Fac. Med. Baghdad*, **4**: 83.

DAMLUJI, S. ET AL. (1976) In: *World Health Organization Conference on Intoxication due to Alkyl mercury Treated Seed, Baghdad, 9–13 November 1974*, Geneva, World Health Organization, p. 11 (Suppl. to *Bull. World Health Organ.*, Vol. 53).

DANZINGER, S. J. & POSSICK, P. A. (1973) *J. occup. Med.*, **15**: 15.

DDR Standards (1968) TGL 22310, Berlin, German Democratic Republic.

DIEHL, J. F. & SCHELLENZ, R. (1974) In: *Proceedings of the International Symposium on the problems of contamination of man and his environment by mercury and cadmium, Luxembourg, 3–5 July 1973*, CEC, Luxembourg, p. 401.

D'ITRI, F. M. (1972) *The environmental mercury problem.* Cleveland, Chemical Rubber Co., p. 69.

DOHERTY, P. E. & DORSETT, R. S. (1971) *Anal. Chem.*, **43**: 1887.

DONOVAN, P. P. (1974) In: *Proceedings of the International Symposium on the problems of contamination of man and his environment by mercury and cadmium, Luxembourg, 3–5 July 1973*, CEC, Luxembourg, p. 573.

DURUM, W. H. ET AL. (1971) *Reconnaissance of minor elements in surface waters of the United States*, Washington, DC, p. 49 (US Geological Survey Circular 643).

DUTKIEWICZ, T. & OGINSKI, M. (1967) *Int. Arch. Gewebepath. Gewebehygiene*, **23**: 197.

EINARSSON, K. ET AL. (1974) *Acta med. scand.*, **195**, 527.

EL-BERGEAMI, M. M. ET AL. (1973) *Fed. Proc.*, **32**: 886.

ENGLESON, G. & HERNER, T. (1952) *Acta pediatr. scand.*, **41**: 289.

ERIKSSON, E. (1967) *Oikos*, Suppl. 9, p. 13.

FAGERSTROM, R. & JERNELOV, A. (1971) *Water Res.*, **5**: 121.

FERNANDEZ, M. DE N. ET AL. (1966) *Rev. de Sanid Hyg. Publica, Madrid*, **40**: 1.

FESENKO, S. (1969) *Vrac. Delo,* **19**: 85.
FILBY, R. H. ET AL. (1970) *Mikrochim. Acta.,* **6**: 1130.
FISHMAN, M. J. (1970) *Anal. Chem.,* **42**: 1462.
FITZHUGH, O. G. ET AL. (1950) *Arch. ind. Hyg. occup. Med.,* **2**: 433.
FLEWELLING, F. J. (1975) In: Krenkel, D. A., ed. *Heavy metals in the aquatic environment,* Oxford, Pergamon.
FONDS, A. W. (1971) *TNO-Nieuws.,* **26**: 375.
FORRESTER, C. R. ET AL. (1972) *J. Fish. Res. Board Can.,* **29**: 1487.
FOWLER, B. A. (1972a) *Science,* **175**: 780.
FOWLER, B. A. (1972b) *Am. J. Pathol.,* **69**: 163.
FRIBERG, L. (1951) *Norw. Hyg. Tidschr.,* **32**: 240.
FRIBERG, L. (1956) *Acta pharmacol. toxicol.,* **12**: 411.
FRIBERG, L. & VOSTAL, J. ed. (1972) *Mercury in the environment—a toxicologicàl and epidemiological appraisal,* Cleveland, The Chemical Rubber Co.
FRIBERG, L. & NORDBERG, G. F. (1972) *Inorganic mercury—relation between exposure and effects.* In: Friberg, L. & Vostal, J. ed. *Mercury in the environment—a toxicological and epidemiological appraisal,* Cleveland, Chemical Rubber Co., p. 7.
FRIBERG, L. & NORDBERG, G. F. (1973) *Inorganic mercury—a toxicological and epidemiological appraisal.* In: Miller, M. W. & Clarkson, T. W., ed. *Mercury, mercurials and mercaptans,* Springfield, C. C. Thomas, p. 5.
FRIBERG, L. ET AL. (1961) *Acta derm. venereol.,* **41**: 4952.
FUJIMURA, Y. (1964) *Jpn. J. Hyg.,* **18**: 10.
FUJITA, M. ET AL. (1968) *Anal. Chem.,* **40**: 2042.
FURUKAWA, K. ET AL. (1969) *Agric. biol. Chem.,* **33**: 128.
GAFAFER, W. M. ed. (1966) *Occupational diseases—a guide to their recognition,* US Department of Health, Education and Welfare, Public Health Service, pp. 175–176 (Public Health Service Publication 1097).
GAGE, J. C. (1961) *Br. J. ind. Med.,* **18**: 287.
GAGE, J. C. (1964) *Br. J. ind. Med.,* **21**: 197.
GAGE, J. C. (1974) *The metabolism of methoxyethyl mercury and phenyl mercury in the rat.* In: Miller, M. W. & Clarkson, T. W., ed. *Mercury, mercurials and mercaptans,* Springfield, C. C. Thomas, p. 346.
GANTHER, H. E. & SUNDE, M. L. (1974) *J. Food Sci.,* **39**: 1.
GANTHER, H. E. ET AL. (1972) *Science,* **175**: 1122.
GAVIS, J. & FERGUSON, J. F. (1973) *Water Res.,* **6**: 989.
GIOVANOLI, T. & BERG, G. G. (1974) *Arch environ Health,* **28**: 139.
GIOVANOLI, T. ET AL. (1974) *Clin. Chem.,* **30** (2): 222.
GLOMME, J. & GUSTAVSON, K. H. (1959) *Acta med. scand.,* **164**: 175.
GOLDWATER, L. J. (1964) *J. R. Inst. publ. health Hyg.,* **27**: 270.
GOLDWATER, L. J. (1973) *Aryl and alkoxyalloyl mercurials.* In: Miller, M. W. & Clarkson, T. W., ed. *Mercury, mercurials and mercaptans,* Springfield, C. C. Thomas, p. 56.
GOLDWATER, L. J. ET AL. (1963) *Arch. environ. Health,* **7**: 586.
GOULD, E. S. (1962) *Inorganic reactions and structures,* New York, Holt, Rinehart and Winston.
GRANT, C. A. (1973) *Pathology of experimental methyl mercury intoxication.* In: Miller, M. W. & Clarkson, T. W., ed. *Mercury, mercurials and mercaptans,* Springfield, C. C. Thomas, p. 294.
GREENWOOD, M. R. & CLARKSON, T. W. (1970) *Am. ind. Hyg. Assoc.,* **31**: 250.
GUTENMANN, W. H. & LISK, D. J. (1960) *J. Agric. Food Chem.,* **8**: 306.
HAQ, I. U. (1963) *Br. med. J.,* **5335**: 1579.
HARADA, Y. (1968) *Congenital (or fetal) Minamata Disease.* In: Katsuna, M., ed. *Minamata Disease,* Kumamoto University, p. 93.
HARADA, Y. ET AL. (1971) *Clinical Study (Rinsho to Kenkyu),* **48**: 1431.
HASANEN, E. (1974) In: *Proceedings of the International Symposium on the problems of contamination of man and his environment by mercury and cadmium, Luxembourg, 3–5 July 1973,* CEC, Luxembourg, p. 109.

124

HATCH, W. R. & OTT, W. L. (1968) *Anal. Chem.*, **40**: 2085.
HAY, W. J. ET AL. (1963) *J. Neurol. Neurosurg. Psychiatr.*, **26**: 199.
HAYES, A. L. & ROTHSTEIN, A. (1962) *J. Pharmacol. exp. Ther.*, **138**: 1.
HEINDRYCKX, R. ET AL. (1974) In: *Proceedings of the International Symposium on the problems of contamination of man and his environment by mercury and cadmium, Luxembourg, 3–5 July 1973*, CEC, Luxembourg, p. 135.
HINKLE, M. E. & LEARNED, R. E. (1969) *U.S. Geol. Survey Prof. Paper*, No. 650, p. 1251.
HOLDEN, A. B. (1972) *Present levels of mercury in man and his environment.* In: *Mercury contamination in man and his environment*, Vienna, IAEA, p. 143.
HOSOHARA, K. (1961) *J. chem. Soc. Jpn.*, **82**: 1107.
HUGHES, W. L. (1957) *Ann. N.Y. Acad. Sci.*, **65**: 454.
HUNTER, D. (1969) *Diseases of occupations*, London, Little, Brown and Co., p. 306.
HUNTER, D. ET AL. (1940) *Quart. J. Med.*, **9**: 193.
HUNTER, D. & RUSSELL, D. S. (1954) *J. Neurol. Neurosurg. Psychiat.*, **17**: 235.
HURSH, J. ET AL. (1976) *Arch. environ. Health* (in press).
INTERNATIONAL ATOMIC ENERGY AGENCY (1972) *Mercury contamination in man and his environment*, Vienna, International Atomic Energy Agency, Technical Report Series No. 137.
INTERNATIONAL LABOUR ORGANISATION (1970) *Permissible levels of toxic substances in the working environment*, Geneva, International Labour Organisation, Occupational Safety and Health Series 20.
JACOBS, M. B. ET AL. (1964) *Arch. environ. Health*, **9**: 454.
JAKUBOWSKI, M. ET AL. (1970) *Toxicol. appl. Pharmacol.*, **16**: 743.
JALILI, M. A. & ABBASI, A. H. (1961) *Br. J. ind. Med.*, **18**: 303.
JENSEN, S. & JERNELOV, A. (1967) *Biocidinformazion*, **10**: 4.
JENSEN, S. & JERNELOV, A. (1968) *Biocidinformazion*, **16**: 3.
JENSEN, S. & JERNELOV, A. (1969) *Nature (Lond.)*, **223**: 753.
JENSEN, S. & JERNELOV, A. (1972) *Behavior of mercury in the environment.* In: *Mercury contamination in man and his environment*, Vienna, International Atomic Energy Agency, Technical Report Series No. 137, p. 43.
JERNELÖV, A. (1968) *Fatten*, **24**: 456.
JERNELÖV, A. (1973) *A new biochemical pathway for the methylation of mercury and some ecological considerations.* In: Miller, M. W. & Clarkson, T. W., ed. *Mercury, mercurials and mercaptans*, Springfield, C. C. Thomas, p. 315.
JOENSUU, O. L. (1971) *Science*, **172**: 1027.
JOHNELS, A. G. (1971) *Observed levels and their dynamics in the environment, results from Sweden.* In: *Proceedings of the Symposium on Mercury in Man's Environment, Ottawa, 15–16 February 1971*, Ottawa, Royal Society of Canada, p. 66.
JOHNELS, A. G. ET AL. (1967) *Oikos*, **18**: 323.
JOHNSON, W. C. & VICKARY, C. (1970) *Analyst.*, **95**: 357.
JOINT FAO/WHO EXPERT COMMITTEE ON FOOD ADDITIVES (1972) *Evaluation of mercury, lead, cadmium and the food additives amaranth, diethylpyrocarbamate and octyl gallate*, WHO Food Additives Series, No. 4, FAO Nutrition Meeting Report Series No. 51a.
JOINT FAO/WHO MEETING (1974) *The use of mercury and alternative compounds as seed dressings*, WHO Technical Report Series, No. 555, 29 pp.
JONASSON, I. R. & BOYLE, R. W. (1971) In: *Proceedings of the Symposium on Mercury in Man's Environment, Ottawa 15–16 February 1971*, Ottawa, Royal Society of Canada, p. 5.
JONSSON, E. ET AL. (1972) *Var Foda*, **24**: 59.
JOSELOW, M. M. & GOLDWATER, L. J. (1967) *Arch. environ. Health*, **15**: 155.
JULIUSBERG, F. (1901) *Arch. Derm. Syph.*, **56**: 5.
KANAZAWA, J. & SATO, R. (1959) *Jpn. Analyst.* **8**: 322.
KATSUNA, M. ed. (1968) *Minamata disease*, Kumamoto University, Japan.
KAZANTZIS, G. ET AL. (1962) *Quart. J. Med.*, **31**: 403.
KAZANTZIS, G. ET AL. (1976) In: *World Health Organization Conference on Intoxication due to Alkyl mercury Treated Seed, Baghdad, 9–13 November 1974*, Geneva, World Health Organization, pp. 37, 49 (Suppl. to *Bull. World Health Organ.*, Vol. 53).

KESIC, B. & HAEUSLER, V. (1951) *Ind. Med. Surg.*, **20**: 485.

KHERA, K. S. (1973) *Toxicol. appl. Pharmacol.*, **24**: 167.

KITAGAWA, T. (1968) *Clinical investigations in Minamata Disease. Rehabilitation in Minamata Disease.* In: Katsuna, M., ed. *Minamata Disease*, Kumamoto University, p. 127.

KLEINERT, S. J. (1972) *Mercury levels in Wisconsin fish*, In: Hartung, R. & Dinman, E. D., ed. *Environmental mercury contamination*, Ann Arbor, Ann Arbor Science Publishers, p. 58.

KLEIN, R. H. & HERMAN, S. P. (1971) *Science*, **172**: 872.

KLEIN, R. ET AL. (1972) *Arch. Pathol.*, **93**: 408.

KLEIN, R. ET AL. (1973) *Arch. Pathol.*, **96**: 83.

KLOKE, A. (1974) In: *Proceedings of the International Symposium on the problems of contamination of man and his environment by mercury and cadmium, Luxembourg, 3–5 July 1973*, CEC Luxembourg, p. 83.

KOEMAN, J. H. ET AL. (1972) *TNO Nieuws*, **27**: 570.

KOEMAN, J. H. ET AL. (1973) *Nature Lond.*, **245**: 385.

KOPP, J. F. ET AL. (1972) *J. Am. Water Works Assoc.*, p. 40.

KORONOWSKI, P. (1973) *Nebenwirkungen von Quecksilberverbindungen auf Mensch und Tier.* Mitt. d. Biolog. Bundesanstalf f. Land-und Forstwirtschaft, No. 153, Berlin.

KORRINGA, P. & HAGEL, P. (1974) In: *Proceedings of the International Symposium on the problems of contamination of man and his environment by mercury and cadmium, Luxembourg 3–5 July 1973*, CEC, Luxembourg, p. 279.

KOSTA, L. ET AL. (1975) In: *Recent advances in the assessment of the health effects of pollution, WHO/EPA/CEC Symposium, Paris 8–11 June 1974*, Paris, p. 245.

KOSTYNIAK, P. J. ET AL. (1975) *J. Pharmacol. exp. Ther.*, **192**: 260.

KRENKEL, D. A., ed. (1975) *Heavy metals in the aquatic environment*, Oxford, Pergamon, pp. 352.

KUDSK, F. N. (1964) *Scand. J. clin. Lav. Invest.*, **16**: 575.

KUDSK, F. N. (1965a) *Acta Pharmacol.*, **23**: 250.

KUDSK, F. N. (1965b) *Acta Pharmacol.*, **23**: 263.

KUDSK, F. N. (1973) *Biological oxidation of elemental mercury*, In: Miller, M. W. and Clarkson, T. W., ed. *Mercury, mercurials and mercaptans*, Springfield, C. C. Thomas, p. 355.

KURLAND, L. T. (1973) *An appraisal of the epidemiology and toxicology of alkyl mercury compounds.* In: Miller, M. W. & Clarkson, T. W., ed. *Mercury, mercurials and mercaptans*, Springfield, C. C. Thomas, p. 23.

LADD, A. C. ET AL. (1964) *Arch. environ. Health*, **9**: 43.

LADD, A. C. ET AL. (1966) *J. occup. Med.*, **8**: 127.

LAMM, C. G. & RUZICKA, J. (1972) *The determination of traces of mercury by spectrophotometry, atomic absorption, radioisotope dilution and other methods.* In: *Mercury contamination in man and his environment*. Vienna, International Atomic Energy Agency, Technical Report Series No. 137, p. 111.

LANDNER, L. (1971) *Nature (Lond.)*, **230**: 452.

LEATHERLAND, T. M. ET AL. (1971) *Nature (Lond.)*, 232: 112.

LEHOTZKY, K. (1972) *Int. Arch. Arbeitsmed.*, **30**: 193.

LEVINE, W. G. (1970) *Heavy metal antagonists.* In: Goodman, L. S. & Gilman, A., ed. *The pharmacological basis of therapeutics*, 4th ed., London, Macmillan, p. 944.

LIDMUS, V. & ULFVARSON, U. (1968) *Acta chem. scand.*, **22**: 2150.

LOFROTH, G. (1969) *Methyl mercury. A review of health hazards and side effects associated with the emission of mercury compounds into natural systems*, Stockholm, Swedish Natural Research Council, Ecol. Research Bull., No. 4.

LOVEJOY, H. B. ET AL. (1974) *J. occup. Med.*, **15**: 590.

LUCIER, G. ET AL. (1971) *Chem. biol. Interactions*, **4**: 265.

LUCIER, G. ET AL. (1972) *Pestic. Biochem. Physiol.*, **2**: 244.

LUCIER, G. ET AL. (1973) *Molecular Pharmacol.*, **9**, 237.

LUNDGREN, K. ET AL. (1967) *Scand. J. clin. Lab. Invest.*, **20**: 164.

MCCARTHY, J. H. (1968) *Mining Eng.*, **20**: 46.

McCarthy, J. H. (1970) *Mercury in the atmosphere.* In: *Mercury in the environment,* Washington, DC, US Government Printing Office, US Geological Survey, Professional Paper No. 713, p. 37.

McDuffie, B. R. (1973) *Discussion.* In: Miller, M. W. & Clarkson, T. W., ed. *Mercury, mercurials and mercaptans,* Springfield, C. C. Thomas, p. 50.

McGill, C. M. et al. (1964) *J. occup. Med.,* **6**: 335.

MAC Committee (1969) *Arch. environ. Health,* **19**: 891.

Magos, L. (1967) *Environ. Res.,* **1**: 323.

Magos, L. (1968) *Br. J. ind. Med.,* **25**: 152.

Magos, L. (1971) *Analyst.,* **96**: 847.

Magos, L. & Butler, W. H. (1972) *Food, Cosmet. Toxicol.,* **10**: 513.

Magos, L. & Clarkson, T. W. (1972) *J. Assoc. Offic. Anal. Chem.,* **56**: 966.

Magos, L. et al. (1964) *Br. J. ind. Med.,* **21**: 294.

Magos, L. et al. (1973) *Toxicol. appl. Pharmacol.,* **26**: 180.

Magos, L. et al. (1974) *Toxicol. appl. Pharmacol.,* **28**: 367.

Marsh, D. O. et al. (1974) In: *Proceedings of the 1st International Congress on Mercury, Barcelona 6–10 May 1974,* Barcelona, p. 235.

Matsumoto, H. et al. (1967) *Life Sci.,* **6**: 2321.

Miettinen, J. K. (1973) *Absorption and elimination of dietary mercury ($Hg^{++}$) and methyl mercury in man.* In: Miller, M. W. & Clarkson, T. W., ed. *Mercury, mercurials and mercaptans,* Springfield, C. C. Thomas, p. 233.

Miller, G. E. et al. (1972) *Science,* **175**: 1121.

Miller, M. W. & Clarkson, T. W. (1973) *Mercury, mercurials and mercaptans,* Springfield, C. C. Thomas.

Morrow, P. et al. (1964) *Health Phys.,* **10**: 543.

Mufti, A. W. et al. (1976) In: *World Health Organization Conference on Intoxication due to Alkylmercury Treated Seed, Baghdad 9–13 November 1974,* Geneva, World Health Organization p. 23 (Suppl. to *Bull. World Health Organ.,* Vol. 53).

Neal, P. A. et al. (1937) *Study of chronic mercurialism in the hatters, furcutting industry.* Washington, DC, USPHS, Public Health Bulletin 234.

Niigata Report (1967) *Report on the cases of mercury poisoning in Niigata,* Tokyo, Ministry of Health and Welfare.

NIOSH (1973) *Criteria for a recommended standard. Occupational exposure to inorganic mercury.* Washington, Public Health Service, US Department of Health, Education and Welfare.

Nordberg, G. F., ed. (1976) Effects and dose-response relationships of toxic metals. Proceedings from an international meeting organized by the Sub-committee on the Toxicology of Metals of the Permanent Commission and International Associations on Occupational Health, Tokyo, 18–23 November 1974. Amsterdam, Elsevier.

Nordberg, G. & Serenius, F. (1969) *Acta Pharmacol. Toxicol.,* **27**: 269.

Nordberg, G. & Skerfving, S. (1972) *Metabolism.* In: Friberg, L. & Vostal, J., ed. *Mercury in the environment,* Cleveland, Chemical Rubber Co., p. 29.

Nordberg, G. & Strangert, P. (1976) In: *Effects and dose–response relationships of toxic metals. A report from an International Meeting arranged by the Sub-committee on the Toxicity of Metals, Tokyo, 18–23 November 1974,* Tokyo, p. 273.

Nordberg, G. et al. (1971) *Proceedings of the International Congress of Occupational Health,* **16**: 234.

Norseth, T. & Clarkson, T. W. (1971) *Arch. environ. Health,* **22**: 568.

Okinaka, S. et al. (1964) *Neurology,* **14**: 69.

Okuno, I. et al. (1972) *J. Assoc. Offic. Anal. Chem.,* **55**: 96.

Ordonez, J. V. et al. (1966) *Bull. Offic. Sanit. Panamer.,* **60**: 510.

Ostlund, K. (1969a) *Norw. Hyg. Tidschr.,* **50**: 82.

Ostlund, K. (1969b) *Acta Pharmacol., Suppl.* **1**: 1.

Paccagnella, R. et al. (1974) In: *Proceedings of the International Symposium on the problems of contamination of man and his environment by mercury and cadmium, Luxembourg 3–5 July 1973,* CEC, Luxembourg, p. 463.

PARAMESHVERA, V. (1967) *Br. J. ind. Med.*, **24**: 73.

PARIZEK, J. & OSTADALOVA, J. (1967) *Experientia*, **23**: 142.

PARIZEK, J. ET AL. (1969) In: Barltrop, D. & Burland, W. L., ed. *Mineral metabolism in paediatrics*, Oxford and Edinburgh, Blackwell Scientific Publications, p. 117.

PARIZEK, J. ET AL. (1971) In: Mertz, W. & Cornatzer, W. E., ed. *Newer trace elements in nutrition*, NY, M. Dekker Inc., pp. 85–122.

PARIZEK, J. ET AL. (1974) In: Hoekstra, W. G., Ganther, H. E. & Mertz, W., ed. *Trace element metabolism in animals*, University Park Press, p. 119.

PASSOW, H. ET AL. (1961) *Pharmacol. Rev.* **13**: 185.

PETERS, R. A. (1963) *Biochemical lesions and lethal synthesis*, New York, Macmillan.

PETERSON, C. L. ET AL. (1973) *Fisheries Bull.*, **71**: 603.

PHILLIPS, R. & CEMBER, H. (1969) *J. occup. Med.*, **11**: 170.

PIERCE, P. E. ET AL. (1972) *J. Am. med. Assoc.*, **220**: 1439.

PILLAY, K. K. ET AL. (1971) *Anal. Chem.*, **43**: 1419.

PIOTROWSKI, J. & BOLANOWSKA, W. (1970) *Med. Pracy.*, **21**: 338.

PIOTROWSKI, J. ET AL. (1969) *Med. Pracy.*, **20**: 589.

PIOTROWSKI, J. ET AL. (1974a) *Arch. Toxicol.*, **32**: 351.

PIOTROWSKI, J. ET AL. (1974b) *Toxicol. appl. Pharmacol.*, **27**: 11.

PIOTROWSKI, J. ET AL. (1975) *Int. Arch. Arbeits. Med.* (in press).

PISCATOR, M. & LIND, B. (1972) *Arch. environ. Health*, **24**: 426.

POLEŠAJEV, N. (1936) *N. Gigena truda i technicka bezopastnos i*, No. 6, p. 86.

POTTER, S. D. & MATRONE, G. (1973) *Fed. Proc.*, **32**: 929.

PRICKETT, C. S. ET AL. (1950) *Proc. Soc. exp. biol. Med.*, **73**: 585.

RAEDER, M. G. & SNEKVIK, E. (1940) *Kgl. Norske Videnskab. Selskabs Forhandl.*, **13**: 169.

RAEDER, M. G. & SNEKVIK, E. (1948) *Kgl. Norske Videnskab. Selskabs Forhandl.*, **21**: 102.

RAHOLA, T. ET AL. (1973) *Ann. clin. Res.*, **5**: 214.

RAMEL, C. (1972) *Genetic effects.* In: Friberg, L. & Vostal, J., ed. *Mercury in the environment —a toxicological and epidemiological appraisal*, Cleveland, Chemical Rubber Co., p. 9.

RAUHUT, A. & WILD, L. (1973) *Metallurgy*, **27**: 933.

REICHERT, J. K. (1973) *Z. Gewasserschultz, Wasser, Abwasser*, **10**: 277.

RESEARCH COMMITTEE ON MINAMATA DISEASE (1975) *Pathological, clinical and epidemiological research about Minamata Disease, 10 years after*, Kumamoto, Kumamoto University.

RIVERS, J. B. ET AL. (1972) *Bull. environ. Contamin. Toxicol.*, **8**: 257.

ROCHOW, E. L. ET AL. (1957) *The chemistry of organometallic compounds*, New York, John Wiley & Sons.

ROTHSTEIN, A. (1973) *Mercaptans, the biological targets for mercurials.* In: Miller, M. W. & Clarkson, T. W., ed. *Mercury, mercurials and mercaptans*, Springfield, C. C. Thomas, p. 68.

ROTHSTEIN, A. & HAYES, A. (1960) *J. Pharmacol. exp. Ther.*, **130**: 166.

SAHAGAIN, B. M. ET AL. (1966) *J. Nutr.*, **90**, 259.

SAHAGAIN, B. M. ET AL. (1967) *J. Nutr.*, **93**: 291.

SAPOTA, A. ET AL. (1974) *Med. Pracy*, **25**: 129.

SCHAMBERG, J. ET AL. (1918) *J. Am. med. Assoc.*, **70**: 142.

SCHRAMEL, P. ET AL. (1973) *Intern. J. environ. Stud.*, **5**: 37.

SEIFERT, P. & NEUDERT, H. (1954) *Zbl. Arbeitsmed.*, **4**: 129.

SELYE, H. (1970) *Science*, **168**: 775.

SERA, K. ET AL. (1962) *Kumamoto med. J.*, **15**: 38.

SERGEEV, YE. A. (1967) *Methodology of mercuriometric investigations.* In: Kransnikov, V. I., ed. *Proceedings of the first all-union conference on geochemical methods of prospecting for ore deposits.* Moscow.

SHAHRISTANI, H. & SHIHAB, K. (1974) *Arch. environ. Health*, **28**: 342.

SHAHRISTANI, H. ET AL. (1976) In: *World Health Organization Conference on Intoxication due to Alkyl mercury Treated Seed, Baghdad 9–13 November 1974*, Geneva, World Health Organization, p. 105 (Suppl. to *Bull. World Health Organ.*, Vol. 53).

SHAIKH, Z. A. ET AL. (1973) *Sequestration of mercury by cadmium-induced metallothioneine.* In: Hemphill, D. D., ed. *Trace Substances in Environmental Health*, Columbia, University of Missouri, Vol. 7, p. 12.

SILLEN, L. G. (1963) *Svensk. Kem. Tidskr.*, **75**: 161.

SJOSTRAND, B. (1964) *Anal. Chem.*, **36**: 814.

SKERFVING, S. (1974a) *Conference on Environmental Effects of Mercury, Brussels, November 1973.* In: Publicaties om het Institut voor Weltenschappen om het haefmilieu, p. 103, Brussels, Vrije Universiteit.

SKERFVING, S. (1974b) *Toxicol.*, **2**: 3.

SKERFVING, S. ET AL. (1974) *Environ. Res.*, **7**: 83.

SKOG, E. & WAHLBERG, J. E. (1964) *J. Invest. Derm.*, **43**: 187.

SMART, N. A. (1968) *Residue Rev.*, **23**: 1.

SMART, N. A. ET AL. (1969) *Analyst.*, **94**: 143.

SMITH, A. D. M. & MILLER, J. W. (1961) *Lancet*, **1**: 640.

SMITH, A. R. & MOSKOWITCH, S. (1948) *Monthly Rev. N.Y. State Dept. Labor, Div. Ind. Hyg. Safety Stand.*, **27**: 45.

SMITH, A. R. ET AL. (1949) *Monthly Rev. N.Y. State Dept. Labor, Div. Ind. Hyg. Safety Stand.*, **28**: 17.

SMITH, R. G. ET AL. (1970) *Am. ind. Hyg. Assoc. J.*, **31**: 687.

SMITH, J. C. ET AL. (1971a) *Nature (Lond.)*, **232**: 393.

SMITH, J. C. ET AL. (1971b) *Fed. Proc.*, **30**: 221.

SNYDER, R. D. (1971) *New Eng. J. Med.*, **284**: 1014.

SOLOMON, J. & UTHE, J. F. (1971) In: Buhler, D. R., ed. *Mercury in the western environment*, Corvallis, Continuing Education Publications, p. 265.

SOMERS, E. (1971) In: *Proceedings of the Symposium on Mercury in Man's Environment, Ottawa 15–16 February 1971*, Ottawa, Royal Society of Canada, p. 99.

SOMJEN, G. G. ET AL. (1973a) *J. Pharmacol. exp. Ther.*, **186**: 579.

SOMJEN, G. G. ET AL. (1973b) *J. Pharmacol. exp. Ther.*, **187**: 602.

SPRAGUE, J. B. & CARSON, W. G. (1970) *Spot-checks of mercury residues in some fishes from the Canadian Atlantic coast*, Ottawa, Fisheries Research Board, Canada, Manuscript Rep. No. 1085, p. 16.

SPYKER, J. M. ET AL. (1972) *Science*, **177**: 621.

STOCK, A. & CUCUEL, F. (1934) *Naturwissenschaften*, **22**: 319.

SUMINO, K. (1968) *Kobe J. med. Sci.*, **14**: 115.

SUMINO, K. (1975) In: Krenkel, D. A., ed. *Heavy metals in the aquatic environment*, Oxford, Pergamon.

SUZUKI, T. & YOSHINO, K. (1969) *Ind. Med.*, **11**: 21.

SUZUKI, T. & TANAKA, A. (1971) *Ind. Med.*, **13**: 52.

SUZUKI, T. ET AL. (1979) *Ind. Health*, **8**: 39.

SUZUKI, T. ET AL. (1971) *Bull. environ. Contam. Toxicol.*, **5**: 502.

SUZUKI, T. ET AL. (1973) In: Miller, M. W. & Clarkson, T. W., ed. *Mercury, mercurials and mercaptans*, Springfield, C. C. Thomas, p. 209.

SWEDISH EXPERT GROUP (1971) *Norw. Hyg. Tidschr.*, Suppl. 4, p. 65.

SWENSSON, A. & ULFVARSON, U. (1968) *Acta Pharmacol. Toxicol.*, **26**: 273.

SWENSSON, A. & ULFVARSON, U. (1969) *Poult. Sci.*, **48**: 1567.

TAKAHATA, N. ET AL. (1970) *Polio Psychiat. Neurol. Jpn*, **24**: 59.

TAKEUCHI, T. (1970) In: *International Conference on Environmental Mercury Contamination, Ann Arbor 3–7 September, 1970*, Ann Arbor, p. 50.

TAKAHASHI, H. & HIRAYAMA, K. (1971) *Nature (Lond.)*, **232**: 201.

TASK GROUP ON LUNG DYNAMICS (1966) *Health Phys.*, **12**: 173.

TASK GROUP ON METAL ACCUMULATION (1973) *Environ. Physiol. Biochem.*, **3**: 65.

TATTON, J. O. (1972) *Identification of mercurial compounds.* In: *Mercury contamination in man and his environment*, Vienna, International Atomic Energy Agency, Technical Report Series 137, p. 131.

TATTON, J. O. & WAGSTAFFE, P. J. (1969) *J. Chromatog.*, **44**: 284.

TEISINGER, J. & FIEROVA-BERGEROVA, V. (1965) *Ind. Med. Surg.*, **34**: 580.
TEJNING, S. (1970) *Mercury contents in blood corpuscles and in blood plasma in non-fisheaters.* Department of Occupational Medicine, University Hospital, Lund, Report No. 700406.
TEJNING, S. & OHMAN, H. (1966) In: *Proceedings of the XVth International Congress of Occupational Health, Vienna 1966*, Vienna, p. 239.
THOMPSON, J. A. J. & McCOMAS, F. T. (1973) *Environ. Lett.*, **5**: 189.
TONOMURA, K. & KANZAKI, E. (1969) *Biochem. Biophys. Acta*, **184**: 227.
TONOMURA, K. ET AL. (1968a) *J. Ferment. Technol.*, **46**: 685.
TONOMURA, K. ET AL. (1968b) *J. Ferment. Technol.*, **46**; 506.
TORIBARA, T. Y. ET AL. (1970) *Talanta*, **17**: 1025.
TRAHTENBERG, I. M. (1969) *Zdorov'e Kiev*, p. 292 (in Russian, German translation).
TROJANOWSKA, B. (1966) *Med. Pracy.*, **17**: 535.
TROJANOWSKA, B. & AZENDZIKOWSKI, S. (in press) *Toxicol. appl. Pharmacol.*
TROJANOWSKA, B. ET AL. (1971) *Toxicol. appl. Pharmacol.*, **18**: 374.
TSUBAKI, T. (1971) In: *Special Symposium on Mercury in Man's Environment, Ottawa 15–16 February 1971*, Ottawa, p. 131.
TSUDA, M. ET AL. (1963) *Yokohama med. Bull.*, **14**: 287.
TURNER, M. D. ET AL. (1974) In: *Proceedings of the 1st International Congress on Mercury, Barcelona 5–10 May 1974*, Barcelona (in press).
UI, J. (1967) *Norw. Hyg. Tidschr.*, **50**, 139.
UKITA, T. ET AL. (1963) *J. Hyg. Chem.*, **9**: 138.
ULFVARSON, U. (1962) *Int. Arch. Gewerbepath.*, **19**: 412.
UTHE, J. P. ET AL. (1970) *J. Fish. Res. Board, Canada*, **27**: 805.
VOEGE, F. A. (1971) In: *Proceedings of the Symposium on Mercury in Man's Environment, Ottawa 15–16 February 1971*, Ottawa, Royal Society of Canada, p. 107.
VON BURG, R. ET AL. (1974) *J. Chromatogr.*, **97**: 65.
VOSTAL, J. (1972) *Transport and transformation of mercury in nature and possible routes of exposure.* In: Friberg, L. & Vostal, J. ed. *Mercury in the environment*, Cleveland, Chemical Rubber Co., p. 3.
VOSTAL, J. (1972) In: *Proceedings of the XVIIth International Congress on Occupational Health, Buenos Aires, October 1972*, Buenos Aires.
VOUK, V. B. ET AL. (1950) *Br. J. Med.*, **7**: 168.
WALLACE, R. A. ET AL. (1971) *Mercury in the environment, the human element*, Oak Ridge, Oak Ridge National Laboratory, ORNL NSF–EP–1.
WATANABE, S. (1971) In: *Proceedings of the XVIth International Congress on Occupational Health, Tokyo, 22–27 September 1969*, Tokyo, p. 553.
WAHLBERG, J. E. (1965) *Arch. environ. Health*, **11**: 201.
WEAVER, J. N. (1973) *Anal. Chem.*, **45**: 1950.
WEBB, J. L. (1966) *Enzyme and metabolic inhibitors*, New York, Academic Press, Vol. 2.
WEISS, H. V. ET AL. (1971) *Science*, **174**: 692.
WERSHAW, R. L. (1970) *Sources and behavior of mercury in surface waters.* In: US Geological Survey, *Mercury in the environment*, Washington, DC, US Government Printing Office, Professional paper No. 713.
WEST, I. & LIM, J. (1968) *J. occup. Med.*, **10**: 697.
WESTERMARK, T. & LJUNGGREN, K. (1972) *The determination of mercury and its compounds by destructive neutron activation analysis.* In: *Mercury contamination in man and his environment*, Vienna, IAEA, p. 99.
WESTÖÖ, G. (1966) *Acta chem. scand.*, **20**: 2131.
WESTÖÖ, G. (1967) *Acta chem. scand.*, **21**: 1790.
WESTÖÖ, G. (1968) *Acta chem. scand.*, **22**: 2277.
WESTÖÖ, G. (1973) *Heavy metals in the aquatic environment, Nashville 4–7 December 1973*, Nashville (in press).
WHITE, M. N. & LISK, D. J. (1970) *J. Assoc. Offic. Anal. Chem.*, **53**: 530.
WHO REGIONAL OFFICE FOR EUROPE (1973) *The hazards to health of persistent substances in water.* Copenhagen, p. 147.
WILLFORD, W. A. ET AL. (1973) *J. Assoc. Offic. Anal. Chem.*, **56**: 1006.

WILLISTON, S. H. (1968) *J. geophys. Res.*, **73**: 7051.

WILLISTON, S. H. (1971) *J. geophys. Res.*, **28**: 1285.

WINTER, D. ET AL. (1968) *Arch. environ. Health*, **16**: 626.

WINTER, J. A. & CLEMENTS, H. A. (1972) *Analyses of mercury in water—a preliminary study of methods.* Cincinnati, US Environmental Protection Agency, EPA-R4-72-303, p. 56.

WISNIEWSKA, J. M. ET AL. (1970) *Toxicol. appl. Pharmacol.*, **16**: 754.

WISNIEWSKA, J. M. ET AL. (1972) *Acta Biochim. Pol.*, **19**: 11.

WOLF, D. ET AL. (1974) *Quecksilbermessungen am Arbeitsplatz bei der Herstellung von Leuchtstoffrohren.* STF report 1–74 Staubforschungsinstitut der gewerblichen Berufsgenossenschaften, Bonn 1974.

WOOD, K. ET AL. (1968) *Nature (Lond.)*, **220**: 173.

WORLD HEALTH ORGANIZATION (1971) *International standards for drinking water*, 3rd ed., Geneva

YOSHINO, Y. ET AL. (1966) *J. Neurochem.*, **13**: 1223.

# ERRATA

Page 29, line 21:

> *For* "gungicide" read "fungicide"

Page 41, line 28–29:

> *Delete* D'Itri et al., 1972
> *Insert* D'Itri, 1972

Page 55, line 28:

> *Delete* environmental layers
> *Insert* environmental levels

# *Environmental Health Criteria* 2

# Polychlorinated Biphenyls and Terphenyls

Published under the joint sponsorship of the United Nations Environment Programme and the World Health Organization

WORLD HEALTH ORGANIZATION GENEVA 1976

*Environmental Health Criteria 2*

# POLYCHLORINATED BIPHENYLS AND TERPHENYLS

Published under the joint sponsorship of
the United Nations Environment Programme
and the World Health Organization

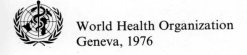

World Health Organization
Geneva, 1976

6201478X

ISBN 92 4 154062 1

PRINTED IN UNITED KINGDOM

# CONTENTS

# NOTE TO READERS OF THE CRITERIA DOCUMENTS

While every effort has been made to present information in the criteria documents as accurately as possible without unduly delaying their publication, mistakes might have occurred and are likely to occur in the future. In the interest of all users of the environmental health criteria documents, readers are kindly requested to communicate any errors found to the Division of Environmental Health, World Health Organization, Geneva, Switzerland, in order that they may be included in corrigenda which will appear in subsequent volumes.

In addition, experts in any particular field dealt with in the criteria documents are kindly requested to make available to the WHO Secretariat any important published information that may have inadvertently been omitted and which may change the evaluation of health risks from exposure to the environmental agent under examination, so that information may be considered in the event of updating and re-evaluating the conclusions contained in the criteria documents.

# WHO TASK GROUP ON ENVIRONMENTAL HEALTH CRITERIA FOR POLYCHLORINATED BIPHENYLS AND TERPHENYLS

*Copenhagen, 20–24 October 1975*

## Participants

*Members*

Dr V. Beneš, Department of Toxicology, Institute of Hygiene and Epidemiology, Prague, Czechoslovakia (*Vice-Chairman*)

Dr H. L. Falk, National Institute for Environment Health Services, Research Triangle Park, NC, USA

Mr L. Gordts, Institute of Hygiene and Epidemiology, Brussels, Belgium

Dr D. L. Grant, Pesticides Section, Toxicology Evaluation Division, Bureau of Chemical Safety, Department of Health and Welfare, Ottawa, Ontario, Canada (*Rapporteur*)

Mr A. V. Holden, Department of Agriculture and Fisheries for Scotland, Freshwater Fisheries Laboratory, Faskally, Pitlochry, Perthshire, Scotland

Dr S. Jensen, Naturvardsverkets Specialanalytiska Laboratorium, Wallenberg-laboratoriet, Lilla Frescati, Stockholm, Sweden

Dr Renate Kimbrough, Center for Disease Control, Toxicology Branch, Atlanta, GA, USA

Professor H. Kuratsune, Department of Public Health, Faculty of Medicine, Kyushu University, Fukuoka, Japan

Dr E. Schulte, Institut für Lebensmittelchemie der Westfälischen Wilhelms-Universität, Münster/Westf., Federal Republic of Germany

Dr J. G. Vos, Laboratory for Pathology, National Institute for Public Health, Bilthoven, Netherlands (*Chairman*)


*Observer*

Dr D. Axelrod, Division of Laboratories and Research, New York State Department of Health, Albany, NY, USA

---

Unable to attend

*a* Dr M. V. Kryžanovskaja, All-Union Institute for Research on Hygiene and Toxicology of Pesticides, Polymers, and Plastics, Kiev, USSR.

*Representatives of other organizations:*

*International Federation of Pharmaceutical Manufacturers Associations*
Professor P. Fabiani, Laboratoire du Chimie et de Toxicologie de l'Hôtel Dieu, Paris, France

*Permanent Commission and International Association on Occupational Health*
Dr Aa. Grut, State Labour Inspection Service, Hellerup, Denmark

*Secretariat:*

Dr J. C. Gage, 21 Lambolle Road, London, England (*Temporary Adviser*)

Dr M. J. Suess, Environmental Pollution Control, WHO Regional Office for Europe, Copenhagen, Denmark

Dr A. H. Wahba, Health Laboratory Services, WHO Regional Office for Europe, Copenhagen, Denmark

Dr G. Vettorazzi, Food Additives Unit, Division of Environmental Health, World Health Organization, Geneva (*Secretary*)

Dr D. C. Villeneuve,[a] Laboratory of Toxicology, National Institute of Public Health, Bilthoven, Netherlands (*Temporary Adviser*)

---

[a] On sabbatical leave from: Biochemical Toxicology Unit, Environmental Toxicology Division, Environmental Health Centre, Department of Health and Welfare, Ottawa, Ontario, Canada.

# ENVIRONMENTAL HEALTH CRITERIA FOR POLYCHLORINATED BIPHENYLS AND TERPHENYLS

A WHO Task Group on Environmental Health Criteria for Polychlorinated Biphenyls (PCBs) and Terphenyls (PCTs) met in Copenhagen from 20-24 October 1975. Dr F. A. Bauhofer, Director of Health Services of the WHO Regional Office for Europe opened the meeting on behalf of the Director-General and the Director of the Regional Office for Europe. The Task Group reviewed and amended the second draft criteria document and made an evaluation of the health risks from exposure to these compounds.

The preparation of the first draft criteria document was based on national reviews of health effects research on polychlorinated biphenyls, received from the national focal points collaborating in the WHO Environmental Health Criteria Programme in Canada, the Federal Republic of Germany, Finland, France, Japan, the Netherlands, New Zealand, Sweden, the United Kingdom, and the USA. Dr J. C. Gage, London, England, prepared the first draft as well as the second draft criteria document which took into account the comments received from the national focal points in Canada, Czechoslovakia, the Federal Republic of Germany, France, Japan, New Zealand, Sweden, the United Kingdom, the USA, and the USSR; from the United Nations Industrial Development Organization (UNIDO), Vienna, the Food and Agriculture Organization of the United Nations (FAO) Rome, and from the United Nations Educational, Scientific and Cultural Organization (UNESCO) Paris; the Organization for Economic Co-operation and Development, Paris, and the Health Protection Directorate of the Commission of the European Communities, Luxembourg.

Comments were also received at the request of the Secretariat, from Dr K. Kojima, Japan, Dr D. S. May, United Kingdom, and Dr V. Zitko, Canada.

The collaboration of these national institutions, international organizations and individual experts is gratefully acknowledged. Without their assistance the document could not have been completed.

This document is based primarily on original publications listed in the reference section. In addition, some recent publications reviewing the environmental and health aspects of polychlorinated biphenyls were also used. These include reviews by the Commission of the European Communities (1974), the US Department of Health, Education and

Welfare (1972), the International Agency for Research on Cancer (1974), the International Council for the Exploration of the Sea (1974), Jensen (1974), Kimbrough (1974), the National Swedish Environment Protection Board (1973), the Panel on Hazardous Trace Substances (1972), the USDA/USDC/EPA/FDA/USDA (1972), and a WHO working group (1973).

Details about the WHO Environmental Health Criteria Programme including the definition of some terms frequently used in the document may be found in the general introduction to the Environmental Health Criteria Programme published together with the Environmental Health Criteria Document on mercury (Environmental Health Criteria 1, Geneva World Health Organization, 1976).

# 1. SUMMARY AND RECOMMENDATIONS FOR FURTHER RESEARCH

## 1.1 Introductory Note

The commercial production of the polychlorinated biphenyls (PCBs) began in 1930, and during the 1930s cases of poisoning were reported among men engaged in their manufacture. The nature of this occupational disease was characterized by a skin affliction with acneiform eruptions; occasionally the liver was involved, in some cases with fatal consequences. Subsequent safety precautions appear largely to have prevented further outbreaks of this disease in connection with the manufacture of PCBs, but from 1953 onwards, cases have been reported in Japanese factories manufacturing condensers.

The distribution of PCBs in the environment was not recognized until Jensen started an investigation in 1964 to ascertain the origins of unknown peaks observed during the gas–liquid chromatographic separation of organochlorine pesticides from wild-life samples. In 1966, he and his colleagues succeeded in attributing these to the presence of PCBs. Since that time, investigations in many parts of the world have revealed the widespread distribution of PCBs in environmental samples.

The serious outbreaks of poisoning in man and in domestic animals from the ingestion of food accidentally contaminated with PCBs have stimulated investigations into the toxic effects of PCBs on animals and on nutritional food chains. This has resulted in limitation of the commercial exploitation of PCBs and polychlorinated terphenyls (PCTs), and in regulations to limit the residues in human and animal food.

The environmental impact of the PCBs and PCTs has been the subject of several reviews, and has been discussed at a number of regional and international meetings. The relevant publications are mentioned in the previous section.

## 1.2 Summary

### 1.2.1 Composition and analytical problems

The commercial production of PCBs and PCTs by the direct chlorination of biphenyl and terphenyl leads to a mixture of components with a range of chlorine contents, the mean percentage chlorine in the product being controlled to give the required technical properties. Most of these components have been separated by gas–liquid chromatography, and the PCBs in the mixtures have been characterized after synthesis of the

11

components by unequivocal routes. Techniques are available to analyse environmental samples for PCBs and PCTs, and experience has shown that interlaboratory collaborative studies are necessary to establish competence to determine residues below the 1 mg/kg level.

The commercial PCB mixtures contain various quantities of impurities, among which chlorinated dibenzofurans and chlorinated naphthalenes have been identified.

### 1.2.2 Sources and pathways in the environment

The estimated cumulative world production of PCBs since 1930 is of the order of 1 million tonnes. Of this, more than one-half has entered dumps and landfills, where it is likely to be stable and released only very slowly. Much of the remainder has entered the environment by the disposal of industrial fluids into rivers and coastal waters, by leakage from nonenclosed systems, or by volatilization into the atmosphere from incineration of PCB-containing material at dumps. The ultimate reservoirs of PCBs and PCTs that enter the environment are mainly sediments of rivers and coastal waters. PCBs and PCTs are stable in the environment, but a small proportion is transformed by biological action and possibly by photolysis.

### 1.2.3 Concentration in the environment

Measured concentrations of PCBs in air range from 50 ng/m$^3$ to less than 1 ng/m$^3$. Nonpolluted fresh waters should contain less than 0.5 ng of PCBs per litre compared with 50 ng per litre in moderately polluted rivers and estuaries, and 500 ng per litre in highly polluted rivers. The concentration in living organisms depends upon the extent of local pollution, the amount of fat in the tissues, and the trophic stage of the organism in food chains. Highest tissue levels were found in marine ecosystems with very high values in top predators from polluted areas, though most of the fish caught for human consumption contains PCBs at levels of less than 0.1 mg/kg in muscle tissue. There is no information available on the environmental distribution of PCTs.

### 1.2.4 Metabolism

PCBs are well absorbed by mammals through the gastrointestinal tract, lungs, and skin. They are stored mainly in adipose tissue and there is some placental transfer. Excretion in mammals is mainly through faeces, where the PCBs appear as phenolic metabolites; they appear unchanged in human milk. In birds, there is a considerable excretion in eggs. The rate of excretion in faeces is dependent on the rate of metabolism and this is much influenced by the number and orientation of the chlorine substituents. As environmental PCBs pass up biological food chains, there is a progressive loss of the lower chlorinated components owing to selective biotransformation, and only traces of PCBs containing less than five chlorine atoms per molecule are found in human fat.

The smaller amount of information available concerning the PCTs indicates that they also are absorbed from the gastrointestinal tract and undergo selective biotransformation, but that the concentration in fat in relation to that in other tissues appears to be less than is observed with the PCBs.

### 1.2.5 The extent of human exposure

Surveys on human adipose tissue in several countries have shown that most samples contain levels of PCBs in the region of 1 mg/kg or less, although higher values have been reported from some countries. Much higher values, up to 700 mg/kg, have been found in fat from men occupationally exposed. Several national surveys give PCB concentrations in the blood in the region of 0.3 μg/100 ml but levels approaching 200 μg/100 ml have been measured in men occupationally exposed, and these may be associated with skin lesions. Most surveys on human milk have shown PCB concentrations in the region of 0.02 mg/litre, although concentrations up to 0.1 mg/litre have been recorded. Results from the very few investigations on PCT concentrations in fat and blood suggest that these may be equal to those of PCBs.

An estimate of the total daily intake of PCBs in air, water, and diet by individuals not occupationally exposed indicates that this falls within the range of 5–100 μg, which may be supplemented by unknown amounts from nondietary sources. This estimate has some support from measurements of concentrations in human milk.

### 1.2.6 Experimental studies on the effects of PCBs and PCTs

Most of the studies on the toxicity of PCBs have been performed with the commercial mixtures. The PCBs are of low acute toxicity but the effects are cumulative with prolonged administration; in mammals, liver enlargement is observed that may progress to liver damage. Non-metastasizing neoplastic liver nodules have been produced in rats and mice, some of which were classified as hepatocellular carcinomas on the basis of histological criteria in one study in rats and one study in mice.

The monkey is much more sensitive to PCBs than the rat, showing effects similar to those seen in human Yusho patients (See section 8, p 65) with a similar order of exposure. Low dose effects on fertility have been seen in both the monkey and the mink, a species that is also relatively sensitive to PCBs.

Other effects of PCBs include porphyria, immunosuppression, and interference with steroid metabolism; some of these may be attributable to the increase in microsomal enzyme activity associated with liver enlargement. Some of the toxic effects can be attributed to impurities in the commercial products.

The toxicity of PCBs to fish is not high by comparison with that of some pesticides, but some aquatic invertebrates are more sensitive. There is little information on the toxicity of the PCTs.

### 1.2.7 Clinical studies of the effects of PCBs in man

Information on the effects of PCBs in man has been obtained from a large-scale incident in Japan (Yusho), in which over 1000 individuals showed signs of poisoning from the ingestion of rice oil contaminated with PCBs from a heat exchanger liquid. The most striking effects were hypersecretion in the eyes, pigmentation and acneiform eruptions of the skin, and disturbances of the respiratory system. Babies born to Yusho mothers were of less than normal size and initially showed skin pigmentation. Over a six-year period, the effects on the skin diminished very gradually, but the nonspecific symptoms tended to become somewhat more prominent. The smallest dose of PCBs calculated to produce an effect was approximately 0.5 g over about 120 days, but as the rice oil contained chlorinated dibenzofurans at a concentration of 5 mg/kg of rice oil in addition to PCBs at 2000–3000 mg/kg it is not certain that the symptoms were due solely to PCBs.

### 1.2.8 Dose–effect relationships

Experimental studies on the dose–effect relationship have shown that no effects on growth, and reproduction are seen in rats receiving PCB levels of 1 mg/kg body weight per day; there may be liver enlargement and a reversible induction of microsome enzyme activity at a level of 1 mg/kg/day but not at 0.1 mg/kg/day. Effects on reproduction are seen in the monkey with PCB levels of about 0.12 mg/kg/day. Symptoms were reported in some Yusho patients ingesting less than 0.1 mg/kg/day.

### 1.3  Recommendations for Further Research

#### 1.3.1  Analytical methods

Collaborative intercalibration studies on the determination of PCBs, PCTs, and chlorinated dibenzofurans should be established between all laboratories engaged in determining these compounds in environmental samples, and adequate standards should be made available for individual chlorinated biphenyls and dibenzofurans.

Improved analytical techniques, including those involving capillary gas–liquid chromatography and mass spectroscopy, should be developed for the determination of PCBs, PCTs, polybrominated biphenyls, polychlorinated dibenzofurans, and naphthalenes, and their metabolites and degradation products.

#### 1.3.2  Environmental pollution

The content of chlorinated dibenzofurans should be studied in a range of commercial PCB mixtures, and in used PCBs from existing or newly designed heat exchangers, capacitors, and hydraulic transmissions. The possibility of the formation of chlorinated dibenzofurans from PCBs in cooking oils before and after use and in other foods during storage or heating requires investigation.

The current production, use patterns, and methods of disposal of PCBs should be carefully examined to gain information on the impact of PCBs on the environment at the present time. The rate of leaching of PCBs from waste dumps and landfills should be studied, and methods of incineration should be investigated to ascertain which of the components survive inefficient combustion, and whether chlorinated dibenzofurans or other compounds are released into the atmosphere.

Information is required on the metabolism and environmental fate of the chlorinated dibenzofurans.

#### 1.3.3  Effects on man

The intake of PCBs and PCTs from all sources by typical populations should be studied, and an attempt should be made to trace the sources of PCBs and PCTs in those items of the diet that make the greatest contribution to the daily intake. Further measurements are required on levels in human body fat, blood, and milk, and an attempt should be made to relate these levels to the daily intake.

Clinical and epidemiological studies are required on individuals exposed to relatively high concentrations of PCBs and PCTs, either occupationally or by virtue of the nature of their diet, and their health status should be correlated with exposure and tissue levels.

### 1.3.4   Experimental studies

Further toxicological and metabolic studies are required on individual polychlorinated biphenyls, dibenzofurans, naphthalenes, and other impurities occurring in commercial products, on a variety of species including primates, in order to assess the nature of the toxic effects, the dose–response relationship, and the threshold of toxic effects. Such investigations should be extended to include dermal and inhalation exposure.

Carcinogenic and cocarcinogenic studies should be undertaken to identify the components in commercial PCBs responsible for neoplastic effects.

### 1.3.5   PCB substitutes

More information should be made available on the production and use patterns of PCTs, polybrominated biphenyls, and of other possible substitutes for PCBs, and when appropriate, these products should be subjected to adequate toxicological investigations.

# 2. PROPERTIES AND ANALYTICAL METHODS

## 2.1   Chemical Composition

The PCBs form a class of chlorinated hydrocarbons and are manufactured commercially by the progressive chlorination of biphenyl in the presence of a suitable catalyst. They are known by a variety of trade names: Aroclor (USA), Phenochlor (France), Clophen (Federal Republic of Germany), Kanechlor (Japan), Fenchlor (Italy), and Sovol (USSR). Their value for industrial applications depends upon their chemical inertness, resistance to heat, non-flammability, low vapour pressure (particularly with the higher chlorinated compounds), and high dielectric constant. There are many different trade names for mixtures of PCBs with other compounds.

Individual manufacturers have their own system of identification for their products. In the Aroclor series, a four digit code is used; biphenyls

16

are generally indicated by 12 in the first two positions, while the last two numbers indicate the percentage by weight of chlorine in the mixture; thus Aroclor 1260 is a polychlorinated-biphenyl mixture containing 60% of chlorine. An exception to this generalization is Aroclor 1016 which is a distillation product of Aroclor 1242 containing only 1% of components with five or more chlorine atoms (Burse et al., 1974). With other commercial products the codes may indicate the approximate mean number of chlorine atoms in the components; thus Clophen A60, Phenochlor DP6, and Kanechlor 600 are biphenyls with an average of about six chlorine atoms per molecule (equivalent to 59.0% chlorine by weight).

In the Aroclor series, terphenyls are indicated by 54 in the first two places of the four digit code. In Japan, the PCTs are coded Kanechlor KC-C.

Individual PCBs have been synthesized for use as reference samples in the identification of gas–liquid chromatographic peaks, for toxicological investigations, and for studying their metabolic fate in living organisms, for which purpose they have been prepared labelled with carbon-14 (Hutzinger et al., 1971; Tas & de Vos, 1971; Webb & McCall, 1972; Moron, et al., 1972; Sundström & Wachtmeister, 1973; Jensen & Sundström, 1974).

The chlorination of biphenyl can lead to the replacement of 1–10 hydrogen atoms by chlorine; the conventional numbering of substituent positions is shown in the diagram. It has been calculated that 210 different

biphenyls of different chlorine content are theoretically possible, although Sissons & Welti (1971) have demonstrated that chlorine substituents in the 3,5- and 2,4,6-positions are not obtained by the direct chlorination of biphenyl. The proportions of PCBs with 1–9 chlorine substituents in the Aroclors are shown in Table 1.

There have been several investigations to identify individual PCBs in commercial products. Sissons & Welti (1971) separated the components of the Aroclors by column and gas–liquid chromatography, and characterized many of the peaks by high-resolution mass spectrometry and nuclear magnetic resonance, and by comparison with 40 synthesized

## Table 1. Approximate composition of Aroclors

| No. of Cl atoms in molecule | % of chlorine weight | Aroclor | | | | |
|---|---|---|---|---|---|---|
| | | 1221[a] | 1242[b] | 1248[b] | 1254[b] | 1260[b] |
| 0 | 0 | 12.7 | | | | |
| 1 | 18.8 | 47.1 | | | | |
| 2 | 31.8 | 32.3 | 3 | 2 | | |
| 3 | 41.3 | | 13 | 18 | | |
| 4 | 48.6 | | 28 | 40 | 11 | |
| 5 | 54.4 | | 30 | 36 | 49 | 12 |
| 6 | 59.0 | | 22 | 4 | 34 | 38 |
| 7 | 62.8 | | 4 | | 6 | 41 |
| 8 | 66.0 | | | | | 8 |
| 9 | 68.8 | | | | | 1 |

[a] Willis & Addison (1972).
[b] Panel on Hazardous Trace Substances (1972).

## Table 2. Polychlorinated biphenyls in Aroclors 1221–1254 (Webb & McCall, 1972)

| Retention time[a] | Synthetic chlorobiphenyl | 1221 | 1232 | 1242 | 1248 | 1254 |
|---|---|---|---|---|---|---|
| 10 | Biphenyl | ×[b] C[c] | × | × | × | |
| 13 | 2 | × C | × | × | × | |
| 16.9 | 3 | × D[d] | | | | |
| 17 | 4 | × C | × | | | |
| 17.7 | | | | × | | |
| 18.5 | 2,2' | × C | × | × C | × | |
| 21.3 | | × | × | × | | |
| 21.6 | | × | × | × | | |
| 23.5 | 2,3' | × C | × | × | × | |
| 24 | 2,4' | × C | × | × C | × | |
| 26 | 2,6,2' | | × | × D | × | |
| 29 | 2,5,2' | | × | × C | × | |
| 29.2 | 2,4,2' | × | × | × C | × | |
| 29.5 | 4,4' | × C | × | × C | | |
| 34 | 2,3,2' | | × | × C[e] | × | |
| 38 | 2,4,3' | | × | × C | × | |
| 38.5 | 2,4,4' | | × | × C | × | |
| 38.7 | 2,5,4' | | × | × C | × | |
| 41.5 | | | | × | × | |
| 42.5 | 3,4,2' | | | × C | × | |
| 43 | | | | × | × | |
| 44 | | | | × | × | |
| 44.5 | | | | | × | |
| 48 | 2,5,2',5' | | | × C | × | × C |
| 49.5 | 2,4,2',5' | | | × C | × | × C |
| 50.5 | 2,4,2',4' | | | × D | × | |
| 51.5 | | | | × | × | |
| 55 | 2,3,2',5' | | | × | × | × C |
| 57 | | | | × | × | |
| 59 | 2,3,6,2',6' | | | × | × | × D |
| 60 | | | | × | × | |
| 64 | | | | × | × | |
| 69 | | | | × | × | |
| 69.5 | | | | × | × | |
| 70 | 2,5,3',4' | | | × C[e] | × | |
| 70.5 | 2,3,6,2',5' | | | | | × C |
| 71 | 2,4,3',4' | | | × C[e] | × | |
| 72 | | | | × | × | |
| 76 | 2,3,6,2',4' | | | | | × C |
| 82 | | | | | × | |
| 83 | 2,3,6,2',3' | | | | × | × C |
| 84 | | | | | × | |
| 85 | 2,4,5,2',5' | | | | × | |
| 87 | 2,4,5,2',4' | | | | × | × C |
| 96 | | | | | | × |
| 99 | 2,4,5,2',3' | | | | × | × C |
| 101 | | | | | × | × |
| 104 | | | | | × | × |
| 107 | 2,3,6,3',4' | | | | × | × |
| 115 | | | | | | × |
| 117 | | | | | × | × |
| 119 | | | | | | × |
| 125 | 2,4,5,2',3',6' | | | | | × C |
| 126 | 2,4,5,3',4' | | | | × | × C |
| 135 | | | | | | × |
| 147 | | | | | | × |
| 148 | 2,4,5,2',4',5' | | | | | × C |
| 149 | | | | | | × |
| 152 | | | | | | × |
| 165 | | | | | | × |
| 178 | | | | | | × |

[a] Relative to p,p'–DDE at 190°C on 100' × 0.02' SCOT SE 30 column.
[b] × indicates a GLC peak in this Aroclor at this retention time.
[c] C means that the synthetic compound in the second column was confirmed by GLC and IR in this Aroclor. Because of the labour involved, most compounds were only confirmed in one Aroclor. For example, biphenyl is probably present in 1232–1248 as well as 1221.
[d] Only GLC data available.
[e] Something else also present.

PCBs. Webb & McCall (1972) identified the gas–liquid chromatographic peaks with those of synthesized compounds by retention times and infrared spectrometry (Table 2). The most exhaustive study is that of Jensen & Sundström (1974). They recognized that conventional gas–liquid chromatography could not separate all the components, so they devised a preliminary fractionation on a charcoal column, which separated the component PCBs according to the number of chlorines in the 2,2′,6 or 6′ positions in the molecule (o-chlorines). They compared the gas–liquid chromatographic peaks with those of 90 synthesized PCBs, and were able to characterize and quantify 60 components of Clophens A50 and A60 (Table 3). Tables 2 and 3 show a considerable overlap between the components of Aroclor 1254 and Clophen A50.

## 2.2 Purity of Products

Commercial PCBs are not sold on a composition specification, but on their physical properties. Different batches may vary somewhat from the compositions shown in Tables 1–3. Impurities known to be present in commercial PCBs are chlorinated dibenzofurans and chlorinated naph-thalenes (Vos et al., 1970; Bowes et al., 1975). Bowes et al. (1975) found chlorinated dibenzofurans at 0.8–2.0 mg/kg in samples of the Aroclor 1248–1260 series, but none in Aroclor 1016, and at levels of 8.4 mg/kg in Clophen A60 and 13.6 mg/kg in Phenoclor DP–6. Roach & Pomerantz (1974) found chlorinated dibenzofurans levels of 1 mg/kg and Nagayama et al. (1976) found 18 mg/kg in different batches of Kanechlor 400, but no chlorinated dibenzodioxins.

## 2.3 Determination of PCB Residues

Reviews have been published on methods used for the determination of organochlorine compounds including PCBs in environmental samples (Holden, 1973a, Panel on Hazardous Trace Substances, 1972). No two laboratories have identical methods, although all have features in common. The techniques appear to be those previously developed for the deter-mination of organochlorine pesticides with appropriate modifications for the presence of PCBs, and the studies on PCBs sometimes form part of a wider programme for monitoring persistent organochlorine compounds in the environment. The major difficulties in the determination of PCBs are to separate them from interfering organochlorine pesticides, and to derive a single quantitative figure from a variable mixture of components.

Table 3. Percentages of polychlorinated biphenyls in Clophen A50 and A60 and in human fat (Jensen & Sundström, 1974)

| Compound No. | Structure[a] | No. of ortho-chlorines | Relative retention time Apiezon L | Relative retention time SF 96 | Percentage in Clophen A50 | Percentage in Clophen A60 | Human tissue |
|---|---|---|---|---|---|---|---|
| 1 | 2,5–2'5' | 2 | 0.25 | 0.30 | 5.0 | | |
| 2 | 2,4–2'5' | 2 | 0.26 | 0.30 | 1.4 | | |
| 3 | 2,3–2'5' | 2 | 0.27 | 0.33 | 1.9 | | 1.1 |
| 4 | (2,–2',3',4') | 2 | 0.30 | 0.43 | 1.2 | | 0.66 |
| 5 | (4–2',3'6') | 2 | 0.32 | 0.43 | 2.1 | | 0.56 |
| 6 | 2,5–2',3',6' | 3 | 0.35 | 0.43 | 4.4[b] | 2.9 | 1.2 |
| 7 | 2,3–2',3',6' | 3 | 0.38 | 0.49 | 2.5[b] | 0.28 | 0.48 |
| 8 | 3,4–2',5' | 1 | 0.41 | 0.42 | 3.9 | | 1.5 |
| 9 | | 1 or 2 | 0.42 | | 2.2[c] | | 2.0[c] |
| 10 | 2,5–2',3',5' | 2 | 0.45 | 0.49 | 2.2 | 1.1 | 1.2 |
| 11 | 2,3,6–2',3',6' | 4 | 0.47 | 0.61 | 0.50 | 1.0[d] | |
| 12 | 2,5–2',4',5' | 2 | 0.48 | 0.50 | 7.0[b] | 5.6 | 4.2 |
| 13 | 2,4–2',4',5' | 2 | 0.51 | 0.51 | 1.8[b] | —[e] | 1.9 |
| 14 | 2,3–2',4',5' | 2 | 0.53 | 0.57 | 1.4[b] | | |
| 15 | 2,5–2',3',4' | 2 | 0.53 | 0.58 | 5.4 | 1.4 | 2.3 |
| 16 | 2,3–2',3',4' | 2 | 0.55 | 0.66 | 1.0 | | |
| 17 | 3,4–2',3',6' | 2 | 0.56 | 0.62 | 7.6[b] | 2.9 | 4.7 |
| 18 | 2,3,5–2',3',6' | 3 | 0.60 | 0.70 | 1.2 | 4.2 | 1.0 |
| 19 | 2,4,5–2',3',6' | 3 | 0.64 | 0.73 | 2.0[b] | 6.5[d] | 0.13 |
| 20 | 2,5–2',3',5',6' | 3 | 0.65 | 0.68 | 1.3 | 3.3 | 0.43 |
| 21 | 2,3–2',3',5',6' | 3 | 0.70 | 0.74 | 0.5 | | 0.05 |
| 22 | 2,3,4–2',3',6' | 3 | 0.72 | 0.83 | 1.8 | 3.2 | 0.15 |
| 23 | | 1 or 2 | 0.76 | | 0.6[c] | | |
| 24 | 2,3,6–2',3',5',6' | 4 | 0.77 | 0.91 | 0.09 | 0.96 | |
| 25 | 3,4–2',4',5' | 1 | 0.79 | 0.74 | 5.0[b] | 1.6 | 5.4 |
| 26 | 2,3,6–2',3'4',6' | 4 | 0.84 | 0.95 | 0.05 | 0.37 | |
| 27 | 2,3,5–2',4',5' | 2 | 0.85 | 0.86 | 0.90 | 2.9 | 2.7 |
| 28 | 3,4–2',3',4' | 1 | 0.87 | 0.86 | 3.6 | | 1.9 |
| 29 | 2,4,5–2',4',5' | 2 | 0.90 | 0.86 | 4.2[b] | 12.9[d] | 21.5 |
| 30 | 2,3,4–2',3',5' | 2 | 0.94 | 0.97 | 1.1 | 1.5 | —[e] |
| 31 | | 1 or 2 | 0.96 | 0.93 | | —[e] | |
| 32 | 2,3,4–2',4',5' | 2 | 1.00 | 1.00 | 5.1 | 11.3[d] | 14.0 |
| 33 | 2,3,5–2',3',5',6' | 3 | 1.01 | 1.06 | 0.04 | 0.49 | 0.90 |
| 34 | | 4 | 1.02 | | 0.05[c] | | |
| 35 | | 1 or 2 | 1.04 | 1.10 | 1.1[c] | 2.0[c] | 1.5[c] |
| 36 | 2,4,5–2',3',5',6' | 3 | 1.08 | 1.13 | 0.39 | 3.3 | 3.5 |

## 2.3.1 Extraction of sample

### Air

Particulate fallout from air has been trapped on 200 µm nylon net coated with silicone oil, and the PCBs then extracted with hexane (Södergren, 1972a). Separate determinations of particulate and vapour phase PCBs in air have been made by the passage of a large volume of air through a filter which was followed by an impinger containing hexane (Hasegawa et al., 1972b), or a polyurethane plug (Bidleman & Olney, 1974) or ceramic saddles coated with OV 17 silicone (Harvey & Steinhauer, 1974) to absorb the vapour.

### Water

PCBs have been extracted from water by passing a sample through a

Table 3.—continued

| Compound | | No. of ortho-chlorines | Relative retention time | | Percentage in | | |
| --- | --- | --- | --- | --- | --- | --- | --- |
| | | | | | Clophen | | Human tissue |
| No. | Structure[a] | | Apiezon L | SF 96 | A50 | A60 | |
| 37 | 2,3,4–2',3',4' | 2 | 1.11 | 1.16 | 1.3 | 2.0 | 0.81 |
| 38 | 2,3,6–2',3',4',5' | 3 | 1.13 | 1.29 | 0.33 | 3.7[d] | —[e] |
| 39 | 2,4,5–2',3',4',6' | 3 | 1.14 | 1.16 | 0.17 | 1.8 | 2.5 |
| 40 | 2,3,4–2',3',5',6' | 3 | 1.19 | 1.32 | 0.27 | 2.1 | 1.3 |
| 41 | 2,3,5,6–2',3',5',6' | 4 | 1.23 | 1.34 | 0.005 | 0.07 | |
| 42 | 2,3,4–2',3',4',6' | 3 | 1.28 | 1.40 | 0.13 | 1.3 | 0.57 |
| 43 | 2,3,4,6–2',3',5',6' | 4 | 1.34 | 1.44 | 0.007 | 0.09 | |
| 44 | 2,4,5–3',4',5' | 1 | 1.41 | 1.19 | 0.47 | 1.0 | 0.49 |
| 45 | 2,3,6–2',3',4',5',6' | 4 | 1.45 | 1.59 | 0.008 | 0.09 | |
| 46 | 2,3,4,6–2',3',4',6' | 4 | 1.46 | 1.51 | —[e] | 0.03 | |
| 47 | 3,4–2',3',4',5' | 1 | 1.55 | 1.37 | 0.81 | 1.5 | 2.0 |
| 48 | 2,3,5–2',3',4'.5' | 2 | 1.59 | 1.49 | 0.23 | 0.90 | 1.2 |
| 49 | 2,4,5–2',3',4',5' | 2 | 1.71 | 1.56 | 0.98 | 7.6[d] | 7.7 |
| 50 | 2,3,4–2',3',4',5' | 2 | 1.88 | 1.82 | 0.72 | 4.1[d] | 5.9 |
| 51 | 2,3,4,5–2',3',5',6' | 3 | 1.94 | 1.99 | 0.08 | 0.74 | 0.77 |
| 52 | | 1 or 2 | 2.02 | 1.96 | 0.23[c] | 1.0[c] | —[e] |
| 53 | 2,3,4,5–2',3',4',6' | 3 | 2.07 | 2.05 | 0.06 | 0.44 | 0.94 |
| 54 | 2,4,5–2',3',4',5',6' | 3 | 2.15 | 2.04 | 0.01 | 0.28 | 0.46 |
| 55 | (2,3,5,6–2',3',4',5',6') | 4 | 2.23 | | | | —[e] |
| 56 | 2,3,4–2',3',4',5',6' | 3 | 2.40 | 2.20 | 0.01 | 0.17 | 0.31 |
| 57 | 2,3,4,6–2',3',4',5',6' | 4 | 2.45 | 2.56 | —[e] | —[e] | —[e] |
| 58 | 3,4,5–2',3',4',5' | 1 | 2.74 | 2.41 | —[e] | —[e] | |
| 59 | 2,3,4,5–2',3',4',5' | 2 | 3.18 | 2.81 | 0.35 | 0.67 | 1.7 |
| 60 | 2,3,4,5,6–2',3',4',5',6' | 4 | 4.12 | | | | 0.62 |
| | | | | Total | 86.7% | 99.4% | 100.0% |
| | p,p'–DDE | | 0.52 | 0.58 | | | |
| | p,p'–DDT | | 0.71 | 0.73 | | | |
| | p,p'–DDT | | 0.90 | 0.97 | | | |

[a] Tentative structures are given in brackets.

[b] Component found present in Aroclor 1254 by Webb & McCall (1972).

[c] Figures calculated using responses of chlorobiphenyls with similar retention times in the same fraction from the charcoal column.

[d] Component found present in Phenoclor DP6 by Tas & Vos (1971).

[e] Present in trace amounts only.

filter of undecane and Carbowax 400 monostearate supported on Chromosorb W (Ahling & Jensen, 1970) or a porous plug of polyurethane coated with a suitable gas–liquid chromatographic stationary phase (Uthe et al., 1972), or Amberlite XAD–2 resin (Harvey, et al., 1973) followed by elution of the PCBs with a solvent. Anhoff & Josefsson (1975) have described a liquid–liquid extraction into cyclohexane.

*Biological samples*

Most analysts have used standard methods developed for organochlorine pesticides, in which the PCBs are extracted together with the fat; the sample is ground with anhydrous sodium sulphate and extracted with petroleum ether or hexane. Porter et al. (1970) studied the optimal conditions for this procedure. A dehydrating solvent may be included to facilitate the breakdown of cell structures; ethanol (Norén & Westöö,

21

1968) and acetone (Jensen et al., 1973) have been used. Rote & Murphy (1971) digested the sample with a mixture of acetic and perchloric acid prior to hexane extraction.

## 2.3.2 Clean-up

Methods for the removal of fat from the extract include solvent partitioning between hexane and acetonitrile or dimethylformamide, or treatment with strong sulfuric acid or ethanolic potassium hydroxide. Gel permeation has also been used (Stalling et al., 1972), and Holden & Marsden (1969) removed fat on dry, partially deactivated alumina columns. Certain pesticides such as dieldrin are destroyed by the sulfuric acid treatment, so this method cannot be used if such pesticides are to be determined together with PCBs (Jensen et al., 1973).

PCBs may be separated from organochlorine pesticides by column chromatography on Florisil (Mulhern et al., 1971), silica gel (Holden & Marsden 1969; Armour & Burke, 1970; Collins et al., 1972) or on charcoal (Berg et al., 1972; Jensen & Sundström, 1974). Several laboratories have reported difficulties in repeating results obtained by other investigators; the ease of separation appears to depend upon the characteristics of the absorbent, of the eluting solvent, and of the sample extract, though there appears to be no difficulty in separating all interfering substances except DDE, a metabolite of DDT. Thin-layer chromatography has been used for separation by Norén & Westöö (1968), Bagley et al. (1970), and Reinke et al. (1973).

In many environmental samples, DDE is present in large excess over the PCBs, and must be removed before the quantitative determination of PCBs. Oxidation procedures have been used to convert DDE to dichlorobenzophenone; recommended oxidants are potassium dichromate and sulfuric acid (Westöö & Norén, 1970b) and chromium (II) oxide and acetic acid (Mulhern et al., 1971). Jensen & Sundström (1974), who were interested in determining DDT/PCB ratios in environmental samples, preferred sodium dichromate in acetic acid with a trace of sulfuric acid. They claim that this does not destroy DDT and its metabolite DDD, which may be present in extracts after clean-up with strong sulfuric acid, and that using this mixture makes possible the quantitative determination of the dichlorobenzophenone from the oxidation of DDE.

Conversion of DDT to DDE may be achieved by treatment with ethanolic potassium hydroxide, which also removes interference from elemental sulfur (Ahling & Jensen, 1970). Sulfur may also be removed by activated Raney nickel (Ahnoff & Josefsson, 1975) or by metallic mercury.

Södergren (1973b) has scaled down the clean-up procedure for small samples, using microlitre volumes.

### 2.3.3 Chromatographic separation of PCBs

*Gas–liquid chromatography*

Most analysts use gas–liquid chromatography with an electron-capture detector for the separation of PCBs from the extract after clean-up. Stationary phases commonly used are silicones or their derivatives, for example, DC 200, SF 96, OV 1, and QF 1, or Apiezon L. Jensen & Sundström (1974) state that with a mixture of SF 96 and QF 1, 14 peaks can be obtained from Clophen A50, but that Apiezon L gives much better resolution. They obtained better peak separation by prior fractionation on a charcoal column, which separated the PCBs according to the number of *o*-chlorine substituents; they regard such refinements as unnecessary in PCB residue analysis, although they may be of value in the study of the selective, environmental degradation of PCBs. Column temperatures used ranged between 170°C and 230°C. Glass capillary columns gave good separation of PCBs from DDT and its metabolites (Schulte & Acker, 1974).

*Thin-layer chromatography*

This has been used in the clean-up stage, but it may also provide semi-quantitative results by visualization of the spots followed by densitometry, or by comparison with spots produced by known amounts of PCBs. Mulhern et al. (1971) separated PCBs on a plate coated with alumina containing silver nitrate, and the spots were developed by exposure to ultraviolet light; the detection limit was in the region of 1 µg. Collins et al. (1972) devised a similar method in which the PCBs remained together in a single spot, and they claimed a limit of detection of about 50 ng. Reversed-phase chromatography on plates coated with kieselguhr treated with liquid paraffin has been used to separate Phenochlor DP6 into several spots with a detection limit of a few micrograms (de Vos & Peet, 1971).

### 2.3.4 Quantification of PCB content

The response of the electron capture detector is not equal for all PCB components, being much affected by the degree of chlorination (Zitko et al., 1971). This does not lead to difficulties when the sample under investigation has been directly contaminated by a commercial PCB mixture, as that mixture can be used as a standard. Difficulties are encountered when the PCBs in the sample have undergone selective environmental degradation (see sections 3 and 5). Several investigators have noted that the pattern of peaks from such samples resembles fairly closely that of one or other of the higher chlorinated PCB mixtures such as Aroclor 1254, and they have compared the total area of the peaks with that of the nearest commercial product in order to determine the amount of PCBs in the sample (Armour & Burke, 1970). Collins et al. (1972) observed that, under their conditions, the area of peaks usually encountered in extracts of tissue samples was closely similar to that of an equivalent amount of DDE, thus DDE could be used for calibration. In order to overcome the uncertainties of these procedures, Rote & Murphy (1971) divided the peaks into groups according to the number of chlorine atoms in the molecule, as determined from mass spectrographic data, and calculated the PCB content of each group from the theoretical response of the detector to chlorine content. Jensen et al. (1973) selected a commercial PCB that included all the peaks from the extract; they determined the PCB content of each peak by combined mass spectrometry and coulo-metry, and determined the total PCBs in the sample by comparing the height of each peak obtained with the extract with those obtained with the reference sample. Simpler methods have been used by Koeman et al. (1969), who compared the height of a single peak obtained with the extract with that of a peak with the same retention time obtained with a commercial PCB mixture; others have averaged out more than one peak for this calculation (Reynolds, 1971; Reinke et al., 1973). Rote & Murphy (1971) have calculated that such procedures may more than double the values obtained by a more accurate method.

A different technique has been recommended by Berg et al. (1972); the PGBs are chlorinated with antimony pentachloride to decachloro-biphenyl, which can then be measured as a single peak.

### 2.3.5 Accuracy of PCB determinations

A group of eight analysts engaged in an investigation of pollution in the North Sea undertook a collaborative study to determine the PCB content of a sample of fish oil, using the methods currently employed in their laboratories (International Council for the Exploration of the Sea, 1974). The PCB values obtained ranged from 1.0 to 3.9 mg/kg with a mean of 1.97 mg/kg and a standard deviation of 0.93 mg/kg. Better agreement was obtained with the same fish oil fortified with PCBs at a concentration of 10 mg/kg; the mean of the results for the fortified sample was 10.0 mg/kg with a standard deviation of 1.1 mg/kg.

A probable source of error is incomplete initial extraction of PCBs from the sample (Holden & Marsden, 1969). Another source of variation between laboratories lies in the method used to quantify gas–liquid chromatographic peaks (section 2.3.4); van Hove Holdrinet (1975) considered this to be the major source of error.

It is evident that caution should be exercised in accepting the analytical results from a laboratory, particularly for samples with a low PCB content, until the competence of that laboratory has been established by an inter-laboratory collaborative study.

### 2.3.6 Confirmation of identity

Since Jensen first identified as PCBs those hitherto unknown substances that interfered in the gas–liquid chromatographic determination of organochlorine pesticides using mass spectrographic data, other investigators have confirmed the presence of PCBs in environmental samples by combining gas–liquid chromatography with mass spectrometry (Bagley et al., 1970) and with coulometry to measure the chlorine content. The conversion of PCBs to bicyclohexyl and decachlorobiphenyl is further confirmation (Berg et al., 1972). The widespread distribution of PCBs is now well established, and, as adequate methods are available to remove interference from organochlorine pesticides, there is no evidence of the presence of other interfering substances in the types of sample that have so far been analysed, down to a limit of detection of around 0.01 mg/kg. This does not necessarily apply to other types of sample, particularly when very low levels are being sought; Ahnoff & Josefsson (1973, 1975) reported a number of unknown interfering substances, when measuring PCBs in water at levels below 1 ng/litre, one of which was subsequently identified as elemental sulfur. They recommend confirmation by mass fragmentography for such samples.

### 2.4 Determination of PCTs

A few methods have been published for the determination of PCTs; extraction and clean-up procedures are similar to those used for PCBs, but the gas–liquid chromatographic details are different because of the lower volatility of the PCTs. Zitko et al. (1972b) used 3% OV 210 as the stationary phase with a column temperature of 200°C. Thomas & Reynolds (1973) also used OV 210 with a column temperature of 250°C and another system with 3% Dexsil as stationary phase at 300°C with a $^{63}$Ni electron capture detector; this was also used by Addison et al. (1972). Sosa-Lucero et al. (1973) used OV 210 and SE 30 at 255°C and Freudental & Greve (1973) used OV 17 with a temperature programmed from 200°C to 285°C. Thomas & Reynolds (1973) confirmed the identity by chlorination to tetradecachloroterphenyl with antimony pentachloride.

A thin-layer chromatographic technique has also been described with a limit of detection of about 1 µg (Addison et al., 1972).

# 3. SOURCES OF ENVIRONMENTAL POLLUTION

## 3.1 Production and Uses of PCBs

Details of the production and uses of PCBs in the USA have been released, and have been summarized by Nisbet & Sarofim (1972). Annual production has increased steadily since 1930 and reached a maximum in 1970 of 33 000 tonnes. During this peak year, 65% of the production was of the 42% chlorinated type, 25% was less chlorinated, and the remainder more chlorinated. After 1970, production sharply decreased owing to voluntary limitation of sales by the Monsanto Company, the sole manufacturer in the USA. According to information collected by the Organization for Economic Co-operation and Development (OECD), the 1971 production in the USA was 18 000 tonnes and the total in OECD countries in that year was 48 000 tonnes. It has been estimated that the cumulative total production of PCBs in North America up to 1971 was 0.5 million tonnes, and in the whole world probably double this figure.

The commercial applications of PCBs have been reviewed in an OECD report (Organization for Economic Co-operation and Development, 1973). From an environmental viewpoint these can be divided into three categories:

*Controllable closed systems.* PCBs used as dielectrics in transformers and large capacitors have a life equal to that of the equipment, and with proper design leakage does not occur. When the equipment is scrapped the quantity of dielectric is sufficiently large to justify regeneration.

*Uncontrollable closed systems.* PCBs are used in heat transfer and hydraulic systems which, although technically closed, permit leakage. The need for frequent replacement of small quantities makes recovery impracticable. PCBs are very widely dispersed in small capacitors, and there are great difficulties in collecting these items for disposal.

*Dissipative uses.* PCBs have been used in the formulation of lubricating and cutting oils, in pesticides, and as plasticizers in paints, copying paper, adhesives, sealants, and plastics. In these applications, the PCBs are in direct contact with the environment, and there is no way of recovering them when the product is scrapped.

The uses of PCBs in the USA in 1970 have been analysed by Nisbet & Sarofim (1972). Of the total 33 000 tonnes, 56% was used as a dielectric, with 36% in capacitors and 20% in transformers. Various plasticizer outlets accounted for 30%, hydraulic fluids and lubricants 12%, and heat transfer liquids 1.5%. Following the restriction of sales for non-dissipative uses, the percentage of PCBs sold as dielectrics rose to 77% in 1971 and the proportion of highly chlorinated products was considerably reduced, Aroclor 1016 replacing Aroclor 1242. In Japan, 44 800 tonnes of PCBs were used from 1962 to 1971, and of this 65.4% was used in the electrical industry, 11.3% in heat exchangers, 17.9% in pressure-sensitive duplicating paper, and 5.4% for other dissipative uses (Ishi, 1972). In Sweden, most of the 600 tonnes imported in 1969 was used in the electrical industry and a large part of the remainder in paints (Jensen, unpublished report 1972).

According to the OECD report, transformers and capacitors provided the major outlets for PCBs in most OECD countries in 1971. In 1972, several countries restricted sales; in Sweden the importation and use of PCBs were restricted by law; in the United Kingdom, as in the USA, sales were voluntarily restricted to the lower chlorinated PCBs for use as dielectrics in enclosed systems; in the Federal Republic of Germany, use in hydraulic and heat transfer fluids was also permitted. In Japan the production and use of PCBs were banned in 1972. Limitations on sales were subsequently introduced in other countries.

No information is available on the scale of production and the uses of PCTs.

### 3.2 Entry of PCBs into the Environment

Surveys of the sources of environmental pollution with PCBs were made before production and use were limited, and the information available may not now apply in North America and elsewhere. Table 4 gives an estimate of the fate of the PCBs produced in 1970 in the USA (Nisbet & Sarofim, 1972). Only 20% of the annual production can be regarded as a net increase in current usage, and the remainder is balanced by a loss to the environment. More than one-half of this entered dumps and landfills and it has been calculated that 0.3 million tonnes of PCBs have accumulated in such locations in North America since 1930. Much of this was originally enclosed in containers such as capacitors or was in plasticized resins and will not be released until the containing medium decays. The diffusion of PCBs from landfills is likely to be slow on account of their low volatility and low water solubility; Carnes et al. (1973) found little leaching from one site they tested.

Table 4. Entry of PCBs into the environment[a]

| Route | Percentage of annual production | PCB type (% chlorination) |
|---|---|---|
| Vaporization from plasticizers | 4.5 | 48–60 |
| Vaporization during incineration | 1 | 42 |
| Leaks and disposal of industrial fluids | 13 | 42–60 |
| Destruction by incineration | 9 | mainly 42 |
| Disposal in dumps and landfills | 52.5 | 42–60 |
| Net increase in current usage | 20 | 42–54 |

[a] From Nisbet & Sarofim, 1972.

Pollution of the environment has occurred mainly from the first three routes mentioned in Table 4. In addition, there are other routes, which although involving relatively small amounts, nevertheless have an influence on the entry of PCBs into food chains. PCBs have been used in the USA in amounts of about 10 tonnes/year in pesticide formulations (Panel on Hazardous Trace Substances, 1972), and the unauthorized use of scrap transformer fluid for this purpose has led to local contamination of milk supplies.

Pressure sensitive duplicating paper containing PCBs has found its way into waste paper supplies and has been recycled into paper and board used as food packaging materials; paints for coating the bottom of ships contained 3–5% of PCBs—about 3% of the annual quantity imported into Sweden has been used for this purpose—and this has been a source of plankton contamination (Jensen et al., 1972a).

### 3.2.1 Release of PCBs into the atmosphere

There appears to be little widespread atmospheric contamination during the manufacture and processing of PCBs, but this can occur during their subsequent use and disposal. Although PCBs have a low volatility, there may be an appreciable loss to the atmosphere during the lifetime of a PCB-plasticized resin, particularly of the lower chlorinated products. Further pollution may occur during the incineration of industrial and municipal waste. Most municipal incinerators are not very effective in destroying PCBs; efficient incinerators can be designed for this purpose (Jensen & Wickberg, unpublished report 1971; Jensen, unpublished report 1972), although the higher chlorinated PCBs are more resistant to pyrolysis. Secondary sources of atmospheric pollution are volatilization from soil, and from the drying of sewage sludge. Laveskog (1973) found that the PCB emissions from municipal incineration and from the drying of sewage sludge each amounted to about 1 kg/year per million inhabitants, a small amount compared with the 2 tonnes deposited yearly from aerial fallout in the south of Sweden.

### 3.2.2 Leakage and disposal of PCBs in industry

The major source of environmental pollution with PCBs, which eventually affects food chains, is the leakage and the disposal of industrial fluids. There have been serious cases of poisoning in man (Kuratsune et al., 1972) and in animals (Panel on Hazardous Trace Substances, 1972) due to leakage from a heat exchanger. Leakages, or the unintentional or deliberate discharge of waste, have contaminated seas, lakes, waterways, and sewers (Duke et al., 1970; Schmidt et al., 1971; Veith, 1972).

Analysis of solids from factory wastes in Japan have revealed a wide variation in PCB content. In most, less than 1 mg/kg was detected, but in some the contamination was very heavy, the highest level recorded being 8.26% from a factory manufacturing electrical equipment (Japanese Environment & Safety Bureau, unpublished report 1972).

## 4. ENVIRONMENTAL TRANSPORT AND TRANSFORMATION

### 4.1 Environmental Transport

Nisbet & Sarofim (1972) emphasize that the available data are insufficient to determine anything but a very crude model of the transfer of PCBs into the environment. Guesses can be made by referring to the distribution of DDT, which resembles the PCBs to some extent in its physical and chemical properties and about which more is known. Much of the following discussion arises from their analysis of the situation in the North American continent.

### 4.1.1 Air transport

By analogy with DDT, it might be expected that PCBs entering the atmosphere in the vapour phase would be adsorbed rapidly on to particles, which would be deposited or washed out in rain at a rate depending on their particle size, the average residence time in the atmosphere being 2–3 days. This has been confirmed by Södergren (1972a), who measured the deposition in southern Sweden of PCBs that had originated from municipal incinerators or had been carried over from Denmark by the prevailing winds, and showed that the amount deposited in central Sweden was much less. Carnes (1973) and Laveskog, (unpublished report 1973) found mainly particulate PCBs in the emission from incinerators. Harvey & Steinhauer (1974), however, considered that the results of their analyses of air in north Atlantic regions indicated that most of the PCBs carried in the air were in the vapour phase.

### 4.1.2  Transport in soil

The PCBs in soil are derived mainly from particulate deposition, estimated at 1000–2000 tonnes annually in North America, most of which is in urban areas. Small amounts have originated from the use of sewage sludge as a fertilizer, from the leaching of landfills, and from the use of PCBs in pesticide formulations. Tucker et al. (1975) found that under experimental conditions, the higher chlorinated PCBs were not leached from soils by percolating water, and those with a lower level of chlorination were removed only slowly, particularly from soils of high clay content. Losses do occur by volatilization and by biotransformation; by analogy with DDT and its metabolites the half-time in soil has been estimated at 5 years. Haque et al. (1974) showed that the rate of evaporation decreased with the clay content of the soil and the degree of chlorination of the biphenyl, and increased with temperature. Biotransformation has also been shown to play a part in the disappearance of the lower chlorinated compounds from soil (Iwata et al., 1973).

The total amount of PCBs distributed over North America, apart from that in dumps and landfills, has been estimated at 20 000 tonnes, of which one quarter has subsequently been transported via the air to the seas. (Nisbet & Sarofim, 1972).

### 4.1.3  Transport in water

The entry of PCBs into water occurs mainly at the points of discharge of industrial and urban wastes into rivers, lakes, and coastal waters. Sewage treatment appears to remove particulate PCBs from water, but not PCBs in solution; these are concentrated in the sludge (Ahling & Jensen, 1970), which may be dumped into rivers and coastal waters. Holden (1970b) found a mean PCB content of 3 mg/kg in liquid sludges from Glasgow, and calculated that PCBs at the rate of 1 tonne/year were released into the Clyde and Thames estuaries; a similar output was calculated from water treatment plants off the California coast (Schmidt et al., 1971). Other localized sources of pollution are leakages or waste disposal from ships. PCBs in water are attached mainly to particulate matter (Södergren, 1973a) and eventually fall to the bottom sediment at a rate that depends on the particle size. PCBs may be leached from the sediment and may reach coastal waters, but Nimmo et al. (1971) noted little change in the PCB content of a sediment at a point downstream from a source of contamination over a period of 9 months. The process may be accelerated by the dumping of dredging spoil.

### 4.1.4  Transport through biota

According to the approximate calculations of Nisbet & Sarofim (1972), less than 1000 tonnes of PCBs are located in living organisms throughout the world, so that biological transport and degradation play little part in the fate of PCBs in the environment, though these factors have great ecotoxicological significance.

## 4.2  Transformation in the Environment

### 4.2.1  Abiotic transformation

The fate of the various PCBs in commercial mixtures depends on their physical and chemical properties. Some fractionation occurs during the volatilization of PCBs, because of a decrease in vapour pressure with increasing chlorination. The PCBs are chemically very stable, and are not likely to be degraded at a significant rate by hydrolytic or similar reactions under environmental conditions. They are, however, fairly easily degraded by photolysis under laboratory conditions; Safe & Hutzinger (1971), and Hutzinger et al. (1972b) have shown that PCBs are dechlorinated in hexane solution at a rate that increases with increasing chlorination. PCBs in aqueous-dioxane suspensions and in thin films give hydroxy and carboxylic acid derivatives on irradiation. The lower chlorinated biphenyls, at a vapour concentration in air of 1.5 mg/m$^3$ are readily destroyed by photolysis under laboratory conditions. There is no direct evidence of the extent of PCB breakdown in the atmosphere under environmental conditions, and nothing is known of the persistence and toxicity of any transformation products.

### 4.2.2  Biotransformation

The biotransformation of PCBs is discussed in section 6.6. Owing to the small proportion of the total environmental PCBs contained in living matter, biotransformation does not significantly influence the overall environmental concentrations of PCBs, though it has a marked influence on PCBs passing through food chains.

### 4.2.3  Metabolism in limited ecosystems

The presence of PCBs in sewage sludge suggests that they are not all readily transformed by microorganisms. Choi et al. (1974) found no evidence of biotransformation of Aroclor 1254 added to water entering an aerated biological water treatment system, although much of it was removed with the sludge; there appeared to be no interference by the PCBs in the performance of the system. Vodden (1973) quotes investigations showing that PCBs with four or fewer chlorine atoms are readily broken down by microorganisms, but that this can be inhibited by the presence of higher chlorinated PCBs. Mono- and dichlorobiphenyl can be transformed by *Achromobacter* isolated from sewage effluent (Ahmed & Focht, 1973), and a culture of lake bacteria can degrade some of the lower chlorinated components of Aroclor 1242 to chlorine-free derivatives (Kaiser & Wong, 1974).

Södergren (1972b) investigated the transport of Clophen A50 added to a model aquatic ecosystem; this was rapidly taken up by an alga (*Chlorella*) and no change was observed in the proportion of the component PCBs in the first consumer fish; however, there was a relative loss of the lower chlorinated PCBs in the perch, the second consumer. No progressive loss of the lower chlorinated PCBs was observed in sediment, plankton, invertebrates, and fish inhabiting a Swedish lake (Södergren, 1973a). Evidence on the metabolism of PCBs by fish is conflicting, but it seems probable that most fish, particularly those in the lower trophic stages of food chains, cannot readily degrade the low chlorinated PCBs.

### 4.3  Biological Accumulation

Although PCB concentrations in living organisms clearly indicate a progressive accumulation in food chains, the factors discussed in the previous sections make it impossible to give any reliable figure for bioaccumulation at each trophic stage. There is also doubt about which tissue level should be used in the calculation, that of the whole body, the fat, or the liver.

There is good evidence that all aquatic organisms studied in aquaria can absorb PCBs directly from water. The accumulation varies with the duration of exposure and the concentration in the ambient water. A diatom exposed to Aroclor 1242 showed an accumulation factor of 1100

(Keil et al., 1971); with Aroclor 1254 the following values have been obtained: pink shrimp, 6600; blue crab, 4600; oyster, 8100; pinfish, 980 (Duke et al., 1970); spot, 37 000 (Hansen et al., 1971); bluegills, up to 71 400 (Stalling & Mayer, 1972). The accumulation factor in scud exposed to Aroclor 1254 reached a maximum of 24 000 within 4 days and thereafter remained fairly constant (Sanders & Chandler, 1972). Similar results were obtained with other invertebrates, although with crayfish the uptake was slower and the accumulation was still increasing after 21 days. However, it is probable that most of the PCBs entering aquatic systems are retained by particulate matter, and the above accumulation factors are not necessarily applicable to natural ecosystems. Nimmo et al. (1971) demonstrated that the fiddler crab could ingest PCB contained in the bottom sediment. When exposed to Aroclor 1254 in water, ciliated protozoa, stated to be the major benthic input to aquatic food chains, had an accumulation factor of 60 (Cooley et al., 1972).

## 5. ENVIRONMENTAL LEVELS AND EXPOSURES

### 5.1  Air

Mean concentrations in air, in several locations in Sweden, ranged from the detection limit of 0.8 ng/m$^3$–3.9 ng/m$^3$. The highest figure recorded was 12.5 ng/m$^3$ (Ekstedt & Odén, 1974). In the USA, PCB concentrations in air ranged from 5 ng/m$^3$ near the north-east coast to 0.05 ng/m$^3$ at a distance of 2000 m out over the Atlantic Ocean (Harvey & Steinhauer, 1974). Results from the United States Environmental Protection Agency indicate a range between 1 and 50 ng/m$^3$ (Panel on Hazardous Trace Substances, 1972), and similar results have been reported from Japan (Tatsukawa & Watanabe, 1972).

### 5.2  Soil and Sediments

In Sweden, PCBs have been found in natural soil at a concentration of 15 µg/kg by Odén & Berggren (1973). The same authors also found 0.006–1.4 mg/kg in sediments from areas in the Baltic Sea with different degrees of pollution. Nimmo et al. (1971) found PCB levels of 1.4–61 mg/kg in sediment from an estuary at a point near the site of an accidental release of PCBs from a factory, and levels of 0.6 mg/kg at a point 16 km downstream. Soil samples from the bank 6.5 km downstream from the source contained 1.4–1.7 mg/kg. Less than 1 mg/kg has been found in Japanese agricultural soil, but as much as 510 mg/kg in soil near a factory making electrical components (Fukada, et al., 1973).

## 5.3   Water

In heavily contaminated waters, PCB concentrations may be several times greater than their solubility, owing to adsorption on suspended particles (Duke et al., 1970). Water in a Swedish river contained 0.5 ng/litre as it entered a water treatment plant, and 0.33 ng/litre in the tap water produced (Ahling & Jensen, 1970). Values of 0.1–0.3 ng/litre have been measured in other Swedish rivers (Ahnoff & Josefsson, 1974). Södergren (1973a) found a seasonal variation in the PCB level in a Swedish lake, with a maximum of 2 ng/litre, the pollution being attributed to aerial fallout. Concentrations of 10–100 ng/litre have been measured in tap water at Kyoto in Japan (Panel on Hazardous Trace Substances, 1972). In a polluted coastal area of Lake Michigan in 1970, PCB concentrations of from 450 to less than 100 ng/litre were measured in 1970, but there was a marked decrease in 1971, possibly due to the limitation on sales of PCBs (Panel on Hazardous Trace Substances, 1972). From the scanty information on PCBs, reinforced by analogy with the more extensive information on DDT, it has been estimated that nonpolluted fresh waters should contain not more than 0.5 ng/litre up to 5 ng/litre for the Great Lakes of North America, 50 ng/litre for moderately polluted rivers and estuaries, and 500 ng/litre for highly polluted rivers.

## 5.4   Living Organisms

There is now considerable information from Canada, Japan, Sweden, the United Kingdom, and the USA, on the accumulation of PCB's in biological material. Analytical measurements on different organisms, and on the same organism from different localities, vary widely and it is necessary to consider the factors that lead to this variation. Differences in analytical techniques may contribute to this (see section 2.3.5) but important influences are exerted by the extent of local pollution, the amount of fat in the organism studied, and its trophic stage in food chains.

### 5.4.1   The influence of local pollution

Most of the fish eaten by man is taken from waters with relatively little pollution. In a collaborative study by seven national laboratories

(International Council for the Exploration of the Sea, 1974), the PCB content of muscle tissue of fish taken from the North Sea was measured. A mean of 0.01 mg/kg was found in cod, herring contained up to 0.48 mg/kg with most samples in the range of 0.1–0.2 mg/kg, and plaice contained 0.1 mg/kg or less. Similar values were reported by Zitko (1971) for fish taken from the North Atlantic.

There are many examples of different PCB levels in similar species collected from areas of high and low pollution. Jensen, et al. (1972b) found five times as much PCBs in herrings caught in waters off industrialized areas near Stockholm, as in herrings from the cleaner waters of the west coast of Sweden. Similarly, levels in plankton harvested along the Swedish archipelago at various distances from Stockholm fell progressively, away from the more polluted areas (Jensen, et al., 1972c); the concentration in pike fell to one-half (Olsson & Jensen, 1974). Koeman et al., (1972b) found PCB levels of up to 88 mg/kg in the blubber of toothed whales caught in the North Sea, but none was detectable in similar species from New Zealand or Surinam. Holden (1973b) found high PCB concentrations in the blubber of seals in polluted coastal areas of the United Kingdom (up to 235 mg/kg), and much lower levels in unpolluted areas (down to 2 mg/kg).

Risebrough & de Lappe (1972) studied the PCB content of extractable lipids from brown pelican eggs collected from areas throughout North and South America, and showed that the content varied from 4 mg/kg up to 266 mg/kg in highly industrialized regions. They also reported levels greater than 3 mg/kg in fish from New York Sound and Tokyo Bay, both very polluted areas. Even higher levels of PCBs have been found in fish from polluted lakes and inland waterways, a level of 20 mg/kg being found in fish from Lake Ontario, and over 200 mg/kg in fish from the Hudson River (Stalling & Mayer, 1972). Similar correlations between pollution and PCB levels have been reported from the United Kingdom in fish (Portmann, 1970), and in mussels (Holdgate, 1971).

The association between high PCB levels and local pollution may be disturbed by the migratory habits of certain species, particularly in birds that may be exposed to PCBs in their wintering areas or on the migration routes. As much as 400 mg/kg has been measured in the fat of a robin entering Sweden, although the normal value is about 16 mg/kg. Many migratory birds start egg-laying on arrival at their summer quarters, so the PCB content of eggs may reflect the bird's previous exposure rather than the local pollution (Odsjö, 1973). A special case concerning the effect of pollution is seen in the use of fish-feed in poultry and fish farming. Kolbye (1972) stated that this may contain PCB levels of 0.6–4.5 mg/kg.

### 5.4.2 The influence of the fat content of tissues

PCBs are mainly stored in body fat (see section 5.5.5.1), and the total PCB content of the body tissues is much influenced by their fat content (Portmann, 1970; Westöö & Norén, 1970a). Jensen, et al. (1969) found PCB levels of 0.27 mg/kg and 0.33 mg/kg respectively, in the muscle tissue of herring and cod from the same area of the Baltic, although the cod is at a higher trophic stage (see section 5.4.3). These two species have 4.4 and 0.32% of extractable fat respectively, and when the PCB level is calculated on the fat content, values of 6.8 mg/kg for the herring and 11 mg/kg for the cod are obtained. Cod liver has a much higher fat content than cod muscle and Jensen (1973) has reported the ratio of PCB concentrations in cod liver and muscle to be over 100, the maximum in liver being 59 mg/kg. Jensen et al. (1969) have remarked that the considerable seasonal variation in the fat content of the herring, rising from 1% in spring to 10% in autumn, influences the tissue level of PCBs. Peakall et al. (1972) noticed a marked rise in tissue levels in starved birds owing to the mobilization of fat, and it is possible that the high levels of PCBs in the livers of birds dying during the "seabird wreck" in the Irish Sea were secondary to emaciation (Holdgate, 1971). De Freitas & Norstrom (1974) showed that, in pigeons, pure PCBs left fatty tissues and accumulated mainly in muscle during the mobilization of fat associated with starvation.

### 5.4.3 The influence of the trophic stage in food

Swedish work on the distribution of PCBs in aquatic ecosystems has been summarized by Jensen et al. (1972b), and Olsson et al. (1973), and it has been largely confirmed by work in other areas (Risebrough et al., 1968; Risebrough & de Lappe, 1972: Holdgate, 1971). Zoo- and phytoplankton readily absorb or adsorb PCBs from their environment; Södergren (1972) has demonstrated the rapid uptake of Clophen A50 by the unicellular alga *Chlorella*. Marine zooplankton may contain PCB levels of 5 mg/kg in extractable lipids in areas of moderate pollution (Jensen et al., 1972c; Williams & Holden, 1973), with somewhat lower values in relatively unpolluted areas. However, results with plankton, particularly where high levels are found, must be regarded with caution as the sample could be contaminated with PCB-rich oil or tar particles. Herring feeding on plankton in areas of moderate pollution contain PCB levels of about 0.5 mg/kg in muscle tissue (10 mg/kg in extractable fat); plaice and flounder, both bottom-feeding fish contain about one-third of this, presumably because the PCB content of benthic organisms is lower. In predatory fish such as the cod and pike, the PCB level in extractable fat is about 10 mg/kg and a mean of 10 mg/kg has been measured in the blubber of seals.

Much higher PCB values have been obtained in fish-eating birds; levels of 18 mg/kg (650 mg/kg in extractable fat), and 17 mg/kg (420 mg/kg in extractable fat) were found in the herring gull and comorant respectively. Lower values were found in birds feeding on invertebrates, such as the long-tailed duck which contained 14 mg/kg in fat; marine invertebrates contain PCB levels in the region of 0.1–0.2 mg/kg. At the top trophic level, 96 mg/kg (9.7 g/kg in extractable fat) has been measured in the eagle owl; the highest recorded PCB values were from eagle owls found dead in the south-east coastal region of Sweden, 260 mg/kg in the brain (3.4 g/kg in extractable fat) and 110 mg/kg in muscle (12 g/kg in extractable fat). (Odsjö, 1973).

Less information is available on terrestrial ecosystems. PCB concentrations in the region of 0.01 mg/kg have been found in fresh tissue in slugs, snakes, and ants, and slightly higher concentrations in earthworms. Tissue levels were generally at the limit of detection (0.01 mg/kg) in herbivorous mammals (Odsjö, 1973). Brüggemann et al. (1974) reported a mean PCB concentration of 0.22 mg/kg in 20 out of 72 measurements in the adipose tissues of the hare and a higher value (2.5 mg/kg) in 1 out of 5 tests on the adipose tissues of the fox. Higher values have been found in the American mink in Sweden with 0.58 mg/kg in muscle (45 mg/kg in fat) presumably because of a fish diet (Odsjö, 1973). Tissue levels in wild birds on a mixed diet are variable and rather low, but those in predatory birds are higher. Prestt et al. (1970) related the PCB concentration in the liver of wild birds to their diet; less than 1 mg/kg was found in insectivorous birds and more than 70 mg/kg in the sparrow hawk. A level of 0.5 mg/kg has been found in the muscle of the eagle owl in central Sweden, but this is much less than the tissue concentrations encountered in this bird in coastal areas (Odsjö, 1973). High tissue concentrations in predatory and marine birds have also been reported from Canada (Gilbertson & Reynolds, 1974), the Netherlands (Koeman et al., 1972a), and from the United Kingdom (Bourne & Bogan, 1972).

### 5.4.4 Indicator organisms

Several of the organisms, that have been shown to accumulate PCBs from the environment, have been suggested as indicators of the extent of local pollution with PCBs. In aquatic systems, the use of plankton as an indicator has the advantage that it is at the lowest trophic stage of food chains but errors may occur in the determination because of the inclusion in the sample of nonplanktonic particles with a high PCB content (Jensen et al., 1972a). The herring, which feeds on plankton, has been suggested as an indicator (Jensen et al., 1972b) and, at higher trophic stages, the pike, which is a stationary fish (Olsson & Jensen, unpublished report, 1974), and seabirds (Jensen et al., 1972c). In fresh waters, the amphipod *Gammarus pulex* has been used as an indicator organism for chlorinated hydrocarbons (Södergren et al., 1972). However, Zitko et al. (1974) claimed that the variation between individual fish was so high that Atlantic herring and yellow perch could be used to detect trends in pollution only if large numbers were taken for analysis, with an interval between measurements of at least 4 years.

A series of monitoring studies has been made by OECD, the species selected covering terrestial, fresh water, and marine environments. The analytical results, and the general problems of selecting species for monitoring, have been discussed by Holden. (1970a, 1973a, 1973b).

## 5.5   The Extent of Human Exposure to PCBs and PCTs

### 5.5.1   Air and water

The maximum concentration of PCBs in air is not likely to exceed 50 ng/m$^3$ (section 5.1). The highest concentration of PCBs reported in domestic tap water is 100 ng/litre in the Kyoto area of Japan (Panel on Hazardous Trace Substances, 1972), but levels more likely to be encountered should not exceed 1 ng/litre (section 5.3).

### 5.5.2   Food

The PCB content of a variety of foods on the Swedish market has been measured by Westöö & Norén (1970a) and Westöö et al. (1971). Less than 0.1 mg/kg was found in samples of butter, margarine, vegetable oils, eggs, beef, lamb, chicken, bread, biscuits, and baby food; one sample of pork out of more than 100 had a PCB content in the range of 0.2– 0.5 mg/kg. As might be expected from the discussion in section 5.4, higher values were found in fish depending on the fat content and the pollution of the fishing area (Westöö & Norén, 1970a; Berglund, 1972). The PCB levels obtained in an extensive study by the US Food and Drug Administration are shown in Table 5.

These values are considerably higher than those reported from Sweden but they are probably biased, as they include samples originating from areas previously suspected of having been subject to local pollution. In a

Canadian survey, PCB levels of less than 0.01 mg/kg were found in eggs (Mes et al., 1974) and a mean of 0.042 mg/kg was found in domestic and imported cheese with a maximum of 0.27 mg/kg (Villeneuve et al., 1973b). No traces of PCTs were found.

In Japan, a similar range of PCB contents for most foods has been reported; however, some high levels have been reported for rice and vegetables harvested in fields polluted with PCBs (Environmental Sanitation Bureau, 1973). The PCB content of most fish on the market was less than 3 mg/kg, although some contained more than this. The PCT content of fish was much lower (Fukano et al., 1974). In the Netherlands, eel has been reported to contain PCT levels of 0.2–0.5 mg/kg and PCB levels of 4.7 mg/kg (Freudenthal & Greve, 1973).

Table 5. PCB levels in food in the USA[a]

| Food | % positive (0.1 mg/kg) | Level in positive samples (mg/kg) | |
|------|------------------------|------|---------|
| | | Mean | Maximum |
| Cheese | 6 | 0.25 | 1.0 |
| Milk | 7 | 2.3 | 27.8 |
| Eggs | 29 | 0.55 | 3.7 |
| Fish | 54 | 1.87 | 35.3 |

[a] From Kolbye (1972).

Samples of butter from the Westphalian area of the Federal Republic of Germany, obtained in the period 1972–74, contained PCB levels of 0.38 mg/kg (range 0.25–0.54 mg/kg) (Claus & Acker, 1975).

Relatively high PCB levels in some packaged foods in Sweden, mainly of imported origin, could be attributed to migration from the packaging material (Westöö et al., 1971). The highest level encountered was 11 mg/kg in a children's breakfast cereal; PCB levels of 70 mg/kg and 700 mg/kg were found in the material of the inner bag containing this product and in the outer cardboard container respectively. Up to 2000 mg/kg was found in cartons of other samples. Villeneuve et al. (1973a) have analysed packaged food in Canada; they found that 66.7% of the samples contained PCB levels of less than 0.01 mg/kg, 30.7% contained between 0.01 and 1 mg/kg, and 2.6% contained more than 1 mg/kg. PCT determinations were also made on these samples; 94.5% of the samples contained less than 0.01 mg/kg and 5.5% contained 0.01–0.05 mg/kg. The highest PCB levels were encountered in a rice sample with 2.1 mg/kg where the packaging material contained 31 mg/kg, and in a dried fruit sample with 4.5 mg/kg in a container containing 76 mg/kg. In a survey of packaging containers, approximately 80% were found to contain PCB and PCT levels of less than 1 mg/kg, while about 4% contained levels higher than 10 mg/kg. The most likely source of PCBs in packaging materials is the recycling of paper waste containing pressure-sensitive duplicating paper (Masuda et al., 1972).

### 5.5.3 Occupational exposure

Occupational exposure does not only occur during the manufacture of PCBs and with their use in the electrical industry. It may also be widespread among mechanics in contact with lubricating oils and hydraulic fluids, among workers exposed to varnishes and paints, and among office workers from contact with pressure-sensitive duplicating paper, some brands of which readily transfer PCBs to skin (Kuratsune & Masuda, 1972b). Studies in Finland showed that whole blood from persons with no special exposure to PCBs contained 0.3–1.2 µg/100 ml, while blood from persons handling PCBs in an analytical laboratory contained 3.6–6.3 µg/100 ml and blood from workers in a capacitor factory had PCB levels of 7.5–190 µg/100 ml in the blood and 30–700 mg/kg in fat. No signs of toxicity were evident in these workers (Karppanen & Kolho, 1973). Similar plasma values were found in workers from Japanese capacitor factories, but here skin lesions were noted (Hasegawa et al., 1972a). This same study reported that air levels of PCBs of 0.01–0.05 mg/m$^3$ were measured in a factory where KC-300 was used in the manufacture of electric condensers. PCB levels in serum in workers ranged from 10 to 65 µg/100 ml.[a] One month after the use of PCBs had been suspended, serum levels still ranged from 9 to 74 µg/100 ml. However, in another factory making electric condensers, serum levels decreased from an average of 80 µg/100 ml to 30 µg/100 ml within three months of the use of PCBs being suspended (Kitamura et al., 1973). According to Hara et al. (1974), the half-time of PCBs in the blood of workers engaged in the manufacture of electric condensers for less than 5 years was several months, while that of workers employed for more than 10 years was 2–3 years. Hammer et al. (1972) found a higher frequency of measurable plasma values in workers working with refuse burners than in a control group.

### 5.5.4 Other sources of exposure

Broadhurst (1972) has reviewed the many technical applications of PCBs that appear in the literature and in patent specifications, and which

---

[a] In this document, the concentrations of PCBs in blood and serum are expressed in µg/100 ml although in some original papers the values are given in µg/100 g. For practical purposes the differences, about 5% and 3% respectively, can be neglected.

indicate the possibility of a widespread nonoccupational low-level exposure to PCBs, other than that deriving from the diet. PCBs are used in the home in ballast capacitors for fluorescent lighting, and exposure deriving from pressure-sensitive copying paper has not been limited to office workers. The valuable properties of PCBs as plasticizers has led to their use in furnishings, interior decoration, and building construction; examples are surface treatment for textiles, adhesive for waterproof wall coatings, paints, and sealant putties. PCBs have been used as plasticizers for plastic materials and in the formulation of printing inks.

### 5.5.5 Biological indices of human exposure

The only surveys of value have been on body fat, blood, and milk.

#### 5.5.5.1 *Body fat*

In a survey of 637 fat samples taken at autopsy or during surgery in the USA, 68.9% contained PCB levels of less than 1 mg/kg, 25.9% contained 1–2 mg/kg, and 5.2% contained more than 2 mg/kg (Yobs, 1972). A similar distribution was found in a smaller survey by Price & Welch (1972). In the Kochi area of Japan, a mean PCB level of 2.86 mg/kg was recorded with an upper limit of 7.5 mg/kg; about double these values were found in the Kyoto area (Nishimoto et al., 1972a, 1972b). Bjerk (1972) reported average PCB levels of 0.9 mg/kg (1.6 mg/kg on a lipid basis) in adipose tissue, taken at 40 autopsies in the Oslo area. Curley et al. (1973b) found PCB levels ranging from 0.30 to 1.48 mg/kg in a total of 241 human adipose samples in Japan. A mean value of 5.7 mg/kg has been reported for 20 samples from the Federal Republic of Germany (Acker & Schulte, 1970); in a more recent study, 282 adipose tissue samples from different areas were found to contain PCB levels of 8.3 mg/kg of adipose tissue on a lipid basis (Acker & Schulte, 1974). In Austria, a range of PCB levels of 0.3–7.3 mg/kg on a lipid weight basis was found in 32 residents in the Vienna metropolitan area; an increase of the PCB concentration with age was not observed (Pesendorfer et al., 1973). Detectable levels of PCBs were found in only a few of 201 human fat samples in the United Kingdom and in these, the level did not exceed 1 mg/kg (Abbott et al., 1972). A survey of 51 human fat samples in New Zealand showed that all samples contained PCB residues with an average of 0.82 mg/kg (Solly & Shanks, 1974).

Doguchi et al. (1974) found an average PCT level of 0.6 mg/kg in human fat with a range of 0.1–2.1 mg/kg. Takizawa & Minagawa (1974) also found PCT levels of 0.02 mg/kg in human liver ($n=6$), 0.01 mg/kg in the kidney ($n=2$), 0.02 mg/kg in the brain ($n=3$) and 0.04 mg/kg in the pancreas ($n=1$). In the Netherlands, PCTs were found in human fat at levels of 0–1 mg/kg (Freudenthal & Greve, 1973).

### 5.5.5.2 *Blood*

43% of blood plasma samples from 723 volunteers in the USA showed the presence of PCBs; the mean value in these was about 0.5 µg/100 ml (Finklea et al., 1972). Studies in Finland on whole blood showed 0.31–1.2 µg/100 ml in persons with no special exposure to PCBs (Karppanen & Kolho, 1973). In Japan, an average PCB level of 0.32 µg/100 ml and a PCT level of 0.5 µg/100 ml have been recorded in the blood of non-occupationally exposed volunteers (Doguchi & Fukano, 1975). In patients with severe weight loss, high levels of PCBs in the blood (up to 10 µg/100 ml) were noted (Hesselberg & Scherr, 1974). This was attributed to the release of PCBs from the mobilization of fat.

### 5.5.5.3 *Human milk*

PCB concentrations measured in whole human milk in Sweden were 0.014 mg/litre in 1967 and 0.025 mg/litre in 1971–72 (Westöö & Norén, 1972), 0.03 mg/litre in Japan (Nishimoto et al., 1972a), 0.103 in the Federal Republic of Germany (Acker & Schulte, 1970) and 0.02 mg/litre in Canada (Musial et al., 1974). A survey in Colorado, USA, revealed 8 positive samples out of 39, within the range of 0.04–0.1 mg/kg (Savage et al., 1973).

### 5.5.6 Estimated daily intake

From the PCB levels encountered in air and drinking water (section 5.5.1), the daily intake from each of these sources is likely to be less than 1 µg.

It has been stated that the major part of the human dietary intake of PCBs is from fish (Berglund, 1972; Hammond, 1972). This may well be true in areas such as Japan or certain localities near the North American Great Lakes, where fish from polluted waters may form a relatively large part of the diet. Several investigators from Japan have measured the daily intake of PCBs in food; the highest mean value recorded was 48 µg/day, of which 90% was from fish (Kobayashi, 1972); the lowest was 8 µg/day (Ushio et al., 1974).

In much of Europe and North America, however, the daily intake of fish is in the region of 30–40 g and most of the fish is taken from waters of low pollution and contains PCB levels of not more than 0.1 mg/kg. Berglund (1972) has estimated that the daily intake of PCBs from fish in Sweden is in the region of 1 µg, though if the fish consumed were solely

Baltic herring, the intake would be about 10 µg. It is difficult to make an assessment of the PCB intake from foods other than fish. Westöö et al. (1971, 1972), in their extensive study of the Swedish diet, reported that most foods contained PCB levels of less than 0.1 mg/kg; it may be concluded that this corresponds to a daily intake of less than 100 µg. The conclusion that foods other than fish may make a greater contribution to the PCB content of the diet can be drawn from the survey in the USA reported by Kolbye (1972), and also from a Swiss study of three different types of home-prepared meals, not containing fish, which were found to contain 6, 41, and 84 µg of PCBs, respectively (Zimmerli & Marek, 1973).

The figures for the PCB content of human milk (section 5.5.5.3) indicate that most nursing mothers excreted about 30 µg/day by this route, and in some areas up to 100 µg/day. It may be assumed that only a portion of the PCBs absorbed was excreted by this route, so that the daily intake could have been more than 100 µg.

It may be concluded that, in the more industrialized countries, the average daily PCB intake from the diet has rarely been less than 5 µg or greater than 100 µg; it is likely that the non-dietary sources of exposure detailed in section 5.5.4 have made a significant contribution, but this cannot be estimated at this time. In any area, the intake depends not only on the diet, but also on social domestic and environmental conditions; the influence of these factors on the daily intake cannot readily be quantified.

# 6. METABOLISM

## 6.1 Absorption

The experiments of Vos & Beems (1971) who applied several commercial PCBs to rabbit skin and found systemic effects (see section 7.1.1.4) indicate that PCBs can penetrate the skin. Early cases of human poisoning from occupational exposure were probably due to a combination of skin absorption and inhalation. Experimental studies on rats by Benthe et al. (1972a) showed that an aerosol containing Aroclor 1242 (particle size 0.5–3.0 µm) was readily absorbed through the lungs.

Although one means of entry of PCBs into aquatic food chains is through the consumption of plankton by fish, aquatic organisms can also absorb PCBs in solution in the ambient water, presumably mainly through the gills (see section 4.3). Salmon eggs can also absorb PCBs from water (Johansson et al., 1970), and Södergren & Svensson 1973 concluded that mayfly nymphs could take in PCBs from water through the gills and the integument.

Recent work with chlorobiphenyl isomers administered orally to rodents at levels up 100 mg/kg of body weight for lower chlorinated compounds and up to 5 mg/kg for the higher chlorinated compounds, showed that 90% of the compounds were rapidly absorbed (Albro & Fishbein, 1972; Berlin et al., 1973; Melvås & Brandt, 1973). Cholestyramine, a basic anion exchange resin, was shown to interfere with intestinal absorption of KC-400 in mice (Tanaka & Araki, 1974).

PCTs have been shown to be absorbed from the gut (Sosa-Lucero et al., 1973) but very little information is available on the rate of absorption.

## 6.2   Tissue Distribution of PCBs

Grant et al. (1971a) demonstrated that 4 days after an oral dose of Aroclor 1254 at 500 mg/kg was given to rats, the concentrations of PCBs in fat, liver, and brain were 996, 116, and 40 mg/kg, respectively. Similar results showing that the highest concentration was in fat, were obtained in rats given Aroclor 1254 in the diet (Curley et al., 1971), in boars (Platonow et al., 1972), cows (Platonow & Chen, 1973), and in pigeons and quail (Bailey & Bunyan, 1972). In the experiments of Curley

Table 6. Tissue distribution of PCBs (mg/kg wet weight) in rats fed Aroclor 1254 (Grant et al., 1974), Aroclor 1242, or Aroclor 1016 (Burse et al., 1974) at 100 mg/kg for about 6 months

| Tissue | PCB levels in tissue (mg/kg) | | |
| | Aroclor 1254 | Aroclor 1242 | Aroclor 1016 |
| --- | --- | --- | --- |
| Blood | 0.40 | 0.53 (plasma) | 0.38 (plasma) |
| Liver | 16 | 4.21 | 7.86 |
| Brain | 3.4 | 1,69 | 2.98 |
| Kidneys | — | 1.89 | 3.21 |
| Heart | 7.3 | — | — |
| Fat | 32.0 | 110 | 236 |
| Urine | — | 0.03 | 0.28 |

et al. (1971), the tissue concentrations initially showed a rapid rise and thereafter a slow increase while the PCB diet was being administered; Grant et al. (1974) fed diets containing Aroclor 1254 at 0.2, 20, and 100 mg/kg to rats for 8 months, during which period the tissue concentrations reached a steady state that was dose-dependent (Table 6). Similar tissue distribution data for Aroclors 1016 and 1242 have been reported by Burse et al. (1974) and for Kanechlor-400 by Yoshimura et al. (1971).

PCB deposition, in general, depends on the fat content of the tissue (section 5.4.2). Residues in trout, receiving Aroclor 1254 in doses of 15 mg/kg in the diet, stabilized after 16 weeks while the absolute quantity continued to increase as the fish grew (Lieb et al., 1974).

More detailed information on the tissue distribution of PCBs and their metabolites has been obtained by the administration of pure $^{14}$C-labelled compounds, using both whole-body autoradiography and scintillation counting of tissue samples. Berlin et al. (1975) demonstrated that after a single oral dose of $^{14}$C-labelled 2,5,2',4',5'-pentachlorobiphenyl, radio-activity rapidly entered the circulation of mice and was distributed in the tissues, particularly in the liver, kidneys, lungs, and adrenals. Subsequently, the radioactivity in the body fat increased, rising to a maximum within 4–24 h. In most other tissues the radioactivity decreased rapidly after dosing, but the authors noted a special affinity for the skin, the bronchiolar epithelium of the lungs, and certain glandular secreting tissue. Soon after administration of the dose, radioactivity appeared in bile and was excreted in the faeces. Similar results were obtained by Melvås & Brandt (1973) with 2,4,2',4'-tetrachlorobiphenyl in the mouse, and with 2,4,2',3'- and 2,4,3',4'-tetrachlorobiphenyls in the quail; in the mouse they found a high affinity for the adrenal cortex, the corpora lutea, and glandular secreting tissue, and in the quail the radioactivity in egg yolk was high, exceeding that in fat. Brandt & Ullberg (unpublished report 1973) found a similar pattern after administration of hexa- and octachlorobiphenyls to mice.

## 6.3    Tissue Distribution of PCTs

Diets containing Aroclor 5460 at levels of 10, 100, and 1000 mg/kg were administered to rats for 7 days (Sosa-Lucero et al., 1973). Table 7 shows the tissue distribution obtained in this study in rats fed with

Table 7. Tissue distribution (mg/kg wet weight) of PCTs (Aroclor 5460) in rats fed dietary levels of 100 mg/kg for 7 days (Sosa-Lucero et al., 1973) and fed PCB (Aroclor 1254) at 100 mg/kg for 9 days (Curley et al., 1971)

| Tissue | Aroclor 5460 | Aroclor 1254 |
|---|---|---|
| Blood | 1.32 | 0.1 |
| Liver | 47 | 6 |
| Brain | 5.1 | 4 |
| Kidneys | 15.1 | 5 |
| Heart | 21.5 | — |
| Fat | — | 180 |

Aroclor 5460 at 100 mg/kg body weight and the values in rats fed with Aroclor 1254 at 100 mg/kg body weight in a similar study (Curley et al., 1971). After oral administration of Aroclor 5460 to the cod, the concentration of PCTs in the liver was more than 100 times that in muscle on a wet weight basis (Addison et al., 1972), a ratio found by Jensen et al. (1973) for PCBs in the cod.

### 6.4 Placental Transport

Aroclors 1221 and 1254 were found to cross the placenta of rabbits, when administered orally to does during gestation. The concentration in fetal tissues was dose-dependent and much less with Aroclor 1221 than with Aroclor 1254; with the latter, the concentration in the fetal liver was greater than that in the maternal liver (Grant et al., 1971b). Curley et al. (1973a) found some placental transport of Aroclor 1254 in the rat. Platonow & Chen (1973) demonstrated that the PCBs in the fetal kidney of a cow dosed with Aroclor 1254 was greater than that in the mother. Placental transfer of polychlorinated biphenyls has also been reported in the mouse (Berlin et al., 1975; Melvås & Brandt, 1973; Brandt & Ullberg, unpublished report 1973).

PCB concentration in human umbilical blood has been shown to be about 25% of that in maternal blood (Taki et al., 1973). Placental transfer of PCBs was observed in Yusho patients (Tsukamoto et al., 1969).

No information is available on the placental transfer of PCTs.

### 6.5 Excretion and Elimination

#### 6.5.1 Milk

Saschenbrecker et al. (1972) found that after oral administration of doses of Aroclor 1254 of 10, and 100 mg/kg in diet to cows, 6.27 and 74.5 mg/litre, respectively, appeared in the milk after 24 hours. These levels were reduced to less than one-half within 3 days but traces remained at 50 days. Cows receiving 200 mg of Aroclor 1254 daily reached a steady state concentration of 61 mg/kg in milk fat and 42 mg/kg in body fat after 10 days (Fries et al., 1973). The PCBs in milk survived processing into dairy products, and most was located in milk fat (Platonow et al., 1971). PCBs have also been found in human milk (see section 5.5.5.3).

#### 6.5.2 Eggs

Several investigations have demonstrated the presence of high levels of PCBs in the eggs of seabirds (Risebrough & de Lappe, 1972). An incident has been reported in which PCB contamination of poultry food was toxic to chickens and decreased the hatchability of eggs which also contained PCB residues (Pichirallo, 1971). In a laboratory study, Scott et al. (1975) fed Aroclor 1248 to chickens at dietary concentrations of 0, 0.5, 1.0, 10, and 20 mg/kg. After 8 weeks, the approximate PCB levels in eggs were 0, 0.22, 0.41, 3.1 and 7.0 mg/kg respectively. Jensen & Sundström (1974) found 33% of an oral dose of 2,4,5,2′,4′,5′-hexachlorobiphenyl administered to quails was excreted in eggs over a period of 10 days; eggs also provided a major route of excretion in the pheasant (Dahlgren et al., 1971).

### 6.5.3  Urine and faeces

All investigators agree that only traces of PCBs can be found in the urine of dosed animals, and that faeces provide a major route of elimination. When the analysis of faeces is limited to the determination of unchanged PCBs, the recovery of the dose administered is incomplete; in boars receiving single or repeated doses of Aroclor 1254, not more than 16% of the dose was recovered from the faeces and less than 1% in urine (Platonow et al., 1972). Better recoveries have been obtained with PCB labelled with radioactive isotopes. Yoshimura et al. (1971) found 70% of the activity from a dose of tritium-labelled Kanechlor 400 in faeces and 2% in urine over a 4-week period. Berlin et al. (1974, 1975) found over 75% of the activity from $^{14}$C-labelled penta- and hexachlorobiphenyls in faeces and less than 2% in urine; most of the faecal excretion consisted of PCB metabolites (see section 6.6.1). Similar results were obtained by Melvås & Brandt (1973) with tetrachlorobiphenyls.

### 6.6  Biotransformation

### 6.6.1  Metabolic degradation

Most investigators studying the tissue distribution of PCBs after administration of commercial mixtures have noted a relative reduction of the gas–liquid chromatographic peaks with shorter retention times, corresponding to the lower chlorinated biphenyls. This has been reported in the rat (Grant et al., 1971a; Curley et al., 1971), the rabbit (Grant et al., 1971b), the cow (Platanow & Chen, 1973), and in pigeons and quails (Koeman et al., 1969; Bailey & Bunyan, 1972). Samples of tissues from animals and man (see table 3, pp. 20–21) that have absorbed PCBs from the environment have shown, on analysis, a pattern of peaks approaching that of PCB mixtures with more than 50% chlorination, although the major manufactured products contain 42% of chlorine or less. This has led to the belief that the rate of metabolic attack on PCBs decreases with increasing chlorination. Studies on single PCBs with 1, 2, 4, or 5 chlorine atoms have shown that these are more readily excreted as metabolites in faeces by mammals and birds and remain for a shorter time in fatty tissues than most PCBs with 6 or more chlorine atoms (Berlin et al., 1975; Hutzinger et al., 1972a; Melvås & Brandt, 1973; Brandt & Ullberg, unpublished report 1973). See also section 6.6.2.

The administration to rats of diets containing Aroclors 1016 or 1242 at a concentration of 100 mg/kg resulted in a steady state in adipose tissue for both compounds in about 4 months. When gas chromatographic traces were compared with those for standard PCB mixtures a difference in the gas–liquid chromatographic pattern, that is disappearance of the peaks with short retention times was noted. After exposure to PCBs was discontinued, a major portion of the PCBs was eliminated from the body in 4 months. However, 20% of the total PCBs in Aroclor 1242 with longer retention times were present after 6 months, and 10% of the total PCBs of Aroclor 1016 with longer retention times were present in adipose tissue after 5 months. (Burse et al., 1974).

The excretion of monohydroxy metabolites of 3,4,3',4'-tetrachloro-biphenyl and 2,4,3',4'-tetrachlorobiphenyl (orally administered) in rats has been demonstrated by Yoshimura & Yamamoto (1973). Yoshimura et al. (1973), Yamamoto & Yoshimura (1973), Yoshimura & Yamamoto (1974), and Yoshimura et al. (1974). They demonstrated that the meta-bolites of the first isomer were 2-hydroxy or 5-hydroxy compounds while the metabolites of the second isomer were 5-hydroxy and 3-hydroxy compounds. All hydroxy metabolites were excreted non-conjugated via the bile and no parent isomers were found in the bile. Yoshimura & Yamamoto (1975) found that unchanged 2,4,3',4'-tetrachlorobiphenyl was excreted through the intestine, when it was intravenously injected in rats with the bile duct ligated, while no metabolite of this isomer was excreted by this route.

Hutzinger et al. (1972a) demonstrated the presence of hydroxylated derivatives of mono-, di-, and tetrachlorobiphenyls in the excreta of rats and pigeons but not of trout and were unable to detect hydroxylated 2,4,5,2',4',5'-hexachlorobiphenyl. Berlin et al. (1975) isolated a hydroxy derivative of 2,5,2',4',5'-pentachlorobiphenyl from mouse faeces, and Jensen & Sundström (1975) have shown that although 2,4,5,2',4',5'-hexachlorobiphenyl is excreted very slowly, a hydroxy derivative could be detected in rat faeces. Hutzinger et al. (1974), however, showed that this compound was also dechlorinated by the rabbit and excreted as the hydroxy derivative of pentachlorobiphenyl. Gardner et al. (1973) detected hydroxylated metabolites in the urine of rabbits dosed with 2,5,2′,5′-tetrachlorobiphenyl together with a dihydroxy derivative which they regarded as evidence for the formation of an arene oxide (epoxide) intermediate (see section 6.5.3). Jansson et al. (1975) have identified up to 26 mono- and dihydroxylated metabolites of PCB in the bile and faeces of wild grey seal and guillemot from the Baltic.

There is little information on the biotransformation of PCTs. Addison et al. (1972), using gas–liquid chromatography, noted a loss of PCTs with a shorter retention time in the excreta of a cod dosed orally with Aroclor 5460; the same loss was observed in rat faeces after the administration of a diet containing Aroclor 5460 (Sosa-Lucero et al., 1973).

### 6.6.2 The effect of structure on retention

While each of the components of mixtures of PCBs has a different pattern of retention and elimination in different species, the measurement of biological half-times for PCB mixtures in tissues has provided useful information. Bailey & Bunyan (1972) found the half-time of PCBs in the fat of quail and pigeon, after cessation of dosing with Aroclor 1242, or with Aroclor 1254, to be 50 days and 125 days, respectively. A half-time of about 200 days was recorded in the fat of rats after feeding with Aroclor 1254 (Grant et al., 1974). Berlin et al. (1975) noted that in mice dosed with a pentachlorobiphenyl there was an initial fairly rapid elimination while the PCB level in the liver was high, followed by a slower elimination when most of the PCB was located in fat. It seems likely that the mobilization of PCBs from fat, and therefore their half-time in the body, depends upon their rates of metabolism. Berlin et al. (1974) investigated the hypothesis that the ability of a PCB to be readily degraded with a half-time of a few days depended upon the presence of two adjacent unsubstituted carbon atoms in the molecule rather than on the number of chlorine atoms, although the presence of such unsubstituted pairs depends to a large extent on the degree of chlorination. They came to the conclusion that this hypothesis probably applied to unsubstituted pairs in the 3,4-position, but that in the 2,3-position, their susceptibility to metabolic degradation was much influenced by the presence of chlorines in the *o*-position of the ring bridge.

Jensen & Sundström (1974) demonstrated that the retention of PCBs in human fat is also influenced by *o*-chlorine substitution (Table 3, pp. 20–21).

# 7. EXPERIMENTAL STUDIES ON THE EFFECTS OF PCBs AND PCTs

## 7.1 Toxic Effects in Different Species

Most of the available information on the toxicity of the PCBs has been obtained from studies on commercial mixtures. The much smaller amount of information available concerning the impurities in commercial products and PCTs, is given in sections 7.2 and 7.3.

### 7.1.1 Mammals

#### 7.1.1.1 *Acute oral and intravenous toxicity*

Earlier work by the Monsanto Company indicated a low acute oral toxicity for the Aroclors, the $LD_{50}$s to rats ranging from 4.0 g/kg for Aroclor 1221 to 11.3 g/kg for Aroclor 1262 (Panel on Hazardous Trace Substances, 1972). More recent work has demonstrated a slightly higher toxicity (Bruckner et al., 1973; Grant & Philips, 1974). With a single intravenous dose the $LD_{50}$ for Aroclor 1254 was 358 mg/kg body weight in adult female Sherman rats (Linder et al., 1974). The acute oral $LD_{50}$ for Aroclor 1254 in the same strain and sex was between 4 and 10 g/kg (Kimbrough et al., 1972). According to Bruckner et al. (1973), severely poisoned animals showed weight loss, ataxia, diarrhoea, and chromo-dacryorrhoea, and they considered progressive dehydration and central nervous depression were the causes of death. In rats, vacuolation in the liver and kidneys was observed (Bruckner et al., 1973) and also ulceration of the gastric and duodenal mucosa (Kimbrough et al., 1972).

Yamamoto & Yoshimura (1973) found the intraperitoneal $LD_{50}$ of 2,4,3′,4′-tetrachlorobiphenyl in mice to be 2.15 g/kg body weight and that of the 5-hydroxy derivative, which is the main *in vivo* metabolite, to be 0.43 g/kg body weight.

#### 7.1.1.2 *Subacute oral toxicity*

After repeated administration, the PCBs have a cumulative toxic action. In a group of rats receiving Aroclor 1254 at a dose of 1 g/kg of diet, deaths occurred between the 28th and 53rd days of feeding (Tucker & Crabtree, 1970), and with Phenochlor DP6 at 2 g/kg of diet, deaths occurred between the 12th and 26th days (Vos & Koeman, 1970). In the latter experiment enlarged livers, small spleens, and a progressive chemically-induced hepatic porphyria were seen at autopsy. Repeated weekly oral administration of 150 mg of Aroclors 1221, 1242 or 1254 to rabbits for

14 weeks produced liver enlargement and damage with Aroclor 1242 and no effect with Aroclor 1221 (Koller & Zinkl, 1973). Allen et al. (1974) administered diets containing Aroclor 1248 at the concentration of 25 mg/kg of diet to 6 female rhesus monkeys for two months; facial oedema, loss of hair, and acne developed after 1 month and one animal died with severe gastritis 2 months after removal from experimental diet. PCB concentrations in the body fat of the animals, after two months of treatment, averaged 127 mg/kg while 8 months later the value declined to 34 mg/kg.

Mink appear to be unusually sensitive to PCBs. Aulerich et al. (1973) administered diets containing PCB levels of 30 mg/kg (10 mg/kg each of Aroclors 1242, 1248, and 1254) to adult mink and demonstrated 100% mortality within 6 months. PCB residues in the brains of these mink averaged about 11 mg/kg and were approximately twice the level observed in other tissues. Four months of feeding Aroclor 1254 at levels of 5 and 10 mg/kg demonstrated a dose-dependent retardation of weight gain of growing female mink. Female mink fed a diet supplemented with Aroclor 1254 at 5 mg/kg for 9 months failed to produce offspring (Ringer et al., 1972).

### 7.1.1.3 *Chronic oral toxicity*

Aroclors 1242, 1254, and 1260 have been administered for 18 months to rats at 1, 10 and 100 mg/kg in the diet (Keplinger et al., 1971). No adverse effects were recorded with the three Aroclors at 10 mg/kg but with Aroclors 1242 and 1254 at 100 mg/kg there was an increase in liver weight and a reduced survival of litters. In similar experiments on dogs, there was a reduced weight gain with Aroclors 1254 and 1260 in the diet at a level of 100 mg/kg (Keplinger et al., 1971). In the experiments reported by Kimbrough et al. (1972), male rats survived Aroclor 1260 in the diet at 1 g/kg (71.4 mg/kg body weight) for 8 months but 8/10 females died at this dose, 2/10 died at 500 mg/kg, and 1/10 at 100 mg/kg (7.2 mg/kg body weight). With Aroclors 1254 and 1260, a dose-dependent increase in liver weight in male rats was significant down to 20 mg/kg (1.4 mg/kg body weight) in the diet; with females the liver enlargement occurred only at diet levels of 500 mg/kg and higher. The livers showed an orange fluorescence, the cells were enlarged and vacuolated with lipid inclusions; there was also much increased smooth endoplasmic reticulum, and what was termed "adenofibrosis" was present, being more marked with Aroclor 1254 (see section 7.8). Grant et al. (1974) fed diets containing Aroclor 1254 at 0, 2, 20, and 100 mg/kg to rats for 246 days followed by 180 days on a PCB-free diet; after 246 days the body weight of the 100 mg/kg rats was significantly less than that of the controls and the liver weight was greater. The most notable change in the livers on histological examination was the appearance of fat microdroplets in the centrilobular region; this effect was dose-dependent and not seen in the rats receiving the 2 mg/kg diet and was reversible when the rats were returned to a normal diet.

Liver enlargement has been described by a number of authors, in a number of species, with different PCB mixtures and pure isomers and it is considered to be due primarily to hypertrophy of the smooth endoplasmic reticulum of the liver cells (Vos & Beems, 1971; Allen et al., 1973; Kimbrough, 1974; Nishizumi, 1970) but with large enough doses (oral intubation of about 150 mg/kg body weight/week for 14 weeks) of Aroclor 1254 and 1242, it progressed to frank liver damage (Koller & Zinkl, 1973). The smooth endoplasmic reticulum may condense in the liver cell and form hyalin inclusions and this may be accompanied by a loss of enzyme activity. Lipid accumulation, pigment deposition, nuclear changes, and necrosis may also occur (Vos & Notenboom-Ram, 1972).

The rhesus monkey is the only species reported to show signs of poisoning similar to those in human Yusho patients (see section 8). The administration of Aroclor 1248 at 2.5 and 5.0 mg/kg of diet for 1 year produced periorbital oedema, alopecia, erythema, and acneiform lesions involving the face and neck within 1–2 months. The effects were less marked in male monkeys. At 25 mg/kg of diet, one out of a group of six died, and at 100 and 300 mg/kg the mortality approached 100% within 2–3 months. Animals more severely affected showed hypertrophic hyperplastic gastritis with ulceration, anaemia, hyperproteinaemia and bone marrow hypoplasia. The survivors still showed signs of poisoning 8 months after exposure (Allen & Norback, 1973; Allen et al., 1974; Allen 1975).

### 7.1.1.4 *Dermal toxicity*

Vos & Beems (1971) have confirmed earlier reports that PCBs damage the follicular epithelium in experimental animals. They applied three commercial 60% chlorinated mixtures, Clophen A60, Phenochlor DP6, and Aroclor 1260 to rabbit skin at a daily dose of 118 mg/50 cm² (5 times per week) for 38 days. After initial reddening, transverse wrinkling developed with hyperplasia and hyperkeratosis of the epidermal and follicular epithelium. These effects were more marked with Clophen and Phenochlor than with the Aroclor.

During these experiments, deaths occurred in the Clophen- and Phenochlor-treated groups but not in the Aroclor group. Kidney lesions were seen in all groups; liver damage was least in the Aroclor group. There was atrophy of the thymus cortex and a reduction of germinal centres of the lymph nodes as well as lymphopenia, and some animals in all groups showed oedema of the abdominal and thoracic cavities, subcutaneous tissue, and pericardium. Faecal excretion of copro- and protoporphyrins was increased by all three PCBs but was lowest with Aroclor 1260. Vos & Beems attributed the greater severity of the effects observed with Clophen and Phenochlor to the presence of toxic impurities (see section 7.2). In another comparative dermal toxicity study in rabbits (Vos & Notenboom-Ram, 1972), skin lesions in Aroclor 1260-treated animals were more severe than in animals treated with 2,4,5,2′,4′,5′-hexachlorobiphenyl. In a Japanese study, no differences in skin lesions were observed in rabbits after dermal applications of Kanechlor 400, Kanechlor 500, or 3,4,3′,4′-tetrachlorobiphenyl (Komatsu & Kikuchi, 1972).

### 7.1.1.5 *Inhalation toxicity*

Only one inhalation study has been reported using Aroclor 1242 and 1254 (Treon et al., 1956). Rats, mice, rabbits, and guinea pigs were exposed to Aroclor 1242 or 1254 vapours for five days a week for several weeks at concentrations ranging from 1.5 to 8.6 mg/m$^3$. At these concentrations Aroclor 1254 produced liver enlargement in rats.

### 7.1.2 Birds

The acute oral toxicity of Aroclors 1242 and 1254 is low for the mallard duck, the $LD_{50}$s being greater than 2 g/kg body weight (Tucker & Crabtree, 1970). Published figures for the lethal dose for birds after repeated administration are very variable; it appears to be dependent on the species and age of the bird, the method of administration, the degree of chlorination of the PCBs, and the presence of impurities (Vos, 1972). Heath et al. (1970) determined the dietary concentrations of Aroclors 1232 to 1262, administered over a 5-day period, that were required to kill 50% of groups of mallards, pheasants, and quails. The concentrations were in the range of 500 to over 5000 mg/kg, with the bobwhite quail the most sensitive and the Japanese quail the least. There was a positive relationship between the percentage of chlorine in a technical Aroclor and its toxicity. The relationship held true for those containing less than 60% chlorine, Aroclor 1260 and 1262 deviating slightly. Mallards were relatively less responsive to chlorine content than the three gallimaceous species. Prestt et al. (1970) found a 50% mortality in Bengalese finches receiving a daily oral dose of Aroclor 1254 at 254 mg/kg body weight for 56 days; cormorants have been killed with a cumulative dose of 5.7 g of Clophen A60, but herons were more resistant (Koeman et al., 1973).

The toxicity to chickens of diets containing Phenochlor DP6, Clophen A60, or Aroclor 1260 at 400 mg/kg has been studied by Vos & Koeman

(1970). The first two produced a 100% mortality with survival times of 24.3 and 20.5 days, but there was only 15% mortality with the Aroclor over the 60-day test period. The authors demonstrated the presence of impurities in the Phenochlor and Clophen (Vos et al., 1970). The results of Flick et al. (1965) and Platonow & Funnel (1971) indicate that chickens can survive 200 mg/kg in the diet for several months, but that deaths may occur at 250 mg/kg, though Rehfeld et al. (1971) found a much higher toxicity, recording 1/30 deaths with Aroclor 1248 at 30 mg/kg of diet over 25 days and 16/30 at 50 mg/kg. Keplinger et al. (1971) found decreased growth in chickens receiving Aroclor 1242 at 10 mg/kg and Aroclor 1254 at 100 mg/kg of diet, but no effects with Aroclor 1260 at 100 mg/kg. Birds severely poisoned with PCBs show tremors, ataxia, and ruffling and loss of feathers. Oedema of the abdominal and peritoneal cavities has been a characteristic sign at autopsy in some experiments (Flick et al., 1965) and this was observed in a very large number of chickens killed in Japan by the contamination of their feed with Kanechlor KC400 (Kohanawa et al., 1969a, 1969b). Enlargement of the kidneys and sometimes of the liver has been reported; some authors have noted a reduction in the size of the spleen, comb, and testes, an enlargement of the adrenals and thyroid, and a pale pancreas. No pathological signs were observed by Dahlgren et al. (1972) that could account for the deaths of pheasants, which occurred with a cumulative intake of about 900 mg of Aroclor 1254 (daily oral administration of 20 or 200 mg to 11-week-old hens); they found a mean concentration of 520 mg/kg in brain tissue at death and it is possible that effects in the central nervous system were a contributory cause.

### 7.1.3  Aquatic organisms

#### 7.1.3.1  *Fish*

Stalling & Mayer (1972) report an oral $LD_{50}$ of more than 1.5 g/kg for rainbow trout with Aroclors 1242 and 1260. A 15 mg/kg oral dose to cod affected their ability to maintain an upright position in rotating water (Lindahl, unpublished report 1974).

The assessment of toxicity to fish by adding PCBs to aquarium water is subject to considerable error on account of the different solubilities of the components. Zitko (1970) claims the aqueous solubility of Aroclor 1221 to be 3.8–5 mg/litre and that of Aroclor 1254 to be 0.3–0.5 mg/litre. Stalling & Mayer (1972) report the 96-hour $LC_{50}$s on the cut-throat trout to be 1.17 mg/litre for Aroclor 1221 and up to 60 mg/litre for Aroclor 1260, where the solubility is clearly exceeded. In the more prolonged experiments of these authors (Table 8), most of the concentrations are

Table 8. Intermittent-flow bioassays of Aroclors against three species of fish[a]

| Aroclor | Species | LC$_{50}$(µg/litre) | | | | | |
| | | 5 days | 10 days | 15 days | 20 days | 25 days | 30 days |
|---|---|---|---|---|---|---|---|
| 1254 | Rainbow trout | 156 | 8 | — | — | — | — |
| 1260 | | — | 240 | 94 | 21 | — | — |
| DDT | | 2.26 | 0.87 | 0.26 | — | — | — |
| 1242 | Bluegills | 154 | 72 | 54 | — | — | — |
| 1248 | | 307 | 160 | 76 | 10 | — | — |
| 1254 | | — | 443 | 204 | 135 | 54 | — |
| 1260 | | — | — | — | 245 | 212 | 151 |
| 1242 | Channel catfish | — | 174 | 107 | — | — | — |
| 1248 | | — | 225 | 127 | — | — | — |
| 1254 | | — | — | 741 | 300 | 113 | — |
| 1260 | | — | — | — | 296 | 166 | 137 |
| 1248[b] | Bluegills | 137 | 76 | — | — | — | — |
| 1248[c] | Channel catfish | — | 94 | 57 | — | — | — |

[a] From Stalling & Mayer (1972).

[b] Temperature, 20°C; alkalinity 260 mg/litre pH 7.4.

[c] Temperature, 27°C.

within the solubility limits and therefore more reliable. In more prolonged experiments, Hansen et al. (1971) found deaths in fish exposed for up to 45 days to Aroclor 1254 at 5 µg/litre, but none at 1 µg/litre. These results show that the effect of PCBs is cumulative in fish and that the toxicity decreases with increasing chlorination.

Young fish appear to be more sensitive to PCBs than adults, 96-hour LC$_{50}$s for newly hatched fathead minnows were 15 and 8 µg/litre, respectively for Aroclors 1242 and 1254. Growth of young fathead minnows and flagfish was affected above 2.2 µg/litre (Nebeker et al., 1974). There is no information on pathological changes in fish that might be related to the lethal action of PCBs; Hansen et al. (1971) found that fish survived exposure to Aroclor 1254 at 5 µg/litre, but died later, after being returned to clean water, with signs of a lowered resistance to infection. Pathological changes were observed in kidney, spleen, and liver of rainbow trout fed Aroclor 1254 at 10 or 100 mg/kg of diet for up to 330 days (Nestel & Budd, 1975).

### 7.1.3.2 Aquatic invertebrates

The results of several investigations into the toxicity of PCBs to aquatic invertebrates have been reported by Stalling & Mayer (1972). The exposure of a variety of scud (Gammarus pseudolimnaeus) to PCBs showed a decreased toxicity with increasing chlorination; 4-day LC$_{50}$s of 10, 52, and 2400 µg/litre were obtained with Aroclors 1242, 1248 and 1254, respectively. The threshold for survival, growth, and reproduction of

*Gammarus pseudolimnaeus* exposed to Aroclor 1248 was about 5 µg/litre and was the same for *Daphnia magna*. With the crayfish, 7-day $LC_{50}$s were 30 µg/litre (Aroclor 1242) and 80 µg/litre (Aroclor 1254) and with the glass shrimp, 3 µg/litre (Aroclor 1254). Wildish (1970) found that mortality in *Gammarus oceanicus* was dependent on the duration of exposure to Aroclor 1254 at levels above 10 µg/litre. Stalling & Mayer (1972) reported the 15-day $LC_{50}$ of Aroclor 1254 for immature pink shrimp to be 0.94 µg/litre.

### 7.1.3.3 *Microorganisms*

The growth of certain marine diatoms is inhibited by Aroclor 1254 at 10–25 µg/litre, but marine and freshwater algae are more resistant, being unaffected by 100 µg/litre (Mosser et al., 1972). Fisher et al. (1972) found that phytoplankton from the Sargasso Sea did not grow in Aroclor 1254 at 10 µg/litre, though phytoplankton from estuarine and coastal waters were not much affected by this concentration. Keil et al. (1971) report that the growth of a diatom was inhibited by Aroclor 1242 at 100 µg/litre with a reduction of RNA synthesis, but that 10 µg/litre had no effect.

The growth of cultures of lake bacteria was not inhibited by concentrations of Aroclors 1221, 1242, and 1254 in excess of solubility, and Aroclors 1221 and 1242 could be utilized as the sole source of carbon and energy (Wong & Kaiser, 1975).

## 7.2   Toxicity of Impurities in Commercial PCBs

Vos & Koeman (1970) observed that Phenochlor DP6 and Clophen A60 were more toxic to chickens than was Aroclor 1260 (see section 7.1.2) and Vos & Beems (1971) showed a similar difference in the dermal toxicity to the rabbit (see section 7.1.1.4). Vos et al. (1970) subdivided Clophen A60 and Phenochlor DP6 into non-polar PCBs and polar fractions and found polar components that were not detectable in a similar fraction from Aroclor 1260. Phenoclor DP6, Clophen A60, and the polar fraction from Clophen A60 produced a high mortality in the chick embryo test but the polar fraction from Aroclor 1260 did not. A difference between the three fractions was seen in the development of skin lesions in the rabbit (Vos & Beems, 1971). Mass spectrographic analysis indicated that the impurities were tetra- and pentachlorodibenzofurans. Additional contaminants were chlorinated naphthalenes. Vos et al. (1970) calculated that the maximum level of chlorinated dibenzofurans in Clophen A60 was 5 mg/kg and in Phenochlor DP6, 20 mg/kg. They calculated that the chlorinated dibenzofurans were approximately one order of magnitude less toxic than the chlorinated benzodioxins, and considered that they were mainly responsible for the toxicity of the polar fraction and for the difference in toxicity between the three commercial PCB mixtures.

Recently, chlorinated dibenzofurans were detected in the PCB-contaminated oil that was responsible for the Yusho disease in Japan (see section 8, p. 65).

In 1961, Bauer et al. demonstrated the toxicity of a mixture of tri- and tetrachlorodibenzofurans; a single oral dose of 0.5–1.0 mg/kg body weight caused severe and often lethal liver necrosis in rabbits. Application to the rabbit ear resulted in hyperplasia and hyperkeratosis. Similar toxic effects were found with 2,3,7,8-tetrachlorodibenzo-$p$-dioxin in doses that were 10 times lower than those found to be toxic in the case of chlorinated dibenzofurans. See also the review by Kimbrough (1974). 2,3,7,8-Tetrachlorodibenzofuran, which has recently been shown to be present in PCBs (Bowes et al., 1975) caused mortality in chickens after 8–15 days when they were dosed orally with 5 µg/kg/day. A single oral dose of 4000 µg/kg was not lethal to mice. Porphyria was not observed in chicks or mice (Goldstein et al., 1975a). In comparison, 8 out of 10 chicks died after 9–19 days when dosed orally with 2,3,7,8-tetrachlorodibenzo-$p$-dioxin at 1 µg/kg body weight/day (Schwetz et al., 1973). The oral $LD_{50}$ of 2,3,7,8-tetrachlorodibenzofuran in guinea pigs is approximately 7 µg/kg body weight, which is less than one order of magnitude higher than that of 2,3,7,8-tetrachlorodibenzo-$p$-dioxin (Moore, 1975).

Using low resolution mass spectrometry, McKinney (1975) found metabolites in the excreta of chickens fed 2,4,6,2',4',6'-hexachlorobiphenyl which had correct masses for chlorodibenzofurans. In addition, by perchlorination, he identified octachlorodibenzofuran again by low resolution mass spectrometry. The 2,4,6,2',4',6'-hexachlorobiphenyl isomer is not found in commercial mixtures.

Zitko et al. (1972b) looked for the presence of chlorinated dibenzo-furans in fish and fish products taken from a contaminated coastal area. The samples included shark tissues, cormorant and herring gull eggs, herring oil, and herring fishmeal. The detection limit of the method was 0.01–0.02 mg/kg in the sample. No chlorinated dibenzofurans were detected, although the authors admit that the limit of detection was not sufficiently low to detect amounts that might exert a significant toxic action. Curley et al. (1975) detected a component which had a 4 Cl isotopic cluster and a mass number of 304 corresponding with that of tetrachloro-dibenzofuran in the urine of rats dosed with PCBs, but a positive identification was not made.

## 7.3    Toxicity of the PCTs

There have not been any systematic studies on the toxicity of the PCTs. Sosa-Lucero et al. (1973) administered diets containing Aroclor 5460 at 0, 10, 100, and 1000 mg/kg to groups of rats for seven days. There were no adverse effects on health or body weight; a significant liver enlargement was recorded at the 1000 mg/kg level. In a test for estrogenic activity involving the stimulation of glycogen response in the immature rat uterus, Aroclor 5460 was inactive, as was Aroclor 1260 but Aroclor 5442 was more active than Arochlor 1242 (Bitman et al., 1972). A dietary level of Aroclor 5460 of 5000 mg/kg for 12 weeks caused decreased body weight and increased liver weight in rhesus monkeys (Allen & Norback, 1973); after 6 weeks, facial oedema, hair loss, and eye discharge were observed as described by the same authors in experiments with Aroclor 1248 (section 7.1.1.2), and similar gastric changes were also reported.

## 7.4    Biochemical Effects

### 7.4.1    Induction of enzymes

Several investigators have observed an increase in the smooth endoplasmic reticulum of liver cells after administration of PCBs (see section 7.1.1.3). This is accompanied by an induction of microsomal mixed function oxidase. Induction of microsomal enzyme activity is, like liver enlargement, more marked with the higher chlorinated PCBs and relatively low with Aroclors 1221 and 1016 (Villeneuve et al., 1971a, 1972; Bickers et al., 1972; Ecobichon & Comeau, 1974). The effect has also been demonstrated with single pure PCBs administered orally (Fujita et al., 1971); more recently this work has been confirmed by Johnstone et al. (1974) whose results are summarized in Table 9 and show a greater degree of enzyme induction with the higher chlorinated compounds.

Grant et al. (1974) found an increase in microsomal enzyme activity in rats receiving Aroclor 1254 at 20 mg/kg of diet for 246 days, but none at 2 mg/kg. Iverson et al. (1975) and Goldstein et al. (1975) compared the microsomal enzyme inducing potential of Aroclors 1242 and 1016. Iverson et al. (1975) found increased hepatic microsomal enzyme activity with both Aroclors in male rats receiving 21 daily oral doses of 1 mg/kg body weight, and with Aroclor 1242 in females at 10 mg/kg of body weight and with Aroclor 1016 at 100 mg/kg. PCBs administered to rats during pregnancy can induce microsomal enzyme activity in the placenta and fetus and this also occurs in the liver of newborn rats suckled by mothers

Table 9. Stimulation of microsomal enzyme activity by single chlorinated biphenyls

| Chlorine substituents | Hepatic microsomal enzyme activity | | | |
|---|---|---|---|---|
| | O-demethylation | N-demethylation | aniline hydroxylation | nitro-reduction |
| 4 | 0 | 0 | 0 | 0 |
| 2,2' | 0 | + | + | 0 |
| 2,4' | 0 | 0 | + | 0 |
| 4,4' | + + | + + | + + | + |
| 2,5,2',5' | 0 | + | + + | + |
| 2,4,2',4' | + + | + + | + + | 0 |
| 2,4,5,2',4',5' | + + | + + | + + | + + |
| 2,3,5,2',3',5' | + + | + + | + | + + |
| 2,4,6,2',4',6' | + + | + + | + + | + + |
| 2,3,4,5,2',3',4',5' | + + | + + | + + | + + |

<sup>a</sup> After Johnstone et al. (1974).
0     no activity.
+     slight activity.
+ +   marked activity.

fed with diets containing PCBs (Alvares & Kappas, 1975). Benthe et al. (1972b) reported that when rats were stressed by food deprivation or cold, the PCB residues in adipose tissue were released during mobilization of the fat and caused increased hepatic microsomal enzyme activity.

### 7.4.2 Porphyria

Hepatic porphyria has been induced by a number of PCBs (Clophen A60, Phenoclor DP6, Aroclors 1016, 1242, 1254, and 1260) in the chicken, rabbit, Japanese quail, and the rat (Vos & Koeman 1970; Vos & Beems, 1971; Iverson et al., 1975; Vos et al., 1971; Goldstein et al., 1974, 1975a, 1975b). Porphyrin induction has been studied more extensively in rats. A dose-dependent increase in liver porphyrins has been observed in females receiving 21 daily oral doses of Aroclor 1242 at 10 and 100 mg/kg of body weight, but not at 1 mg/kg. Female rats were more sensitive than males, and Aroclor 1016 had less effect (Iverson et al., 1975). In rats receiving Aroclor 1254 at 100 mg/kg in the diet, the increase was usually delayed 2–4 months after the start of dosing (Goldstein, 1974) and was characterized by high hepatic and urinary levels of uroporphyrin. Disturbance of porphyrin biosynthesis had been connected with an increase of the rate limiting enzyme 2.3.1.37-$\delta$-aminolaevulinate synthase (Vos et al., 1971; Goldstein 1974, 1975).

Kawanishi et al. (1973, 1974) have shown that the administration, in the diet, of Kanechlors KC-300 and KC-500 to rats at 500 mg/kg produced a marked increase in urinary excretion of copro- and uroporphyrins, and in faecal excretion of protoporphyrin, but no increase was observed with KC-400. Experimental administration of pure tetrachloro-biphenyls did not produce porphyria. However, porphyria did result from the repeated subcutaneous injection of KC-400 (total dose 1.8 g) to rabbits for 55 days (Miura et al., 1973).

### 7.4.3 Effects on steroid metabolism

PCBs have been shown to stimulate the activity of enzymes responsible for metabolizing steroids such as estrodiol and androsterone more effectively than does DDT or DDE (Risebrough et al., 1968; Lincer & Peakall, 1970). It has been suggested that effects on reproduction (see section 7.4) may be attributed to the induction of steroid-metabolizing enzymes (Kihlström et al., 1973). Long-term administration of daily oral doses of 0.025 mg Clophen A60 to 23 female NMRI strain mice caused a lengthening of the estrous cycle and a reduction in the frequency of implantation of ova (Örberg & Kihlström, 1973).

### 7.4.4 Other biochemical effects

Hepatic vitamin A has been reported by Villeneuve et al. (1971a) to be reduced to half the normal values in pregnant rabbits by Aroclor 1254. Similar observations have been made by Cecil et al. (1973) in male and female Japanese quails and in rats after feeding Aroclor 1242 at 100 mg/kg of diet. The 50% decrease in hepatic vitamin A found in male and female rats was also found in male and female quail provided that the latter were kept in the dark to prevent egg laying.

A lowering of hepatic vitamin A in rats, fed a 0.1% PCB diet (PCB not specified), was described by Innami et al. (1974). The rats given vitamin A supplement (3400 IU) plus PCB for 6 weeks showed better growth than those given PCB alone, while those given PCB and a vitamin A deficient diet showed significant growth retardation. On the PCB diet, the hepatic vitamin A level decreased to 20% of the normal level during the experimental period; supplementing the diet with 1000 IU did not re-establish the vitamin A level in the liver. Administration of 3000 IU of vitamin A with the PCB diet, however, allowed better than normal hepatic vitamin A levels to be re-established. The authors concluded that vitamin A may play a role in the detoxification of PCB rather than that PCB plays a role in the destructive metabolism of vitamin A.

Aroclor 1254 administered intraperitoneally to rats at 25 mg/kg body weight daily for 4 days caused a 4- to 5-fold increase in the biliary excretion of thyroxine during a 3-hour period. Biliary clearance was greatly elevated. Hypobilirubinaemia has been produced in rats by Bastomsky et al. (1975) who investigated the mechanism by administering daily intra-peritoneal injections of Aroclor 1254 (25 mg/kg body weight in corn oil) to female rats for 4 days, then measuring bilirubin glucuronide formation by hepatic microsomes *in vitro*. PCB treatment was not effective in increasing UDPglucuronosyltransferase (2.4.1.7) activity. Serum bilirubin levels were also significantly decreased by PCB treatment of Gunn rats, which are genetically deficient in UDPglucuronosyltransferase (2.4.1.7) activity.

### 7.4.5 Potentiation and antagonism by PCBs

As PCBs can stimulate microsomal enzyme activity, it is to be expected that they may potentiate the action of those chemicals that undergo microsomal activation, and decrease the action of those that are detoxified. Villeneuve et al. (1973) demonstrated the antagonistic effect by the reduction of phenobarbital sleeping time in rats receiving Aroclors 1242, 1254, and 1260 in their diet, but not in those receiving Aroclor 1221. Johnstone et al. (1974) have confirmed this with a series of single PCBs. Tanaka & Komatsu (1972) found that the hexobarbital-induced sleeping time in female rats was reduced to 49% of the control value by daily oral doses of Kaneclor 500 of 2 mg/kg for 3 days (total 6 mg/kg). When a daily dose of 0.4 mg/kg was given for 15 days (total 6 mg/kg), no reduction in sleeping time was observed. When this small dose was continued for 45 and 53 days, the reduction remained at 12–13%. Phillips et al. (1972) did not find any potentiation of the cholinesterase-inhibitory action of parathion in rats dosed with Aroclors 1221 and 1254; this does not necessarily imply that there was no enhanced activation of parathion, as a stimulation of detoxication may have occurred concurrently. A stimula-tion of parathion detoxication but not of activation has been demonstrated in rabbit microsomes (Villeneuve et al., 1971a). Lichtenstein (1972) reports a potentiation by PCBs of the toxicity of parathion to flies.

Cecil et al. (1975) have shown that the ability of PCTs to decrease phenobarbital sleeping time in quails is rather less than that of the PCBs.

Aroclor 1254 at 160 mg/kg of diet fed to 5-week old male and female Fischer-344 rats for 8 weeks reduced mortality due to feeding hexachloro-phene at a concentration of 600 mg/kg of diet from 77% to 7% and com-pletely prevented the paralysis that was observed in all animals on the hexachlorophene diet alone. However, in the animals on the combined treatment, histological changes in the brain characteristic of hexachloro-phene were still apparent and the possibility of delayed toxicity beyond the 8 weeks of the experiment could not be eliminated. The protective effect of Aroclor 1254 was explained by its capacity to enhance detoxifica-tion by means of hepatic microsomal enzyme induction (Jones et al., 1974).

## 7.5 Cytotoxic Effects

Peakall et al. (1972) found a significant increase in chromosome abnormalities in the embryos of ring doves when the parents were fed on a diet containing Aroclor 1254 at 10 mg/kg. Nilsson & Ramel (1974) found no chromosome breakage in *Drosophila melanogaster* when either Clophen A30 or A50 was added to the substrate. Green et al. (1975) administered Aroclor 1242 orally to rats in single doses of 1250, 2500, or 5000 mg/kg or as a repeated dose of 500 mg/kg/day for 4 days. Aroclor 1254 was also administered for 5 days at doses of 75, 150, or 300 mg/kg/day; they found no evidence of a mutagenic potential as assessed by cytogenic analysis of bone marrow and spermatogonia.

## 7.6 Immunosuppressive Effects

Administration of PCBs leads to an atrophy of lymphoid tissue in chickens (Flick et al., 1965; Vos & Koeman, 1970), in pheasants (Dahlgren et al., 1972), and in rabbits (Vos & Beems, 1971). Vos & de Roij (1972) and Vos & van Driel-Grootenhuis (1972) came to the conclusion that these effects could be attributed to an immunosuppressive effect of PCBs. They found that when guinea pigs fed on diets containing Clophen A60 or Aroclor 1260 at 50 mg/kg were stimulated with tetanus toxoid, a lower antitoxin titre and a lower count of antitoxin-producing cells was obtained than in control guinea pigs, resulting in a significant reduction of immunoglobulins. The skin reaction after tuberculination in animals immunized with Freund's complete adjuvant (as a parameter of cell-mediated immunity) was also depressed at the 50 mg/kg of diet level. Vos & de Roij (1972) suggested that the ability of PCBs to increase the susceptibility of ducklings to duck hepatitis virus (Friend & Trainer, 1970) and of fish to fungal disease (Hansen et al., 1971) could be attributed to this immunosuppressive effect. Kimuru & Baba (1973) found an increased incidence of pneumonia, and lung and intracranial abscesses in rats on a diet containing Kanechlor 400 and suggested that this was due to a lowered resistance to infection.

## 7.7 Effects on Reproduction

In a one-generation reproduction study, rats were fed with a diet containing Aroclor 1242, 1254, or 1260 at levels of 1, 10, and 100 mg/kg. Decreased survival of pups with Aroclors 1242 and 1254 at 100 mg/kg was noted. No effect on reproduction was detected with Aroclor 1260 (Keplinger et al., 1971). In a two-generation reproduction study (Linder et al., 1974), rats were fed a diet containing Aroclor 1254 at levels of 0, 1, 5, 20, and 100 mg/kg and Aroclor 1260 at levels of 0, 5, 20, and 100 mg/kg. Rats exposed to Aroclor 1254 at dietary levels of 20 mg/kg or more had fewer pups per litter. Aroclor 1260 had no effect on reproduction, even at levels of 100 mg/kg. Kihlström et al. (1973) showed that a daily oral dose of 25 μg of Clophen A60 to female mice for 62 days significantly increased the length of the estrus cycle and decreased the frequency of implanted ova. In order to study the effect of PCBs on the development of sexual functions in the early postnatal period they also mated mice that had been suckled by mothers dosed with Clophen A60 during the lactation period. A decrease in the frequency of implanted ova was noted when both parents of the couple had been suckled with milk containing PCBs.

In the rhesus monkey, Allen (1975) reports a lower fertility and a diminished weight of the young at birth, after the administration of a diet containing Aroclor 1248 at 2.5 mg/kg for several months.

Studies by Aulerich et al. (1971), and Ringer et al. (1972) showed that Aroclor 1254 fed to mink at 5 mg/kg severely affected reproduction.

There have been several reports that PCBs can adversely affect egg production and hatchability in birds. Contamination of feed by PCBs from a heat exchanger was shown to be the cause of reduced hatchability of hen eggs at a large poultry hatchery in the USA (Kolbye, 1972), and this has been confirmed in laboratory experiments. Keplinger et al. (1971) found poor hatchability of eggs from hens receiving diets containing Aroclor 1242 at 10 mg/kg or Aroclor 1254 at 100 mg/kg, but not with hens receiving Aroclor 1260 at 100 mg/kg. An adverse effect on eggs has been noted with Aroclor 1248 at 10 mg/kg (Panel on Hazardous Trace Substances, 1972). Peakall et al. (1972) fed ring doves on a diet containing Aroclor 1254 at 10 mg/kg for 3 months and found a marked reduction in egg hatchability 6 months later, due to embryo mortality.

The viability of salmon eggs bears some relation to their PCB content; Johansson et al. (1970) found a 12% mortality in eggs containing PCBs at the rate of 9.2 mg/kg in extractable fat, and a 100% mortality with 34 mg/kg. Adverse effects on the reproduction of aquatic invertebrates have occurred at water concentrations of PCBs in the region of 5 μg/litre.

The potential teratogenic effect of the PCBs has been studied by dosing pregnant females during the gestation period. No fetal abnormalities were produced in the rat by daily doses of Aroclors 1242, 1254, or 1260 at 10 and 30 mg/kg (Keplinger et al., 1971), or Aroclor 1254 at 100 mg/kg (Villeneuve et al., 1971b), or in the rabbit dosed with Aroclor 1254 at 10 and 50 mg/kg (Villeneuve et al., 1971b). Injection of PCBs into eggs has been reported to produce beak abnormalities in chicks (McLaughlin et al., 1963). Cecil et al. (1974) have claimed that the administration of Aroclors 1232 and 1254 to hens at 20 mg/kg in the diet causes teratogenic effects and a reduction in hatchability of fertile eggs.

### 7.8 Neoplasia and Adenofibrosis

The liver is the only organ where tumours have been reported following the ingestion of PCBs. Ito et al. (1973) fed groups of 12 male dd mice with diets containing 500, 250, 100 and 0 mg/kg of Kanechlor 500, 400, and 300, respectively. After 1 year, 7/12 mice developed neoplastic nodules (hyperplastic nodules) and 5/12 developed hepatocellular carcinomas, all in mice from the group fed with 500 mg/kg Kanechlor 500. Metastases were not observed. In a second study on mice, combined exposure to Kanechlor 500 and either $\alpha$- or $\beta$-hexachlorocyclohexane isomers enhanced the development of neoplastic nodules (hyperplastic nodules) and hepato-cellular carcinomas. A combination of Kanechlor 500 and $\gamma$-hexachlorocyclohexane did not produce tumours. Dosing with Kanechlor 500 alone at a dietary level of 250 mg/kg, and $\beta$- or $\gamma$-hexachlorocyclohexane at dietary levels of 250, 100 or 50 mg/kg did not produce tumours. However, $\alpha$-hexachlorocyclohexane produced 8/30 hepatacellular carcinomas and 23/30 hyperplastic nodules.

Groups of 50 BALB/cj male inbred mice were fed dietary concentrations of Aroclor 1254 of 0 or 300 mg/kg (49.8 mg/kg body weight) for 6 or 11 months respectively. No liver tumours were noted in a total of 58 surviving controls. A total of 10 hepatomas (neoplastic or hyperplastic nodules) were noted in 9/22 surviving mice fed Aroclor 1254 for 11 months and in 1/24 surviving mice fed Aroclor 1254 for 6 months. In addition adenofibrosis (cholangiofibrosis) was observed in all 22 livers of mice fed Aroclor 1254 for 11 months, but not in the other groups (Kimbrough & Linder 1974).

In another study (Kimbrough et al., 1975), 200 female Sherman strain COBS random bred rats (descendants of the Osborne Mendel strain) were given a diet containing Aroclor 1260 at 100 mg/kg (11.6–4.3 mg/kg body weight) for approximately 21 months; 200 female rats were kept as controls. The rats were sacrificed when 23 months old. Hepatocellular carcinomas were found in 26/184 of the experimental groups and in 1/173 of the control rats. None of the controls, but 144/184 experimental rats had neoplastic nodules (hyperplastic nodules). Areas of hepatocellular alteration were noted in 28/173 controls and 182/184 experimental rats. No effect of the Aroclor on the incidence of tumours in other organs and no metastases from the liver tumours were observed. In this and two earlier studies, adenofibrosis of the liver was observed in male and female rats fed either Aroclor 1260 or Aroclor 1254 (Kimbrough et al., 1972; 1973); 1975). Adenofibrosis of the liver is a persistent progressive lesion that consists of a marked proliferation of fibrous tissue and epithelial glandular cells which are well differentiated in the mouse but appear atypical in the rat.

In a preliminary study, liver tumours (multiple adenomatous nodules) were induced by Kanechlor 400 in 6/10 female, but not in male Donryu rats. The dietary exposure varied throughout the study and the number of animals used was small (10 experimental and 5 control rats for each sex) (Kimura & Baba, 1973). Makiura et al. (1974) reported that PCBs (Kanechlor 500) inhibited the induction of liver tumours in male Sprague-Dawley rats by the known carcinogens 3′ methyl-*p*-dimethylaminoazo-benzene, N-2-fluosenyl acetamide, and diethylnitrosamine.


## 8. EFFECTS OF PCBs AND PCTs ON MAN—
## EPIDEMIOLOGICAL AND CLINICAL STUDIES

In June 1968, patients appeared at the Dermatology Clinic of Kyushu University Hospital, Fukuoka, Japan suffering from chloracne. A group at the University undertook intensive clinical, chemical, and epidemio-logical investigations and found that the disease originated from the consumption of a batch of rice oil supplied in February 1968; the disease was called Yusho (rice oil disease) (Katsuki, 1969). This batch of rice oil was found to be contaminated with Kanechlor 400, a 48% chlorinated biphenyl, at 2000–3000 mg/kg which entered the oil through a leak in a heat exchanger (Tsukamoto et al., 1969). The symptoms and signs of Yusho were described by Goto & Higuchi (1969) and by Okumura & Kat-suki (1969). The earliest signs were enlargement and hypersecretion of the Meibomian glands of the eyes, swelling of the eylids, and pigmentation of the nails and mucous membranes, occasionally associated with fatigue, nausea, and vomiting. This was usually followed by hyperkeratosis and darkening of the skin with follicular enlargement and acneform eruptions, frequently with a secondary staphylococcal infection. These skin changes were most often seen on the neck and upper chest, but in severe cases extended to the whole body. Biopsy skin samples showed hyperkeratosis, dilation of the follicles, and an accumulation of melanin in the basal cells of the epidermis; melanin granules have also been observed in biopsy samples of the conjunctiva. Oedema of the arms and legs was seen in some patients. There were no definite signs of liver enlargement or liver disorders (Okumura & Katsuki 1969), but slight rises in serum transaminases and in alkaline phosphatase were detected, and a liver sample from a Yusho patient showed an increase in the smooth endoplasmic reticulum (Hirayama et al., 1969). The majority of the patients were found to have respiratory symptoms, and suffered from a chronic bronchitis-like disturbance that persisted for several years (Shigematsu et al., 1971, 1974).

Yusho patients did not appear to suffer from central nervous effects, but some complained of numbness of the arms and legs. Murai & Kuroiwa (1971) found a decrease in the conduction velocity in peripheral sensory nerves.

Yoshimura (1971) reported diminished growth in boys but not in girls, who consumed the oil. Babies born to Yusho mothers were smaller than normal. Newborn babies showed a dark brown skin pigmentation, which disappeared after a few months (Yagamuchi et al., 1971; Taki et al., 1969). Funatsu et al. (1972) found spotted and sporadic ossification of the skull and facial oedema with exophthalmia in four babies, but there was no evidence of any teratogenic action.

Determinations of PCB concentrations in the tissue of Yusho patients were made several months after the ingestion of the oil, apparently by an X-ray fluorescence method for organic chlorine (Goto & Higuchi, 1969). Abdominal fat contained 13.1 mg/kg, subcutaneous fat 75.5 mg/kg, and nails 59 mg/kg. The mesenteric adipose tissue in six Yusho patients, analysed by gas–liquid chromatography 1–3 years after the occurrence of intoxication, contained PCB levels of 2.5 mg/kg on average, which was considerably higher than the normal value. (Masuda et al., 1974a). The mean blood level of PCBs in patients was 0.6 or 0.7 µg/100 ml (0.3 µg/100 ml for the general population) five years after exposure (Masuda et al., 1974b; Takamatsu et al., 1974). These authors also noted a specific gas–liquid chromatographic pattern, peculiar to Yusho patients, which is still observed.

Hirayama et al. (1974) also reported that the serum bilirubin level of patients was significantly lower than the normal level and was negatively correlated with the blood level of PCBs and the serum triglyceride level.

A considerable number of patients had elevated serum triglyceride levels, up to four times the normal values, although this was not correlated with the severity of the symptoms; these high values were maintained for three years in many patients (Uzawa, 1972). There were no marked abnormalities in serum cholesterol and phospholipid levels (Okumura & Katsuki, 1969; Uzawa et al., 1969). Nagai et al. (1969) reported an increase in urinary 17-ketosteroid excretion. Kusuda (1971) also observed changes in the menstrual cycle in approximately 60% of 81 female Yusho patients as compared with their cycles prior to exposure. Okumura et al. (1974) examined the relationship between the blood levels of triglycerides and PCBs in 42 patients and observed a positive correlation. Uzawa et al. (1972) showed that high values of serum triglycerides were maintained for 3 years in many patients. Shigematsu et al. (1971) examined serum immunoglobulin levels in 38 patients, 2 years after onset, and observed a decrease in IgA and IgM and an increase in IgG. Saito et al. (1972) reported lower IgM levels in patients showing chloracne.

Urabe (1974) reported that the total number of Yusho patients had reached 1200 by 13 September 1973 and that 22 of them had died. Mucocutaneous signs had decreased year by year, but neurological and respiratory signs and symptoms and various complaints such as general fatigue, anorexia, abdominal pain, and headache had become more prominent among the patients. The smallest amount of oil that produced symptoms when ingested over approximately 120 days, contained approximately 0.5 g of PCBs, or approximately 0.07 mg/kg body weight per day (Kuratsune, 1972a). Recently chlorinated dibenzofurans at 5 mg/kg were found in three samples of the toxic rice oil that contained PCB levels of about 1000 mg/kg (Nagayama et al., 1975).

Symptoms similar to those of Yusho have been observed in workers in a Japanese condenser factory, including pigmentation of the fingers and, nails, and acneiform eruptions on the jaw, back, and thighs. It was thought that these effects arose from local contact with PCBs; when the use of PCBs ceased, the symptoms disappeared (Hasegawa et al., 1972b).

## 9. EVALUATION OF HEALTH RISKS TO MAN FROM EXPOSURE TO PCBs AND PCTs

### 9.1 Species Variation

The data in sections 7 and 8 indicate that man appears to be the species most sensitive to PCBs, the consumption of relatively small amounts having resulted in a severe disease (Yusho) in 1200 persons in Japan. The monkey is the only experimental species in which effects qualitatively and quantitatively approaching those in man have been observed; Allen (1975) attributed this to metabolic differences leading to a slower elimination than that observed in other species tested.

Conclusions concerning the specific effects of PCBs on different species are confused by uncertainty arising from the presence of toxic impurities.

The rice oil that caused the outbreak of Yusho was contaminated with PCBs containing relatively high amounts of tetrachlorodibenzofuran (see section 8), but the sample used in the monkey experiments had a low content of these impurities, so it is not clear whether PCBs alone were responsible for the Yusho incident. Further uncertainty arises from reports from Finland of high PCB concentrations in blood and body fat of occupationally exposed workers with no indication of adverse effects, while at similar tissue concentrations Japanese workers showed skin lesions typical of Yusho (see section 5.5.3).

A species-specific toxic manifestation that can probably be attributed to toxic impurities, is the abdominal oedema and hydropericardium seen in birds affected by some commercial PCB mixtures.

Mink is another species showing a high sensitivity to PCBs. Deaths have been produced with diets containing PCB levels of 30 mg/kg; no information is available on any species-specific metabolic pathway in the mink that would account for this susceptibility.

## 9.2   Dose–Effect Relationships

The following is a summary of the data in Sections 7 and 8 concerning the relationship between mammalian toxicity and dose. Approximate calculations of the daily dose in mg/kg body weight derived from the dietary concentration are given in parentheses.[a]

### 9.2.1   Body weight

Body weight was reduced in rats after 8 months of dietary intake of Aroclor 1254 at 100 mg/kg (corresponding to 5 mg/kg body weight); no effects were observed at 20 mg/kg in the diet (corresponding to 1 mg/kg body weight).

Dose-dependent retardation of weight gain was observed in mink after 4 months of dietary intake of Aroclor 1254 at 5 and 10 mg/kg (corresponding to 0.5 and 1.1 mg/kg body weight respectively).

### 9.2.2   Effects on liver

*Liver weight*

Dose-dependent increase in liver weight was observed in rats receiving Aroclors 1242, 1254 and 1260 at concentrations of more than 20 mg/kg

---

[a] When no food consumption figures were available from the experimental studies, the following factors were used to transform mg/kg in the diet to mg/kg body weight: mouse (7), rat (20), guinea-pig (25), mink (10), rabbit (33), monkey (25).

in the diet (corresponding to $> 1.4$ mg/kg body weight). Male rats were more sensitive than female rats; no effects were observed with Aroclors 1254 and 1260 at concentrations lower than 20 mg/kg in the diet (corresponding to $< 1.4$ mg/kg body weight). Effects were less marked with the lower chlorinated PCBs.

### Liver changes

Smooth endoplasmic reticulum proliferation with fat droplet inclusions were observed in the liver tissue of rats after 8 months of dietary intake of Aroclor 1254 at 20 mg/kg (corresponding to 1.4 mg/kg body weight).

Liver damage was observed with Aroclors 1242 and 1254 in rabbits receiving 14 weekly oral doses of 150 mg/kg body weight; no effect was observed with Aroclor 1221.

### Liver enzyme activity[a]

Increase in microsomal enzyme activity was observed in male rats after 8 months of dietary intake of Aroclor 1254 of 20 mg/kg (corresponding to 1 mg/kg body weight). No effect was observed at 2 mg/kg in the diet (corresponding to 0.1 mg/kg body weight). Effects were less marked in female rats.

Increased activity was also observed with Aroclors 1242 and 1016 in male rats receiving 21 daily oral doses of 1 mg/kg body weight.

### Liver porphyria

Effects were observed in rats after several months of dietary intake of Aroclor 1254 at 100 mg/kg (corresponding to 5 mg/kg body weight); dose-dependent effects were observed in female rats after 21 daily oral doses of Aroclor 1242 at 10 and 100 mg/kg; no effects were noted at 1 mg/kg body weight.

### Liver vitamin A

Reduction of hepatic vitamin A was observed in rats receiving Aroclor 1242 at the rate of 100 mg/kg in the diet (corresponding to 5 mg/kg body weight).

### Liver tumours

Hepatocellular carcinomas were observed in mice after one year of dietary intake of Kanechlor 500 at 500 mg/kg (corresponding to 75 mg/kg

---

[a] According to Litterst, et al. (1972) the dose producing an effect on nitroreductase activity in the rat corresponds to 0.5 mg/kg in the diet (corresponding to 0.3 mg/kg body weight).

body weight); no carcinomas were observed with Kaneclor 500 at 250 mg/kg in the diet (corresponding to 37.5 mg/kg body weight), or with Kaneclor 300 and 400 at 500 mg/kg in the diet (corresponding to 75 mg/kg body weight).

Hepatomas were observed in mice after 10 months of daily intake of Aroclor 1254 at 300 mg/kg in the diet (corresponding to 49.8 mg/kg body weight).

Hepatocellular carcinomas were observed in rats after 21 months of daily intake of Aroclor 1260 at 100 mg/kg in the diet (corresponding to 11.6–4.3 mg/kg body weight).

### 9.2.3  Reproduction

Effects on reproduction were observed in the mouse at a daily oral dose of 0.025 mg Clophen A60; in the rat at a dietary level of Aroclor 1254 of 20 mg/kg (corresponding to 1 mg/kg body weight) with the effects decreasing with higher chlorinated PCBs; in the mink at a dietary level of Aroclor 1254 of 5 mg/kg (corresponding to 0.5 mg/kg body weight); and in the monkey at a dietary level of Aroclor 1248 of 2.5 mg/kg (corresponding to 0.1 mg/kg body weight).

### 9.2.4  Immunosuppression

Immunosuppressive effects were observed in the guinea-pig at a dietary level of Clophen A60 or Aroclor 1260 of 50 mg/kg (corresponding to 2 mg/kg body weight).

### 9.2.5  Skin effects

In man, symptoms of Yusho disease were observed at a dietary level of 4.2 mg/day of PCBs (corresponding to 0.07 mg/kg body weight/day for a 60-kg person). A value of 0.50 g was estimated as the quantity of PCBs consumed over approximately 120 days above which toxic symptoms were evident. Similar effects were observed in the monkey at a dietary level of Aroclor 1248 of 2.5 mg/kg (corresponding to 0.1 mg/kg body weight) after several months.

### 9.3  Nondetected effect levels

The assessment of non detected effect levels for toxic effects is complicated by the different activities of the component PCBs and by the presence of impurities, in addition to the influence of inter- and intraspecies variation, age, sex, and length of exposure. Moreover, many of the available experimental studies do not include a nondetected effect level.

The most sensitive species appears to be man, and effects have been observed at intake rates of 0.07 mg/kg body weight/day. This may have been influenced by the intake of impurities more toxic than PCBs, but similar effects have been produced in the monkey, at the same order of dosage, with a product containing little of these impurities. At this dosage level, no effects may be expected on growth, liver enlargement, and liver enzyme activity in less sensitive species such as the rat. Although nondetected effect levels are not available for effects on immunosuppression and reproduction, and for certain biochemical effects on the liver, it seems unlikely that these effects would be apparent at intake rates of 0.1 mg/kg body weight/day. Carcinogenic effects have been observed in rats and mice at doses two orders of magnitude greater than this, but there is no epidemiological evidence to suggest that PCBs cause tumours in man. According to Grant et al. (1974) rats fed on a diet containing Aroclor at the rate of 2 mg/kg (equivalent to about 0.1 mg/kg body weight) showed PCB levels of 8 µg/100 ml in blood and 26.1 mg/kg in body fat. However, values much higher than these have been observed in men occupationally exposed to PCBs without evidence of any toxic effects (see section 5.5.3).

It is not possible at present to resolve this conflict in the evidence on the toxicity of PCBs to man.

# REFERENCES

ABBOT, D. C., COLLINS, G. B. & GOULDING, R. (1972) Organochlorine pesticide residues in human fat in the United Kingdom 1969–71. *Br. med. J.*, **2**: 553–556.

ACKER, L. & SCHULTE, E. (1970) Über das Vorkommen von chlorierten Biphenylen und Hexachlorbenzol neben chlorierten Insektiziden in Humanmilch und menschlichem Fettgewebe. *Naturwissenschaften*, **57**: 497 (in German).

ACKER, L. & SCHULTE, E. (1974) Chlorkohlenwasserstoffe in menschlichem Fettgewebe, *Naturwissenschaften*, **61**: 1–4 (in German).

ADDISON, R. F., FLETCHER, G. L., RAY, S. & DOANE, J. (1972) Analysis of a chlorinated terphenyl (Aroclor 5460) and its deposition in tissues of the cod (*Gadus morhua*). *Bull. environ. Contam. Toxicol.*, **8**: 52–60.

AHLING, B. & JENSEN, S. (1970) Reversed liquid–liquid partition in determination of poly-chlorinated biphenyl (PCB) and chlorinated pesticides in water. *Anal. Chem.*, **42**: 1483–1486.

AHMED, M. & FOCHT, D. D. (1973) Degradation of polychlorinated biphenyls by two species of *Achromobacter*. *Can. J. Microbiol.*, **19**: 47–52.

AHNOFF, M. & JOSEFSSON, B. (1973) Confirmation studies on polychlorinated biphenyls (PCB) from river waters using mass fragmentography. *Anal. Lett.*, **6**: 1083–1093.

AHNOFF, M. & JOSEFSSON, B. (1974) Simple apparatus for on-site continuous liquid–liquid extraction of organic compounds from natural waters. *Anal. Chem.*, **46**: 658–663.

AHNOFF, M. & JOSEFSSON, B. (1975) Clean-up procedures for PCB analysis on river water extracts. *Bull. environ. Contam. Toxicol.*, **13**: 159–166.

ALBRO, P. W. & FISHBEIN, L. (1972) Intestinal absorption of polychlorinated biphenyls in rats. *Bull. environ. Contam. Toxicol.*, **8**: 26–31.

ALLEN, J. R. (1975) Response of the non-human primate to polychlorinated biphenyl exposure. *Fed. Proc.*, **34**: 1675–1679.

ALLEN, J. R., ABRAHAMSON, L. J. & NORBACK, D. H. (1973) Biological effects of polychlori-nated biphenyls and triphenyls on the subhuman primate. *Environ. Res.*, **6**: 344–354.

ALLEN, J. R. & NORBACK, D. A. (1973) Polychlorinated biphenyl- and triphenyl-induced mucosal hyperplasia in primates. *Science*, **179**: 498–499.

ALLEN, J. R., CARSTENS, L. A. & BARSOTTI, D. A. (1974) Residual effects of short-term, low-level exposure of non-human primates to polychlorinated biphenyls. *Toxicol. appl. Pharmacol.*, **30**: 440–451.

ALVARES, A. P. & KAPPAS, A. (1975) Induction of aryl hydrocarbon hydroxylase by chlori-nated biphenyls in the foeto-placental unit and in neonatal livers during lactation. *FEBS Lett.*, **50**: 172–174.

ARMOUR, J. A. & BURKE, J. A. (1970) Method for separating polychlorinated biphenyls from DDT and its analogs. *J. Assoc. Off. Anal. Chem.*, **53**: 761–768.

AULERICH, R. J., RINGER, R. K., SEAGRAN, H. L. & YOUATT, W. G. (1971) Effects of feeding coho salmon and other Great Lakes fish on mink reproduction. *Can. J. Zool.*, **49**: 611–616.

AULERICH, R. J., RINGER, R. K. & IWAMOTO, S. (1973) Reproductive failure and mortality in mink fed on Great Lake fish. *J. Reprod. Fertil.*, **19** (Suppl.): 365–376.

BAGLEY, G. E., REICHEL, W. L. & CROMARTIE, E. (1970) Identification of polychlorinated biphenyls in two bald eagles by combined gas–liquid chromatography–mass spectro-scopy. *J. Assoc. Off. Anal. Chem.*, **53**: 251–261.

BAILEY, S. & BUNYAN, P. J. (1972) Interpretation of persistence and effects of polychlorinated biphenyls in birds. *Nature (Lond.)*, **236**: 34–36.

BASTOMSKY, C. H., SOLYMOSS, B., ZSIGMOND, G. & WYSE, J. M. (1975) On the mechanism of polychlorinated biphenyl-induced hypobilirubinaemia. *Clin. chim. Acta*, **61**: 171–174.

BAUER, H., SCHULZ, K. H. & SPIEGELBERG, U. (1961) Berufliche Vergiftungen bei der

Herstellung von Chlorphenol-Verbindungen. *Arch. Gewerbepathol. Gewerbehyg.*, **18**: 538–555.

BENTHE, H. F., KNOP, J. & SCHMOLDT, A. (1972a) Absorption and distribution of polychlorinated biphenyls (PCB) after inhalatory application. *Arch. Toxikol.*, **29**: 85–95 (in German).

BENTHE, H. F., SCHMOLDT, A. & SCHMIDT, H. (1972b) Induction of microsomal liver enzymes after polychlorinated biphenyls (PCB) and following stress. *Arch. Toxikol.*, **29**; 97–106 (in German).

BERG, O. W., DIOSADY, P. L. & REES, G. A. V. (1972) Column chromatographic separation of polychlorinated biphenyls from chlorinated hydrocarbon pesticides and their subsequent gas chromatographic quantitation in terms of derivatives. *Bull. environ. Contam. Toxicol.*, **7**: 338–347.

BERGLUND, F. (1972) Levels of polychlorinated biphenyls in foods in Sweden. *Environ. Health Perspect*, **1**: 67–69.

BERLIN, M., GAGE, J. C. & HOLM, S. (1973) *The metabolism and distribution of 2,4,5,2',5'-pentachlorobiphenyl in the mouse.* In: *PCB Conference II.* National Swedish Environment Protection Board Publications: 4E, pp. 101–107.

BERLIN, M., GAGE, J. C. & HOLM, S. (1974) *Distribution and metabolism of polychlorobiphenyls.* In: *Proceedings of the International Symposium on Recent Advances in Environmental Pollution, Paris, 24–28 June.* pp. 8.

BERLIN, M., GAGE, J. & HOLM, S. (1975) The distribution and metabolism of 2,4,5,2',5'-pentachlorobiphenyl. *Arch. environ. Health*, **30**: 141–147.

BICKERS, D. R., HARBER, L. C., KAPPAS, A. & ALVARES, A. P. (1972) Polychlorinated biphenyls: comparative effects of high and low chlorine containing Aroclors on hepatic mixed function oxidase. *Res, Comm. Chem. Pathol. Pharmacol.*, **3**: 505–511.

BIDLEMAN, T. F. & OLNEY, C. E. (1974) High-volume collection of atmospheric polychlorinated biphenyls. *Bull. environ. Contam. Toxicol.*, **11**: 442–450.

BITMAN, J., CECIL, H. C. & HARRIS, S. J. (1972) Biological effects of polychlorinated biphenyl in rats and quail. *Environ. Health Perspec.*, **1**: 145–149.

BJERK, J. E. (1972) Rester av DDT og polyklorerte bifenyler i Norsk humant materiale. *Tidsskr. Nor-Laegeforen.*, **92**: 15–19 (cited by Kimbrough, 1974).

BOURNE, W. P. P. & BOGEN, A. A. (1972) Polychlorinated biphenyls in North Atlantic seabirds, *Mar. Pollut. Bull.*, **3** (11): 172–175.

BOWES, G. W., MULVIHILL, M. J., SIMONEIT, B. R. T., BURLINGAME, A. L. & RISEBROUGH, R. W. (1975) Identification of chlorinated dibenzofurans in American polychlorinated biphenyls. *Nature (Lond.)*, **256**, 305–307.

BROADHURST, M. G. (1972) Use and replaceability of polychlorinated biphenyls. *Environ. Health Perspect.*, **2**: 81–102.

BRUCKNER, J. V., KHANNA, K. L. & CORNISH, H. H. (1973) Biological responses of the rat to polychlorinated biphenyls. *Toxicol. appl. Pharmacol.*, **24**: 434–448.

BRUGGEMAN, VON J., BUSCH, L., DRESCHER-KADAN, U., EISELI, W. & HOPPE, P. (1974) Pesticid—und PCB—Ruckstunde in Organen von Wildtieren als Indikatoren für Umweltkontamination. *Z. Jagdwiss.*, **20**: 70–74.

BURSE, V. W., KIMBROUGH, R. D., VILLANUEVAS, E. C., JENNINGS, R. W., LINDER, R. E. & SOVOCOOL, G. W. (1974) Polychlorinated biphenyls. Storage, distribution, excretion and recovery: liver morphology after prolonged dietary ingestion. *Arch. environ. Health*, **29**: 301–307.

CARNES, R. A., DOERGER, J. U. & SPARKS, H. L. (1973) Polychlorinated biphenyls in solid waste and solid-waste-related materials. *Arch. environ. Contam. Toxicol.*, **1**: 27–35.

CECIL, H. C., HARRIS, S. J., BITMAN, J. & FRIES, G. F. (1973) Polychlorinated biphenyl-induced decrease in liver vitamin A in Japanese quail and rats. *Bull. environ. Contam. Toxicol.*, **9**: 179–185.

CECIL, H. C., BITMAN, J., LILLIE, R. J., FRIES, G. F. & VERRETT, J. (1974) Embryotoxic and teratogenic effects in unhatched fertile eggs from hens fed polychlorinated biphenyls (PCBs). *Bull. environ. Contam. Toxicol.*, **11**: 489–495.

CECIL, H. C., HARRIS, S. J. & BITMAN, J. (1975) Effects of polychlorinated biphenyls and terphenyls and polybrominated biphenyls on pentobarbital sleeping times of Japanese quail. *Arch. environ. Contam. Toxicol.,* **3**, 183–192.

CHOI, P. S. K., NACK, H. & FLINN, J. E. (1974) Distribution of polychlorinated biphenyls in an aerated biological oxidation wastewater treatment system. *Bull. environ. Contam. Toxicol.,* **11**: 12–17.

CLAUS, B. & ACKER, L. (1975) Zur Kontamination von Milch und Milcherzeugnissen mit chlorierten Kohlenwasserstoffen im Westphälischen Raum II Ergebnisse und Diskussion. *Zeitschrift fur Lebensmitteluntersuchung und forschung,* **159** (3): 129–137.

COLLINS, G. B., HOLMES, D. C. & JACKSON, F. J. (1972) The estimation of polychlorobiphenyls. *J. Chromatogr.,* **71**: 443–449.

COMMISSION OF THE EUROPEAN COMMUNITIES (1974) *European Colloquium. Problems raised by the contamination of man and his environment by persistent pesticides and organohalogenated compounds.* Luxembourg, 14–16 May 1964.

COOLEY, N. R., KELTNER, J. M. & FORESTER, J. (1972) Mirex and Aroclor 1254: effect on accumulation by *Tetrahymena pyriformis* strain W. *J. Protozool.,* **19**: 636–638.

CURLEY, A., BURSE, V. W., GRIM, M. E., JENNINGS, R. W. & LINDER, R. E. (1971) Polychlorinated biphenyls: distribution and storage in body fluids and tissues of Sherman rats. *Environ. Res.,* **4**: 481–495.

CURLEY, A., BURSE, V. W. & GRIM, M. E. (1973a) Polychlorinated biphenyls: evidence of transplacental passage in the Sherman rat. *Fd. cosmet. Toxicol.,* **11**: 471–476.

CURLEY, A., BURSE, V. W., JENNINGS, R. W., VILLANUEVA, E. C., TOMATIS, L. & AKAZAKI, K. (1973b) Chlorinated hydrocarbon pesticides and related compounds in adipose tissue from people of Japan. *Nature (Lond),* **242**: 338–340.

CURLEY, A., BURSE, W. V., JENNINGS, R. W., VILLANUEVA, E. C. & KIMBROUGH, R. D. (1975) Evidence of tetrachlorodibenzofuran (TCDF) in Aroclor 1254 and the urine of rats following dietary exposure to Aroclor 1254. *Bull. environ. Contam. Toxicol.,* **14**: 153–158.

DAHLGREN, R. B., GREICHUS, Y. A. & LINDER, R. L. (1971) Storage and excretion of polychlorinated biphenyls in the pheasant. *J. Wildlife Manag.,* **35**: 823–828.

DAHLGREN, R. B., LINDER, R. L. & CARLSON, C. W. (1972) Polychlorinated biphenyls: their effects on penned pheasants. *Environ. Health Perspect.,* **1**: 89–101.

DE FREITAS, A. S. & NORSTROM, R. J. (1974) Turnover and metabolism of polychlorinated biphenyls in relation to their chemical structure and the movement of lipids in the pigeon. *Can. J. Physiol. Pharmacol.,* **52**: 1080–1094.

DEPARTMENT OF HEALTH EDUCATION AND WELFARE (1972) *Environ. Health Perspect.* Experimental Issue 1, April 1972. DHEW, Bethesda, MD, USA. *Publ. No. (NIH) 72–218.*

DE VOS, R. H. & PEET, E. W. (1971) Thin-layer chromatography of polychlorinated biphenyls. *Bull. environ. Contam. Toxicol.,* **6**: 164–170.

DOGUCHI, M. & FUKANO, S. (1975) Residue levels of polychlorinated terphenyls, polychlorinated biphenyls and DDT in human blood. *Bull. environ. Contam. Toxicol.,* **13**: 57–63.

DOGUCHI, M., FUKANO, S. & USHIO, F. (1974) Polychlorinated terphenyls in the human fat. *Bull. environ. Contam. Toxicol.,* **11** (2): 157–158.

DUKE, T. W., LOWE, J. I. & WILSON, A. J. JR (1970) A polychlorinated biphenyl (Aroclor 1254) in the water, sediment and biota of Escambia Bay, Florida. *Bull. environ. Contam. Toxicol.,* **5**: 171–180.

ECOBICHON, D. J. & COMEAU, A. M. (1974) Comparative effects of commercial Aroclors on rat liver enzyme activities. *Chem.-Biol. Interactions,* **9**: 341–350.

EKSTEDT, J. & ODEN, S. (1974) *Chlorinated hydrocarbons in the lower atmosphere in Sweden.* Report from Department of Soil Science, Royal Agricultural College, S-75007, Uppsala 7. pp. 1–16.

ENVIRONMENTAL SANITATION BUREAU, MINISTRY OF HEALTH & WELFARE (1973) *Survey on foods contamination.* Report of comprehensive investigation on the prevention of pollution by PCBs, 191–210 (in Japanese).

FINKLEA, J., PRIESTER, L. E., CREASON, J. P., HAUSER, T., HINNERS, T. & HAMMER, D. I.

74

(1972) Polychlorinated biphenyl residues in human plasma expose a major urban pollution problem. *Am. J. pub. Health*, **62**: 645–651.

FISHER, N. S., GRAHAM, L. B., CARPENTER, E. J. & WURSTER, C. F. (1972) Geographic differences in phytoplankton sensitivity to PCBs. *Nature (Lond.)*, **241**: 548–549.

FLICK, D. F., O'DELL, R. G. & CHILDS, V. A. (1965) Studies of the chick edema disease. 3. Similarity of symptoms produced by feeding chlorinated biphenyl. *Poult. Sci.*, **44**: 1460–1465.

FREUDENTHAL, J. & GREVE, P. A. (1973) Polychlorinated terphenyls in the environment. *Bull. environ. Contam. Toxicol.*, **10**: 108–111.

FRIEND, M. & TRAINER, D. O. (1970) Polychlorinated biphenyl: interaction with duck hepatitis virus. *Science*, **170**: 1314–1316.

FRIES, G. F., MARROW, G. S. JR & GORDON, C. H. (1973) Long-term studies of residue retention and excretion by cows fed a polychlorinated biphenyl (Aroclor 1254). *J. agr. Food Chem.*, **21**: 117–121.

FUJITA, S., TZUJI, H., KATO, K., SAEKI, S. & TSUKAMOTO, H. (1971) Effect of biphenyl chlorides on rat liver microsomes. *Fukuoka med. Acta.*, **62**: 30–34.

FUKADA, K., INUYAMA, Y., TAKESHITA, T. & YAMAMOTO, S. (1973) Present state of environmental pollution by PCB in Shimane Prefecture. *Shimare Igaku*, **5**: 1–25 (in Japanese).

FUKANO, S., USHIO, F. & DOGUCHI, M. (1974) PCB, PCT and pesticides residues in fish collected from the Tama River. *Ann. Rep. Tokyo Metr. Res. Lab. P.H.*, **25**: 297–305 (in Japanese).

FUNATSU, I., YAMASHITA, F., ITO, Y., TZUGAWA, S., FUNATSU, T., YOSHIKANE, T., HAYASHI, M., KATO, T., YAKUSHIJI, M., OKAMOTO, G., YAMASAKI, S., ARIMA, T., KUNO, T., IDE, H. & IBE, I. (1972) PCB induced fetopathy. I. Clinical observation. *Kurume med. J.*, **19**: 43–51 (in English).

GARDNER, A. M., CHEN, J. F., ROACH, J. A. G. & RAEGLIS, E. P. (1973) Polychlorinated biphenyls. Hydroxylated urinary metabolites of 2,5,2′,5′-tetrachlorobiphenyl identified in rabbits. *Biochem. biophys. res. Commun.*, **55**: 1377–1384.

GILBERTSON, M. & REYNOLDS, L. (1974) *DDE and PCB in Canadian wild birds*. In: *Occasional paper, Canadian Wildlife Service, Environment*, Canada, Ottawa, pp. 1–16.

GOLDSTEIN, J. A., HICKMAN, P. & JUE, D. L. (1974) Experimental hepatic porphyria induced by polychlorinated biphenyls. *Toxicol. appl. Pharmacol.*, **27**: 437–448.

GOLDSTEIN, J. A., McKINNEY, J. D., LUCIEN, G. W., MOORE, J. A., HICKMAN, P. & BERGMAN, H. (1975a) Effects of hexachlorobiphenyl isomers and 2,3,7,8-tetrachlorodibenzofuran (TCDF) on hepatic drug metabolism and porphyria accumulation. *Pharmacologist*, **16**: 239 (Abstract No. 278).

GOLDSTEIN, J. A., HICKMAN, P., BURSE, V. W. & BERGMAN, H. (1975b) A comparative study of two polychlorinated biphenyl mixture (Aroclor 1242 and 1016) containing 42% chlorine on induction of hepatic porphyria and drug metabolizing enzymes. *Toxicol. appl. Pharmacol.*, **32**: 461–473.

GOTO, M. & HIGUCHI, K. (1969) The symptomatology of Yusho (chlorobiphenyls poisoning) in dermatology. *Fuoka Act. Med.*, **60**: 409–431.

GRANT, D. L., MOODIE, C. A. & PHILLIPS, W. E. J. (1974) Toxicodynamics of Aroclor 1254 in the male rat. *Environ. physiol. Biochem.*, **4**: 214–225.

GRANT, D. L. & PHILLIPS, W. E. J. (1974) The effect of age and sex on the toxicity of Aroclor 1254, a polychlorinated biphenyl, in the rat. *Bull. environ. Contam. Toxicol.*, **12**: 145–152.

GRANT, D. L., PHILLIPS, W. E. J. & VILLENEUVE, D. C. (1971a) Metabolism of a polychlorinated biphenyl (Aroclor 1254) mixture in the rat. *Bull. environ. Contam. Toxicol.*, **6**: 102–112.

GRANT, D. L., VILLENEUVE, D. C., McCULLY, K. A. & PHILLIPS, W. E. J. (1971b) Placental transfer of polychlorinated biphenyls in the rabbit. *Environ. Physiol.*, **1**: 61–66.

GREEN, S. G., CARR, J. V., PALMER, K. A. & OSWALD, E. J. (1975) Lack of cytogenetic effects in bone marrow and spermatagonial cells in rats treated with polychlorinated biphenyls (Aroclor 1242 and 1254). *Bull. environ. Contam. Toxicol.*, **13**: 14–22.

HAMMER, D. I., FINKLEA, J. F., PRIESTER, L. E., KEIL, J. E., SANDIFER, S. H. & BRIDBORN, K.

(1972) Polychlorinated biphenyl residues in the plasma and hair of refuse workers. *Environ. Health Perspect.,* **1** (15): 83.

HAMMOND, A. L. (1972) Chemical pollution: polychlorinated biphenyls. *Science,* **175**: 155–156.

HANSEN, D. J., PARRISH, P. R., LOWE, J. I., WILSON, A. J. JR & WILSON, P. D. (1971) Chronic toxicity, uptake and retention of Aroclor 1254 in two estuarine fishes. *Bull. environ. Contam. Toxicol.,* **6**: 113–119.

HAQUE, R., SCHMEDDING, D. W. & FREED, V. H. (1974) Aqueous solubility, absorption and vapour behaviour of polychlorinated biphenyl Aroclor 1254. *Environ. Sci. Technol.,* **8**: 139–142.

HARA, I., HARADA, H., KIMURA, S., ENDO, T. & KAWANO, K. (1974) Follow up health examination in an electric condenser factory after cessation of PCBs usage (1st report) *Jpn. J. Ind. Health,* **16**: 365–366 (in Japanese).

HARVEY, G. R. & STEINHAUER, W. G. (1974) Atmospheric transport of polychlorobiphenyls to the North Atlantic. *Atmos. Environ.,* **8**: 777–782.

HARVEY, G. R., STEINHAUER, W. G. & TEAL, J. M. (1973) Polychlorobiphenyls in North Atlantic Ocean water. *Science,* **180**: 643–644.

HASEGAWA, H., SATO, M. & TSURUTA, H. (1972a) PCBs concentration in the blood of workers handling PCB. *Occup. Health,* **10**: 50–55 (in Japanese).

HASEGAWA, H., SATO, M. & TSURUTA, H. (1972b) *PCB concentration in air of PCB-using plants and health examination of workers* (in Japanese). In: *Report on Special Research on Prevention of Environmental Pollution by PCB-like Substances.* Tokyo, Research Co-ordination Bureau, Science and Technology Agency, pp. 141–149.

HEATH, R. G., SPANN, J. W., KREITZER, J. F. & VANCE, C. (1970) *Effects of polychlorinated biphenyls on birds* In: *Proceedings of the 15th Congress of International. Ornithology, The Hague,* pp. 475–485.

HESSELBERG, R. J. & SCHERR, D. D. (1974) PCBs and *p,p'* DDE in the blood of cachectic patients. *Bull. environ. Contam. Toxicol.,* **11**: 202–205.

HIRAYAMA, C., IRISA, T. & YAMAMOTO, T. (1969) Fine structural changes of the liver in a patient with chlorobiphenyls intoxication. *Fukuoka Acta Med.,* **60**: 455–461 (in Japanese).

HIRAYAMA, C., OKUMURA, M., NAGAI, J. & MASUDA, Y. (1974) Hypobilirubinemia in patients with polychlorinated biphenyls poisoning. *Clin. chim. Acta.,* **55**: 97–100.

HOLDGATE, M. W. (1971) *The sea bird wreck in the Irish Sea, Autumn 1969.* Natural Environmental Research Council (Publication Series C4).

HOLDEN, A. V. (1970a) International co-operative study of organochlorine pesticide residues in terrestial and aquatic wildlife 1967/1968. *Pest. monit. J.,* **4**: 117–135.

HOLDEN, A. V. (1970b) Source of polychlorinated biphenyl contamination in the marine environment. *Nature (Lond.),* **228**: 1220–1221.

HOLDEN, A. V. (1973a) International co-operative study of organochlorine and mercury residues in wildlife, 1969–71. *Pest. monit. J.,* **7**: 37–52.

HOLDEN, A. V. (1973b) *Monitoring PCBs in water and wildlife.* In: *PCB Conference II,* Stockholm, *Swedish Environment Protection Board,* Publication: **4E**, pp. 23–33.

HOLDEN, A. V. & MARSDEN, K. (1969) Single-stage clean-up of animal tissue extracts for organochlorine residue analysis. *J. Chromatogr.,* **44**: 481–492.

HUTZINGER, O., JAMIESON, W. D., SAFE, S., PAULMANN, L. & AM¹ , R. (1974) Identification of metabolic dechlorination of highly chlorinated bipheny ʳʰbit. *Nature (Lond.),* **252**: 698–699.

HUTZINGER, O., NASH, D. M., SAFE, S., DE FREITAS, A. S. W., NORSTROM, R. J., WILDISH, D. J. & ZITKO, V. (1972a) Polychlorinated biphenyls: metabolic behaviour of pure isomers in pigeons, rats and brook trout. *Science,* **178**: 312–313.

HUTZINGER, O., SAFE, S. & ZITKO, V. (1971) Polychlorinated biphenyls: Synthesis of some individual chlorobiphenyls. *Bull. environ. Contam. Toxicol.,* **6**: 209–219.

HUTZINGER, O., SAFE, S. & ZITKO, V. (1972b) Photochemical degradation of chlorobiphenyls (PCBs). *Environ. Health Perspect.,* **1**: 15–20.

INNAMI, S. *et al.* (1974) PCB toxicity and nutrition II. PCB toxicity and Vitamin A (2). *J. Nutr. Sci-Vitaminol.,* **20**: 363–370.

INTERNATIONAL AGENCY FOR RESEARCH ON CANCER (1974) Polychlorinated biphenyls in IARC monographs on the evaluation of carcinogenic risk of chemicals to man, *Vol. 7*, 261–289.

INTERNATIONAL COUNCIL FOR THE EXPLORATION OF THE SEA (1974) *Report of a Working Group for the international study of the pollution of the North Sea and its effects on living resources and their exploitation.* Charlottenlund, Denmark (Co-operative Research Report No. 39).

ISHI, H. (1972) PCB Pollution in Japan. *Environ. Health Rep. No. 14, Jpn. Publ. Health Assoc.,* pp. 13–28.

ITO, N., NAGASAKI, H., ARAI, M., MAKIURA, S., SUGIHARA, S. & HIRAO, K. (1973) Histopathologic studies on liver tumorigenesis induced in mice by technical polychlorinated biphenyls and its promoting effect on liver tumours induced by benzene hexachloride. *J. nat. Cancer Inst.,* **51**: 1637–1646.

IVERSON, F., VILLENEUVE, D. C., GRANT, D. L. & HATINA, G. V. (1975) Effect of Aroclor 1016 and 1242 on selected enzyme systems in the rat. *Bull. environ. Contam. Toxicol.,* **13**: 456–463.

IWATA, Y., WESTLAKE, W. E. & GUNTHER, F. A. (1973) Varying persistence of polychlorinated biphenyls in six California soils under laboratory conditions. *Bull. environ. Contam. Toxicol.,* **9**: 204–211.

JANSSON, B., JENSEN, S., OLSSON, M., RENBERG, L., SUNDSTRÖM, G. & VAZ, R. (1975) Identification by GC-MS of phenolic metabolites of PCB and *p,p'*-DDE isolated from Baltic guillemot and seal. *Ambio.,* **4**: 93–97.

JENSEN, S. (1973) (no title) Report to Swedish National Environment Protection Board, Stockholm, 12 October 1973.

JENSEN, S. (1974) *Identification of some organic substances potentially harmful to the environment.* National Swedish Environment Protection Board, University of Stockholm, Kollenberg Laboratory (*Publication SNV-PM 520*).

JENSEN, S. & SUNDSTRÖM, G. (1974) Structures and levels of most chlorobiphenyls in two technical PCB products and in human adipose tissue. *Ambio,* **3**: 70–76.

JENSEN, S. & SUNDSTRÖM, G. (1975) Metabolic hydroxylation of a chlorobiphenyl containing only isolated unsubstituted positions—2,2′,4,4′,5,5′-hexachlorobiphenyl. *Nature (Lond.)* (in press).

JENSEN, S., JOHNELS, A. G., OLSSEN, M., OTTERLIND, G. (1969) DDT and PCB in marine animals from Swedish waters. *Nature (Lond.),* **224**: 247–250.

JENSEN, S., JOHNELS, A. G., OLSSON, M. & OTTERLIND, G. (1972b) DDT and PCB in herring and cod from the Baltic, the Kattegat and the Skaggerrak. *Ambio spec. Rep.,* **No. 1**: 71–85.

JENSEN, S., JOHNELS, A. G., OLSSEN, M. & WESTERMARK, T. (1972c) *The avifaune of Sweden as indicators of environmental contamination with mercury and chlorinated hydrocarbons.* In: *Proceedings of the 15th International Ornithology Congress, Leiden,* pp. 455–465.

JENSEN, S., RENBERG, L. & OLSSON, M. (1972a) PCB contamination from boat bottom paint and levels of PCB in plankton outside a polluted area. *Nature (London),* **240**: 358–360.

JENSEN, S., RENBERG, L. & VAZ, R. (1973) *Problems in quantification of PCB in biological material.* In: *PCB Conference II.* National Swedish Environment Protection Board. Publications: **4E**, pp. 7–13.

JOHANSSON, N., JENSEN, S. & OLSSON, M. (1970) *PCB indications of effects on fish.* In: *PCB Conference,* National Swedish Environment Protection Board, pp. 59–67.

JOHNSTONE, G. J., ECOBICHON, D. J. & HUTZINGER, O. (1974) The influence of pure polychlorinated biphenyl compounds on hepatic function in the rat. *Toxicol. appl. Pharmacol.,* **28**: 66–81.

JONES, D. C. L., DAVIS, W. E., JR, NEWELL, G. W., SASMORE, D. P. & ROSEN, V. J. (1974) Modification of hexachlorophene toxicity by dieldrin and Aroclor 1254. *Toxicology,* **2**: 309–318.

KAISER, K. L. E. & WONG, P. T. S. (1974) Bacterial degradation of polychlorinated biphenyls.

I. Identification of some metabolic products from Aroclor 1242. *Bull. environ. Contam. Toxicol.*, **11**: 291–296.

KARPPANEN, E. & KOLHO, L. (1973) *The concentration of PCB in human blood and adipose tissue in three different research groups.* In: *PCB Conference II.* National Swedish Environment Protection Board Publications: **4E**: pp. 124–127.

KATSUKI, S. (1969) Foreword. *Fukuoka Acta Med.,* **60**: 403–407.

KAWANISHI, S., SANO, S., MIZUTANI, T. & MATSUMOTO, M. (1973) Experimental porphyria induced by polychlorinated biphenyls (in Japanese). *Jpn. J. Hyg.,* **28**: 84.

KAWANISHI, S., SANO, S., MIZUTANI, T. & MATSUMOTO, M. (1974) Experimental studies on toxicity of synthetic tetrachlorobiphenyl isomers (in Japanese). *Jpn. J. Hyg.,* **29**: 81.

KEIL, J. E., PRIESTER, L. E. & SANDIFER, S. H. (1971) Polychlorinated biphenyl (Aroclor 1242): effects of uptake on growth, nucleic acids and chlorophyll of a marine diatom. *Bull. environ. Contam. Toxicol.,* **6**: 156–159.

KEPLINGER, M. L., FANCHER, O. E. & CALANDRA, J. C. (1971) Toxicologic studies with polychlorinated biphenyls. *Toxicol. appl. Pharmacol.,* **19**: 402–403.

KIHLSTROM, J. E., ORBERG, J., LUNDBERG, C., DANIELSSON, P. O. & SYDHOFF, J. (1973) *Effects of PCB on mammalian reproduction.* In: *PCB Conference II.* National Swedish Environment Protection Board Publications **4E**, pp. 109–111.

KIMBROUGH, R. D. (1974) The toxicity of polychlorinated polycyclic compounds and related chemicals. *Crit. Rev. Toxicol.,* **2**: 445–498.

KIMBROUGH, R. D. & LINDER, R. E. (1974) Induction of adenofibrosis and hepatomas of the liver in BALB/cJ mice by polychlorinated biphenyls (Aroclor 1254). *J. nat. Cancer Inst.,* **53**: 547–549.

KIMBROUGH, R. D., LINDER, R. E. & GAINES, T. B. (1972) Morphological changes in livers of rats fed polychlorinated biphenyls. *Arch. environ. Health,* **25**: 354–364.

KIMBROUGH, R. D., LINDER, R. E., BURSE, V. W., JENNINGS, R. W. & GA, C. (1973) Adenofibrosis in the rat liver with persistence of polychlorinated biphenyls in adipose tissue. *Arch. environ. Health,* **27**: 389–395.

KIMBROUGH, R. D., SQUIRE, R. A., LINDER, R. E., STRANDBERG, J. D., MONTALI, R. J. & BURSE, V. W. (1975) Induction of liver tumours in Sherman strain female rats by polychlorinated biphenyl Aroclor 1260. *J. nat. Cancer Inst.,* **55**: 1453–1459.

KIMURA, N. T. & BABA, T. (1973) Neoplastic changes in the rat liver induced by polychlorinated biphenyl. *GANN,* **64**: 105–108.

KITAMURA, M., TSUKAMOTO, T., SUMINO, K., HAYAKAWA, K., SHIBITA, T. & HIRANO, I. (1973) The PCB levels in the blood of workers employed in a condenser factory. *Jpn. J. indust. Health,* **47**: 354–355 (in Japanese).

KOBAYASHI, Y. (1972) Answer Report to the Questionary paper on "Regulation of Residual level in Foods". *Biol. Pollut.,* **4**: 93–116 (in Japanese).

KOEMAN, J. H., TEN NOEVER DE BRAUW & DE VOS, R. H. (1969) Chlorinated biphenyls in fish, mussels and birds from the river Rhine and the Netherlands coastal area. *Nature (Lond),* **221**: 1126–1128.

KOEMAN, J. H., BOTHOF, TH., DE VRIES, R., VAN VELZEN-BLAD, H. & VOS, J. G. (1972a) The impact of persistent pollutants on piscivorous and molluscivorous birds. *TNO-nieuws,* **27**: 561–569.

KOEMAN, J. H., PEETERS, W. H. M., SMIT, C. J., TJIOE, P. S. & DE GOEIJ, J. J. M. (1972b) Persistent chemicals in marine mammals. *TNO-nieuws,* **27**: 570–578.

KOEMAN, J. H., VAN VELZEN-BLAD, H. C. W., DE VRIES, R. & VOS, J. G. (1973) Effects of PCB and DDE in cormorants and evaluation of PCB residues from an experimental study. *J. Reprod. Fertil.,* **19** (Suppl.): 353–364.

KOHANAWA, M., SHOYA, S., OGURA, Y., MORIWAKI, M. & KAWASAKI, M. (1969a) Poisoning due to an oily by-product of rice-bran similar to chick edema disease. I. Occurrence and toxicity test. *Nat. Inst. anim. Health Quart.,* **9**: 213–219.

KOHANAWA, M., SHOYA, S., YONEMURA, T., NISHIMURA, K. & TSUSHIO, Y. (1969b) Poisoning due to an oily by-product of rice-bran similar to chick edema disease. II. Tetrachlorodiphenyl as toxic substance. *Nat. Inst. anim. Health Quart.,* **9**: 220–228.

KOLBYE, A. C. JR (1972) Food exposures to polychlorinated biphenyls. *Environ. Health Perspect.*, **1**: 85–88.

KOLLER, L. D. & ZINKL, J. G. (1973) Pathology of polychlorinated biphenyls in rabbits. *Am. J. Pathol.*, **70**: 363–373.

KOMATSU, F. & KIKUCHI, M. (1972) Skin lesions by 3,4,3′,4′-tetrachlorobiphenyl in rabbits. *Fukuoka Acta Med.*, **63**: 384–386 (in Japanese).

KURATSUNE, M. (1972a) PCB pollution. *Kosei no Shihyo*, **19**: 11–18 (in Japanese).

KURATSUNE, M. & MASUDA, Y. (1972b) Polychlorinated biphenyls in non-carbon copypaper. *Environ. Health Perspect.*, **1**: 61–62.

KURATSUNE, M., YOSHIMURA, T. & MATSUZAKA, J. (1972) Epidemiologic study on Yusho, a poisoning caused by ingestion of rice oil contaminated with a commercial brand of polychlorinated biphenyls. *Environ. Health Perspect.*, **1**: 119–128.

KUSUDA, M. (1971) Study on the female sexual function suffering from the chlorobiphenyls poisoning. *Sanka to Fujinka*, **4**: 1063–1072 (in Japanese).

LINCER, J. L. & PEAKALL, D. B. (1970) Metabolic effects of polychlorinated biphenyls in the American kestrel. *Nature (Lond.)*, **228**: 783–784.

LICHTENSTEIN, E. P. (1972) PCBs and interactions with insecticides. *Environ. Health Perspect.*, **1**: 151–153.

LIEB, A. J., BILLS, D. B. & SINNHUBER, R. O. (1974) Accumulation of dietary polychlorinated biphenyls (Aroclor 1254) by rainbow trout (*Salmo gairdneri*). *J. agr. Food Chem.*, **22**: 638–642.

LINDER, R. E., GAINES, T. B. & KIMBROUGH, R. D. (1974) The effect of polychlorinated biphenyls on rat reproduction. *Food cosmet. Toxicol.*, **12**: 63–77.

LITTERST, C. L., FARBER, T. M., BAKER, A. M. & VAN LOON, E. J. (1972) Effect of polychlorinated biphenyls on hepatic microsomal enzymes in the rat. *Toxicol. appl. Pharmacol.*, **23**: 112–122.

MAKIURA, S., AOE, H., SUGIHARA, S., HIRAO, K., ARAI, M. & ITO, N. (1974) Inhibitory effect of polychlorinated biphenyls on liver tumorigenesis in rats treated with 3′-methyl-4-dimethylaminoazobenzene, N-2-fluorenylacetamide, and diethylnitrosamine. *J. nat. Cancer Inst.*, **53**: 1253–1257.

MASUDA, Y., KAGAWA, R. & KURATSUNE, M. (1972) Polychlorinated biphenyls in carbonless copying paper. *Nature (Lond.)*, **237**: 41–42.

MASUDA, Y., KAGAWA, R. & KURATSUNE, M. (1974a) Polychlorinated biphenyls in Yusho patients and ordinary persons. *Fukuoka Acta Med.*, **65**: 17–24 (in Japanese).

MASUDA, Y., KAGAWA, R., SHIMAMURA, K., TAKADA, M. & KURATSUNE, M. (1974b) Polychlorinated biphenyls in the blood of Yusho patients and ordinary persons. *Fukuoka Acta Med.*, **65**: 25–27 (in Japanese).

MCKINNEY, J. D. (in press) *Toxicology of selected symmetrical hexachlorobiphenyl isomers: correlating biological effects with chemical structure.* In: *Proceedings of the National Conference on Polychlorinated Biphenyls, Chicago, 19–21 November, 1975.*

MCLAUGHLIN, J. JR., MARLIAC, G. P., VERRETT, M. J., MUTCHLER, M. K. & FITZHUGH, O. G. (1963) The injection of chemicals into the yolk sac of fertile eggs prior to incubation as toxicity test. *Toxicol. appl. Pharmacol.*, **5**: 760–771.

MELVÅS, B. & BRANDT, I. (1973) *The distribution and metabolism of labelled polychlorinated biphenyls in mice and quails.* In: *PCB Conference II.* National Swedish Environment Protection Board Publications: **4E**, pp. 87–90.

MES, J., COFFIN, D. E. & CAMPBELL, D. (1974) Polychlorinated biphenyl and organochlorine pesticide residues in Canadian chicken eggs. *Pestic. Monit. J.*, **8**: 8–11.

MIURA, H., OMORI, S., KATOH, M. (1973) Experimental porphyria in rabbits induced by PCB. *Jpn. J. Hyg.*, **28**: 83 (in Japanese).

MOORE, J. A. (in press) *Toxicity of 2,3,7,8-tetrachlorodibenzofuran: preliminary results. Proceedings of the National Conference on Polychlorinated Biphenyls, Chicago, 19–21 November.*

MOSSER, J. L., FISHER, N. S., TENG, T. & WURSTER, C. F. (1972) Polychlorinated biphenyls: toxicity to certain phytoplankters. *Science*, **175**: 191–192.

MULHERN, B. M., CROMARTIE, E., REICHEL, W. L. & BELISLE, A. A. (1971) Semiquantitative determination of polychlorinated biphenyls in tissue samples by thin layer chromatography. *J. Assoc. Off. Agric. Chem.*, **54**: 548–550.

MURAI, Y. & KUROIWA, Y. (1971) Peripheral neuropathy in chlorobiphenyl poisoning. *Neurol.*, **21**: 1173–1176.

MUSIAL, C. J., HUTZINGER, O., ZITKO, V. & CROCKER, J. (1974) Presence of PCB, DDE and DDT in human milk in the provinces of New Brunswick and Nova Scotia, Canada. *Bull. environ. Contam. Toxicol.*, **12**: 258–267.

NAGAI, J., FURUKAWA, M., YAE, Y. & IKEDA, Y. (1969) Clinicochemical investigation of chlorobiphenyls poisoning. Especially on the serum lipid analysis of the patients. *Fukuoka Acta Med.*, **60**: 475–488 (in Japanese).

NAGAYAMA, J., KURATSUNE, M. & MASUDA, Y. (1976) Determination of chlorinated dibenzofurans in Kanechlors and "Yusho Oil". *Bull. environ. Contam. Toxicol.*, **15**(1): 9–13.

NAGAYAMA, J., MASUDA, Y. & KURATSUNE, M. (1975) Chlorinated dibenzofurans in Kanechlors and rice oils used by patients with Yusho, *Fukuoka Acta Med.*, **66**: 593–599.

NATIONAL SWEDISH ENVIRONMENT PROTECTION BOARD (1973) PCB Conference II. *Publication 1973*: **4E**.

NEBEKER, A. V., PUGLISI, F. A. & DEFOE, D. L. (1974) Effect of polychlorinated biphenyl compounds on survival and reproduction of the fathead minnow and flagfish. *Trans. Am. Fish Soc.*, No. 3, 562–568.

NESTEL, H. & BUDD, J. (1975) Chronic oral exposure of rainbow trout (*Salmo Gairdneri*) to a PCB (Aroclor 1254): pathological effects. *Can. J. comp. Med.*, **39**: 208–215.

NILSSON, B. & RAMEL, C. (1974) Genetic tests on *Drosophila melanogaster* with polychlorinated biphenyls (PCB). *Heriditas*, **77**: 319–322.

NIMMO, D. R., WILSON, P. D., BLACKMAN, R. R. & WILSON, A. J. JR (1971) Polychlorinated biphenyl absorbed from sediments by fiddler crabs and pink shrimp. *Nature (Lond.)*, **231**: 50–52.

NISBET, I. C. T. & SAROFIM, A. F. (1972) Rates and routes of transport of PCBs in the environment. *Environ. Health Perspect.* **1**: 21–38.

NISHIMOTO, T., UEDAM, M., TAUL, S. & CHIKAZAWA, K. (1972a) Organochlorine pesticide residues and PCB in breast milk. *Igaku no Ayumi (Proc. Med. Sci.)*, **82**: 574–575 (in Japanese).

NISHIMOTO, T., UETA, M., TAUL, S., CHIKAZAURA, K. NISHIUCHI, I. & KONDO, K. (1972b) Deposition of organochlorine pesticide residues and PCB in human body fat. *Igaku no Ayumi (Proc. med. Sci.)*, **82**: 515–516.

NISHIZUMI, M. (1970) Light and electron microscope study of chlorobiphenyl poisoning in mouse and monkey liver. *Arch. environ. Health*, **21**: 620–632.

NORÉN, K. & WESTÖÖ, G. (1968) Determination of some chlorinated pesticides in vegetable oils, margarine, butter, milk, eggs, meat and fish by gas chromatography and thin-layer chromatography. *Acta chem. scand.*, **22**: 2289–2293.

ODÉN, S. & BERGGREN, B. (1973) *PCB and DDT in Baltic sediments*. Report Dept. Soil Science, Agricultural College, Uppsala, pp. 1–10.

ODSJÖ, T. (1973) *PCB in some Swedish terrestrial organisms*. In: *PCB Conference II*. National Swedish Environment Protection Board Publications: **4E**, pp. 45–58.

OKUMURA, M. & KATSUKI, S. (1969) Clinical observation on Yusho (chlorobiphenyls poisoning). *Fukuoka Acta Med.*, **60**: 440–446 (in Japanese).

OKUMURA, M., MASUDA, Y. & NAKAMUTA, S. (1974) Correlation between blood PCB and serum triglyceride levels in patients with PCB poisoning. *Fukuoka Acta Med.*, **65**: 84–87 (in Japanese).

OLSSON, M., JENSEN, S. & RENBERG, L. (1973) *PCB in coastal areas of the Baltic. PCB Conference II*. National Swedish Environment Protection Board, Publications: **4E**, pp. 59–68.

ÖRBERG, J. & KIHLSTRÖM, J. E. (1973) Effects of long-term feeding of polychlorinated biphenyls (PCB, Clopehn A60) on the length of the oestrous cycle and on the frequency of implanted ova in the mouse. *Environ. Res.*, **6**: 176–179.

ORGANIZATION FOR ECONOMIC CO-OPERATION AND DEVELOPMENT (1973) *Polychlorinated biphenyls, their use and control.* Environmental Directorate, Paris, OECD.

PANEL ON HAZARDOUS TRACE SUBSTANCES (1972) PCBs—Environmental Impact. *Environ. Res.* **5**: 249–362.

PEAKALL, D. B. LIMIAR, J. L. & BLOOM, S. E. (1972) Embryonic mortality and chromosomal alterations caused by Aroclor 1254 in ring doves. *Environ. Health Perspect.*, **1**: 103–104.

PESENDORFER, VON H., EICHLER, I. & GLOFKE, E. (1973) Informative analyses or organochlorine pesticide and PCB residues in human adipose tissue (from the area of Vienna). *Wiener klinische Wochenschr.*, **85**: 218–222 (summary in English).

PHILLIPS, W. E. J., HATINA, G., VILLENEUVE, D. C. & GRANT, D. L. (1972) Effect of parathion administration in rats following long-term feeding with PCBs. *Environ. physiol. Biochem.*, **2**: 165–169.

PJCHIRALLO, J. (1971) PCBs: leaks of toxic substances raises issue of effects, regulations. *Science*, **173**: 899–902.

PLATONOW, N. S. & FUNNELL, H. S. (1971) Anti-androgenic-like effect of polychlorinated biphenyls in cockerels. *Vet. Rec.*, **88**: 109–110.

PLATONOW, N. & CHEN, N. Y. (1973) Transplacental transfer of polychlorinated biphenyls (Aroclor 1254) in a cow. *Vet. Rec.*, **92**: 69–70.

PLATONOW, N. S., FUNNELL, H. S., BULLOCK, D. H., ARNOTT, D. R., SASCHENBRECKER, P. W. & GRIEVE, D. G. (1971) Fate of polychlorinated biphenyls in dairy products processed from the milk of exposed cows. *J. Dairy Sci.*, **54**: 1305–1308.

PLATONOW, N. S., LIPTRAP, R. M. & GEISSINGER, H. D. (1972) The distribution and excretion of polychlorinated biphenyls (Aroclor 1254) and their effect on urinary gonadal steroid levels in the boar. *Bull. environ. Contam. Toxicol.*, **7**: 358–365.

PORTER, M. L., YOUNG, S. J. V. & BURKE, J. A. (1970) A method for the analysis of fish, poultry and animal tissue for chlorinated pesticide residues. *J. Am. Off. Anal. Chem.*, **53**: 1300–1303.

PORTMANN, J. E. (1970) *Monitoring of organochlorine residues in fish from around England and Wales, with special reference to polychlorinated biphenyls.* Report CM 1970/E: 9 International Council for the Exploration of the Sea, Charlottenlund Slot, DK-2920 Charlottenlund, Denmark.

PRESTT, I., JEFFERIES, D. J. & MOORE, N. W. (1970) Polychlorinated biphenyls in wild birds in Britain and their avian toxicity. *Environ. Pollut.*, **1**: 3–26.

PRICE, H. A. & WELCH, R. L. (1972) Occurrence of polychlorinated biphenyls in humans, *Environ. Health Perspect.*, **1**: 73–78.

REHFELD, B. M., BRADLEY, R. L. JR & SUNDE, M. L. (1971) Toxicity studies on polychlorinated biphenyls in the chick. 1. Toxicity and symptoms. *Poult. Sci.*, **50**: 1090–1096.

REINKE, J., UTHE, J. F. & O'BRODOVICH, H. (1973) Determination of polychlorinated biphenyls in the presence of organochlorine pesticides by thin-layer chromatography. *Environ. Lett.*, **4**: 201–210.

REYNOLDS, L. M. (1971) Pesticide residue analysis in the presence of polychlorobiphenyls (PCBs). *Residue Rev.*, **34**: 27–45.

RINGER, R. K., AULERICH, R. J. & ZABIK, M. (1972) Effect of dietary polychlorinated biphenyls on growth and reproduction of mink. *Proc. Amer. chem. Soc.*, 164th A.C.S. Meeting, New York, pp. 149–154.

RISEBROUGH, R. W. & DE LAPPE, B. (1972) Accumulation of polychlorinated biphenyls in ecosystems. *Environ. Health Perspect.*, **1**: 39–45.

RISEBROUGH, R. W., RIECHE, P., PEAKALL, D. B., HERMAN, S. G. & KIRVEN, M. N. (1968) Polychlorinated biphenyls in the global ecosystem. *Nature (Lond.)*, **220**: 1098–1102.

ROACH, J. A. G. & POMERANTZ, I. H. (1974) The finding of chlorinated dibenzofurans in a Japanese polychlorinated biphenyl sample. *Bull. environ. Contam. Toxicol.*, **12**: 338–342.

ROTE, J. W. & MURPHY, P. G. (1971) A method for the quantitation of polychlorinated biphenyl (PCB) isomers. *Bull environ. Contam. Toxicol.*, **6**: 377–384.

SAFE, S. & HUTZINGER, O. (1971) Polychlorinated biphenyls: photolysis of 2,4,6,2′,4′,6′-hexachlorobiphenyl. *Nature (Lond.)*, **232**: 641–642.

SAITO, R., SHIGEMATSU, N. & ISHIMARU, S. (1972) Immunoglobulin levels in serum and sputum of patients with PCB poisoning. *Fukuoka Acta Med.*, 63: 408–411 (in Japanese).

SANDERS, H. O. & CHANDLER, J. H. (1972) Biological magnification of a polychlorinated biphenyl (Aroclor 1254) from water by aquatic invertebrates. *Bull. environ. Contam. Toxicol.*, 7: 257–263.

SASCHENBRECKER, P. W., FUNNELL, H. S. & PLATONOW, N. S. (1972) Persistence of polychlorinated biphenyls in the milk of exposed cows. *Vet. Rec.*, 90: 100–102.

SAVAGE, E. P., TESSARI, J. D., MALBERG, J. W., WHEELER, H. W. & BAGBY, J. R. (1973) A search for polychlorinated biphenyls in human milk in rural Colorado. *Bull. environ. Contam. Toxicol.*, 9: 222–226.

SCHMIDT, T. T., RISEBROUGH, R. W. & GRESS, F. (1971) Input of polychlorinated biphenyls into California coastal waters from urban sewage outfalls. *Bull. environ. Contam. Toxicol.*, 6: 235–243.

SCHULTE, E. & ACKER, L. (1974) Gas chromatographic unit Glascapillaran bei Temperaturan bis zu 320°C. *Z. Anal. Chem.*, 268: 261–267.

SCOTT, M. L., ZIMMERMAN, J. R., MARINSKY, S. & MULLENHOFF, P. A. (1975) Effects of PCBs, DDT and mercury compounds upon egg production, hatchability and shell quality in chickens and Japanese quail. *Poult. Sci.*, 54: 350–368.

SCHWETZ, B. A., NORRIS, J. M., SPARSCHU, G. L., ROWE, U. K., GEHRING, P. J., EMERSON, J. L. & GERBIG, C. G. (1973) Toxicology of chlorinated dibenzo-*p*-dioxin. *Environ. Health Perspect.*, 5: 87–99.

SHIGEMATSU, N., NORIMATSU, Y., ISHIBASHI, T., YOSHIDA, M., SUETSUGU, S., KAWATSU, T., IKEDA, T., SAITO, R., ISHIMARU, S., SHIRAKISA, T., KIDO, M., EMORI, K. & TOSHIMITSU, H. (1971) Clinical and experimental studies on respiratory involvement in chlorobiphenyls poisoning. *Fukuoka Acta Med.*, 62: 150–156 (in Japanese).

SHIGEMATSU, N., ISHIMARU, S., HIROSE, T., IKEDA, T., EMORI, I. & MIYAZAKI, N. (1974) Clinical and experimental studies on respiratory involvement in PCB poisoning. *Fukuoka Acta Med.*, 65: 88–95 (in Japanese).

SISSONS, D. & WELTI, D. (1971) Structural identification of polychlorinated biphenyls in commercial mixtures by gas–liquid chromatography, nuclear magnetic resonance and mass spectrometry. *J. Chromatogr.*, 60: 15–32.

SÖDERGREN, A. (1972a) Chlorinated hydrocarbon residues in airborne fallout. *Nature (Lond.)*, 236: 395–397.

SÖDERGREN, A. (1972b) Transport, distribution and degradation of chlorinated hydrocarbon residues in aquatic model ecosystems. *Oikos*, 23: 30–41.

SÖDERGREN, A. (1973a) Transport, distribution and degradation of DDT and PCB in a south Swedish lake ecosystem. *Vatten*, 2: 90–108.

SÖDERGREN, A. (1973b) A simplified clean-up technique for organochlorine residues at the microliter level. *Bull. environ. Contam. Toxicol.*, 10: 116–119.

SÖDERGREN, A. & SVENSSON, Bj. (1973) Uptake and accumulation of DDT and PCB by *Ephemera danica* (Ephemeroptera) in continuous-flow systems. *Bull. environ. Contam. Toxicol.*, 9: 345–350.

SÖDERGREN, A., SVENSSON, Bj. & ULFSTRAND, S. (1972) DDT and PCB in South Swedish streams. *Environ. Pollut.*, 3: 25–36.

SOLLY, S. R. B. & SHANKS, V. (1974) Polychlorinated biphenyls and organochlorine pesticides in human fat in New Zealand. *N.Z. J. Sci.*, 17: 535–544.

SOSA-LUCERO, J. C. & DE LA IGLESIA, F. A. (1973) Distribution of a polychlorinated terphenyl (PCT) (Aroclor 5460) in rat tissues and effect on hepatic microsomal mixed function oxidases. *Bull. environ. Contam. Toxicol.*, 10: 248–256.

STALLING, D. L. & MAYER, F. L. JR (1972) Toxicities of PCBs to fish and environmental residues. *Environ. Health Perspect.*, 1: 159–164.

STALLING, D. L., TRINDLE, R. C. & JOHNSON, J. L. (1972) Clean up of pesticides and polychlorobiphenyl residues in fish extracts by gel permeation chromatography. *J. Assoc. Off. Anal. Chem.*, 55: 32–45.

TAKAMATSU, M., INOUE, Y. & ABE, S. (1974) Diagnostic meaning of the blood PCB. *Fukuoka Acta Med.*, 65: 28–31 (in Japanese).

TAKI, I., HISANAGA, S. & AMAGESE, Y. (1969) Report on Yusho (chlorobiphenyl poisoning). Especially further study of its dermatological findings. *Fukuoka Acta Med.*, 62: 132–138 (in Japanese).

TAKI, I., KURATSUNE, M., MASUDA, Y. (1973) *Studies on the transmission to the fetus through placenta and the health of mother and baby. Special Research Report on the effects of PCBs on human health such as chronic toxicity for prevention of pollution by PCBs.* Research Co-ordination Bureau, Science and Technology Agency, pp. 87–95 (in Japanese).

TAKIZAWA, Y. & MINAGAWA, K. (1974) Studies on environmental accumulation and bio-accumulation of organochloric compounds. Mainly mentioned PCT. *Studies on the body effect of degradation-registration substances*, pp. 29–50 (in Japanese).

TANAKA, K. & ARAKI, Y. (1974) Inhibitory effect of cholestyramine on the intestinal absorption of PCB. *Fukuoka Acta Med.*, 65: 53–57.

TANAKA, K. & KOMATSU, F. (1972) Shortening of hexobarbital sleeping time after small doses of PCB in rats. *Fukuoka Acta Med.*, 63: 360–366 (in Japanese).

TAS, A. C. & DE VOS, R. H. (1971) Characterization of four major components in a technical polychlorinated biphenyl mixture. *Environ. Sci. Technol.*, 5: 1216–1218.

TATSUKAWA, R. & WATANABE, I. (1972) Air pollution by PCBs. *Shoku No Kagaku*, 8: 55–63.

THOMAS, G. H. & REYNOLDS, L. M. (1973) Polychlorinated terphenyls in paperboard samples. *Bull. environ. Contam. Toxicol.*, 10: 37–41.

TREON, J. F., CLEVELAND, F. P., CAPPEL, J. W. & ATCHLEY, R. W. (1956) The toxicity of the vapours of Aroclor 1242 and Aroclor 1254. *Am. Ind. Hyg. Assoc. Quart.*, 17: 204–213.

TSUKAMOTO, H. ET AL. (1969) The chemical studies on detection of toxic compounds in the rice bran oils used by the patients of Yusho. *Fukuoka Acta Med.*, 60: 496–512 (in Japanese).

TUCKER, R. K. & CRABTREE, D. G. (1970) *Handbook of toxicity of pesticides to wildlife.* Bureau of Sport, Fisheries and Wildlife Resources Pub. No. 84, Washington, DC, US Dept. of Interior, pp. 130.

TUCKER, E. S., LITSCHGI, W. S. & MEES, W. M. (1975) Migration of polychlorinated biphenyls in soil induced by percolating water. *Bull. environ. Contam. Toxicol.*, 13: 86–93.

URABE, H. (1974) Foreword. *Fukuoka Acta Med.*, 65: 1–4 (in Japanese).

USDA/USDC/EPA/FDA/USDI/(1972) *Polychlorinated biphenyls and the environment.* Washington DC, Interdepartmental Task Force on PCBs. (Report ITF-PCB-72-1).

USHIO, F., FUKANO, S., NISHIDA, K., KANI, T. & DOGUCHI, M. (1974) Some attempt to estimate the total daily intake of pesticides and PCB residues and trace heavy metals. *Ann. Rep. Tokyo Metr. Res. Lab. P.H.*, 25: 307–312 (in Japanese).

UTHE, J. F., REINKE, J. & GESSER, H. (1972) Extraction of organochlorine pesticides from water by porous polyurethane coated with selective absorbent. *Environ. Lett.*, 3: 117–135.

UZAWA, H., ITO, Y., NOTOMI, A. & KATSUKI, S. (1969) Hyperglyceridemia resulting from intake of rice oil contaminated with chlorinated biphenyls. *Fukuoka Acta Med.*, 60: 449–454 (in Japanese).

UZAWA, H., NOTOMI, A., NAKAMUTA, S. & IKEURA, Y. (1972) Consecutive three year follow up study of serum triglyceride concentrations of 82 subjects with PCB poisoning. *Fukuoka Acta Med.*, 63: 401–404 (in Japanese).

VAN HOVE HOLDRINET, M. (1975) Preliminary results of an interlaboratory PCB check sample programme. *J. Environ. Qual.* (in press).

VEITH, G. D. (1972) Recent fluctuations of chlorobiphenyls (PCBs) in the Green Bay, Wisconsin, region. *Environ. Health Perspect.*, 1: 51–54.

VILLENEUVE, D. C., GRANT, D. L., PHILLIPS, W. E. J., CLARK, M. L. & CLEGG, D. J. (1971a) Effects of PCB administration on microsomal enzyme activity in pregnant rabbits. *Bull. environ. Contam. Toxicol.*, 6: 120–128.

VILLENEUVE, D. C., GRANT, D. L., KHERA, K., CLEGG, D. J., BAER, H. & PHILLIPS, W. E. J. (1971b) The fetotoxicity of a polychlorinated biphenyl mixture (Aroclor 1254) in the rabbit and in the rat. *Environ. Physiol.*, 1: 67–71.

VILLENEUVE, D. C., GRANT, D. L. & PHILLIPS, W. E. J. (1972) Modification of pentobarbital sleeping times in rats following chronic PCB ingestion. *Bull. environ. Contam. Toxicol.,* 7: 264–269.

VILLENEUVE, D. C., REYNOLDS, L. M., THOMAS, G. H. & PHILLIPS, W. E. J. (1973a) Polychlorinated biphenyls and polychlorinated terphenyls in Canadian food packaging materials. *J. Assoc. Offic. Anal. Chem.,* 56: 999–1001.

VILLENEUVE, D. C., REYNOLDS, L. M. & PHILLIPS, W. E. J. (1973b) Residues of PCBs and PCTs in Canadian and imported European cheeses, Canada-1972. *Pestic. Monit. J.,* 7: 95–96.

VODDEN, H. A. (1973) *Discussion* In: *PCB Conference II.* National Swedish Environment Protection Board, Publications: 4E, p. 118.

VOS, J. G. (1972) Toxicology of PCBs for mammals and for birds. *Env. Health Perspect.,* 1: 105–117.

VOS, J. G. & BEEMS, R. B. (1971) Dermal toxicity studies of technical polychlorinated biphenyls and fractions thereof in rabbits. *Toxicol. appl. Pharmacol.,* 19: 617–633.

VOS, J. G. & DE ROIJ, Th. (1972) Immunosuppressive activity of a polychlorinated biphenyl preparation on the humoral immune response in guinea pigs. *Toxicol. appl. Pharmacol.,* 21: 549–555.

VOS, J. G. & KOEMAN, J. H. (1970) Comparative toxicologic study with polychlorinated biphenyls in chickens with special reference to porphyria, edema formation, liver necrosis and tissue residues. *Toxicol. appl. Pharmacol.,* 17: 656–668.

VOS, J. G. & NOTENBOOM-RAM, E. (1972) Comparative toxicity study of 2,4,5,2′,4′,5′-hexachlorobiphenyl and a polychlorinated biphenyl mixture in rabbits. *Toxicol. appl. Pharmacol.,* 23: 563–578.

VOS, J. G. & VAN DRIEL-GROOTENHUIS, L. (1972) PCB-induced suppression of the humoral and cell-mediated immunity in guinea pigs. *Sci. Total Environ.,* 1: 289–302.

VOS, J. G., KOEMAN, J. H., VAN DER MAAS, H. L,, TEN NOEVER DE BRAUW, M. G. & DE VOS, R. E. (1970) Identification and toxicological evaluation of chlorinated dibensofuran and chlorinated naphthalene in two commercial polychlorinated biphenyls. *Fd. cosmet. Toxicol.,* 8: 625–633.

VOS, J. G., STRIK, J. J. T. W. A., VAN HOLSTEYN, C. M. W. & PENNINGS, J. H. (1971) Polychlorinated biphenyls as inducers of hepatic porphyria in Japanese quail, with special reference to δ-aminolevulinic acid synthetase activity, fluorescence and residues in the liver. *Toxicol. appl. Pharmacol.,* 20: 232–240.

WEBB, R. G. & MCCALL, A. C. (1972) Identification of polychlorinated biphenyl isomers in aroclors. *J. Am. Off. Anal. Chem.,* 55: 746–752.

WESTÖÖ, G. & NORÉN, K. (1970a) Levels of organochlorine pesticides and polychlorinated biphenyls in fish caught in Swedish water areas or kept for sale in Sweden, 1967–1970. *Var Föda,* 3: 93–146 (summary in English).

WESTÖÖ, G. & NORÉN, K. (1970b) Determination of organochlorine pesticides and polychlorinated biphenyls in animal foods. *Acta Chem. Scand.,* 24: 1639–1644.

WESTÖÖ, G. & NORÉN, K. (1972) Levels of organochlorine pesticides and polychlorinated biphenyls in Swedish human milk. *Var Föda,* 24: 41–54 (summary in English).

WESTÖÖ, G., NORÉN, K. & ANDERSSON, M. (1971) Levels of organochlorine pesticides and polychlorinated biphenyls in some cereal products. *Var Föda,* 10: 341–360 (summary in English).

WHO WORKING GROUP (1973) *The hazzards to health and ecological effects of persistent substances in the environment—polychlorinated biphenyls.* Report of a Working Group convened by the World Health Organization, Regional Office for Europe, EURO, 3109(2).

WILDISH, D. J. (1970) The toxicity of polychlorinated biphenyls (PCB) in sea water to *Grammarus oceanicus. Bull. environ. Contam. Toxicol.,* 5: 202–204.

WILLIAMS, R. & HOLDEN, A. V. (1973) Organochlorine residues from plankton. *Mar. Pollut. Bull.,* 4, 109–111.

WILLIS, D. E. & ADDISON, R. F. (1972) Identification and estimation of the major components

84

of a commercial polychlorinated biphenyl mixture, *Aroclor 1221. J. Fish. Res. Board Can.*, **29**(5): 592–595.

WONG, P. T. S. & KAISER, K. L. E. (1975) Bacterial degradation of polychlorinated biphenyls. II. Rate studies. *Bull. environ. Contam. Toxicol.*, **13**: 249–256.

YAGAMUCHI, A., YOSHIMURA, T. & KURATSUNE, M. (1971) A survey on pregnant women having consumed rice oil contaminated with chlorobiphenyls and their babies. *Fukuoka Acta Med.*, **62**: 112–117 (in Japanese).

YAMAMOTO, H. & YOSHIMURA, H. (1973) Metabolic studies on polychlorinated biphenyls. III. Complete structure and acute toxicity of the metabolites of 2,4,3′,4′-tetrachlorobiphenyl. *Chem. pharm. Bull.*, **21** (10): 2237–2242 (in English).

YOBS, A. R. (1972) Levels of polychlorinated biphenyls in adipose tissue of the general population of the nation. *Environ. Health Perspect.*, **1**: 79–81.

YOSHIMURA, T. (1971) Epidemiological analysis of "Yusho" patients with special reɩꞇ.ence to sex, age, clinical grades and oil consumption. *Fukuoka Acta Med.*, **62**: 109–116 (in Japanese).

YOSHIMURA, H. & YAMAMOTO, H. (1973) Metabolic studies on polychlorinated biphenyls. I. Metabolic fate of 3,4,3′,4′-tetrachlorobiphenyl in rats. *Chem. pharm. Bull.*, **21**: 1168–1169 (in English).

YOSHIMURA, H. & YAMAMOTO, H. (1974) Metabolic studies on polychlorinated biphenyls. IV. Biotransformation of 3,4,3′,4′-tetrachlorobiphenyl, one of the major components of Kanechlor-400. *Fukuoka Acta Med.*, **65**: 5–11 (in Japanese).

YOSHIMURA, H. & YAMAMOTO, H. (1975) A novel route of excretion of 2,4,3′,4′-tetrachlorobiphenyl in rats. *Bull. environ. Contam. Toxicol.*, **13**: 681–687.

YOSHIMURA, H., YAMAMOTO, H., NAGAI, J., YAE, Y., UZAWA, H., ITO, Y., NOTOMI, A., MIMAKAMI, S., ITO, A., KATO, K. & TSUJI, H. (1971) Studies on the tissue distribution and the urinary and faecal excretion of ³H-Kanechlor (chlorobiphenyls) in rats. *Fukuoka Acta Med.*, **62**: 12–19 (in Japanese).

YOSHIMURA, H., YAMAMOTO, H. & SAEKI, S. (1973) Metabolic studies on polychlorinated biphenyls. II. Metabolic fate of 2,4,3′4′-tetrachlorobiphenyl in rats. *Chem. pharm. Bull.*, **21**: 2231–2236 (in English).

YOSHIMURA, H., YAMAMOTO, H. & KINOSHITA, H. (1974) Metabolic studies on polychlorinated biphenyls. V. Biliary excretion of 5-hydroxy-2,4,3′,4′,-tetrachlorobiphenyl, a major metabolite of 2,4,3′,4′-tetrachlorobiphenyl. *Fukuoka Acta Med.*, **65**: 12–16 (in Japanese).

ZIMMERLI, B. & MAREK, B. (1973) Die Belastung der Schweizerischen Bevölkerung mit Pestiziden. *Mitt. Lebensm. Unters. Hyg.*, **64**: 459–479.

ZITKO, V. (1970) Polychlorinated biphenyls (PCB) solubilized in water by nonionic surfactants for studies of toxicity to aquatic animals. *Bull. environ. Contam. Toxicol.*, **5**: 279–285.

ZITKO, V. (1971) Polychlorinated biphenyls and organochlorine pesticides in some freshwater and marine fishes. *Bull. environ. Contam. Toxicol.*, **6**: 464–470.

ZITKO, V., HUTZINGER, O. & SAFE, S. (1971) Retention times and electron-capture detector responses of some individual chlorobiphenyls. *Bull. environ. Contam. Toxicol.*, **6**: 160–163.

ZITKO, V., HUTZINGER, O., JAMIESON, W. D. & CHOI, P. M. K. (1972a) Polychlorinated terphenyls in the environment. *Bull. environ. Contam. Toxicol.*, **7**: 200–201.

ZITKO, V., HUTZINGER, O. & CHOI, P. M. K. (1972b) Contamination of the Bay of Fundy— Gulf of Maine area with polychlorinated biphenyls, polychlorinated terphenyls, chlorinated dibenzodioxins and dibenzofurans. *Environ. Health Perspect.*, **1**: 47–50.

ZITKO, V., CHOI, P. M. K., WILDISH, D. J., MONAGHAN, C. F. & LISTER, N. A. (1974) Distribution of PCBs and *p,p′*-DDE residues in Atlantic herring and yellow perch in eastern Canada, 1972. *Post. Mon. J.*, **8**: 105–109.

# WORLD HEALTH ORGANIZATION PUBLICATIONS

## SUBSCRIPTIONS AND PRICES 1977

### Global Subscription

The global subscription covers all WHO publications, i.e., the combined subscription IV and, in addition, the Monograph Series and non-serial publications, including the books, but not the slides, of the International Histological Classification of Tumour series and not the publications of the International Agency for Research on Cancer.     US $440.00     Sw. fr. 1,100.—

### Combined Subscriptions

Special prices are offered for combined subscriptions to certain publications as follows:

*Subscription*

| | | |
|---|---|---|
| I *Bulletin, Chronicle, Technical Report Series* and *Public Health Papers* | US $96.00 | Sw. fr. 240.— |
| II *World Health Statistics Report* and *World Health Statistics Annual* | US $104.00 | Sw. fr. 260.— |
| III *World Health Statistics Report, World Health Statistics Annual* and *Weekly Epidemiological Record* | US $140.00 | Sw. fr. 350.— |
| IV *Bulletin, Chronicle, Technical Report Series, Public Health Papers, WHO Offset Publications, Official Records, International Digest of Health Legislation, World Health Statistics Report, World Health Statistics Annual, Weekly Epidemiological Record, WHO Regional Publications* and *World Health* | US $340.00 | Sw. fr. 850.— |

The World Health Organization will be pleased to submit a quotation for any other type of combined subscription desired.

### Individual Subscriptions

| | | |
|---|---|---|
| *Bulletin*, vol. 55 (one volume only—six numbers) . . . . . . . . . . . | US $36.00 | Sw. fr. 90.— |
| *Chronicle*, vol. 31 (12 numbers) . . . . . . . . . . . . . . . . . . . | US $18.00 | Sw. fr. 45.— |
| *International Digest of Health Legislation*, vol. 28 (4 numbers) . . . . | US $34.00 | Sw. fr. 85.— |
| *Technical Report Series* . . . . . . . . . . . . . . . . . . . . . . . | US $40.00 | Sw. fr. 100.— |
| *Official Records* . . . . . . . . . . . . . . . . . . . . . . . . . . | US $40.00 | Sw. fr. 100.— |
| *World Health Statistics Report*, vol. 30 . . . . . . . . . . . . | US $40.00 | Sw. fr. 100.— |
| *Weekly Epidemiological Record*, 52nd year (52 numbers) . . . . . . . | US $44.00 | Sw. fr. 110.— |
| *Vaccination Certificate Requirements for International Travel* . . . . | US $7.20 | Sw. fr. 18.— |
| *WHO Regional Publications* . . . . . . . . . . . . . . . . . . . . . | US $20.00 | Sw. fr. 50.— |
| *World Health*, vol. 30 . . . . . . . . . . . . . . . . . . . . . . . . | US $10.00 | Sw. fr. 25.— |

### International Agency for Research on Cancer Subscription

| | | |
|---|---|---|
| *Annual Report, IARC Monographs on the evaluation of Carcinogenic Risk of Chemicals to Man,* and *IARC Scientific Publications* | US $120.00 | Sw. fr. 300.— |

Subscriptions can be obtained from WHO sales agents for the calendar year only (*January to December*).   Prices are subject to change without notice.

\* \*   \*

*Specimen numbers of periodicals and a catalogue will be sent free of charge on request.*

# *Environmental Health Criteria | 3*

# Lead

Published under the joint sponsorship of the United Nations Environment Programme and the World Health Organization

WORLD HEALTH ORGANIZATION GENEVA 1977

*Environmental Health Criteria 3*

# LEAD

Published under the joint sponsorship of
the United Nations Environment Programme
and the World Health Organization

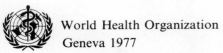

World Health Organization
Geneva 1977

62014791

ISBN 92 4 154063 X

PRINTED IN UNITED KINGDOM

# CONTENTS

# NOTE TO READERS OF THE CRITERIA DOCUMENTS

While every effort has been made to present information in the criteria documents as accurately as possible without unduly delaying their publication, mistakes might have occurred and are likely to occur in the future. In the interest of all users of the environmental health criteria documents, readers are kindly requested to communicate any errors found to the Division of Environmental Health, World Health Organization, 1211 Geneva 27, Switzerland, in order that they may be included in corrigenda which will appear in subsequent volumes.

In addition, experts in any particular field dealt with in the criteria documents are kindly requested to make available to the WHO Secretariat any important published information that may have inadvertently been omitted and which may change the evaluation of health risks from exposure to the environmental agent under examination, so that that information may be considered in the event of updating and re-evaluating the conclusions contained in the criteria documents.

## WHO TASK GROUP ON ENVIRONMENTAL HEALTH CRITERIA FOR LEAD

*Geneva, 29 April–5 May 1975*

## Participants

Professor M. Berlin, Department of Environmental Health, University of Lund, Sweden

Professor A. David, Centre of Industrial Hygiene and Occupational Diseases, Institute of Hygiene and Epidemiology, Prague, Czechoslovakia (*Vice-Chairman*)

Dr F. A. Fairweather, Division of Chemical Contamination of Food and Environmental Pollution, Department of Health and Social Security, London, England

Professor R. A. Goyer, Department of Pathology, University of Western Ontario, London, Ontario, Canada (*Chairman*)

Dr L. Graovac-Leposavić, Institute of Occupational and Radiological Health, Belgrade, Yugoslavia

Dr R. J. M. Horton, Environmental Protection Agency, Research Triangle Park, NC, USA

Dr C. H. Nordman, Institute of Occupational Health, Helsinki, Finland (*Rapporteur*)

Dr H. Sakabe, Department of Industrial Physiology, National Institute of Industrial Hygiene, Kawasaki, Japan

Professor H. W. Schlipköter, Institute of Air Hygiene and Silicosis, Düsseldorf, Federal Republic of Germany

Professor N. Ju. Tarasenko, First Moscow Medical Institute, Moscow, USSR

Professor R. L. Zielhuis, Coronel Laboratory, Faculty of Medicine, University of Amsterdam, Amsterdam, Netherlands

*Representatives of other agencies*

Dr A. Berlin, Health Protection Directorate, Commission of the European Communities, Centre Louvigny, Luxembourg

Professor R. Bourdon, International Union of Pure and Applied Chemistry, Commission on Toxicology, Laboratoire de Biochimie-Toxicologie, Centre Anti-Poison de l'Hôpital Fernand Widal, Paris, France

Dr D. Djordjević, Occupational Safety and Health Branch, International Labour Office, Geneva, Switzerland

Dr R. Morf, International Union of Pure and Applied Chemistry, Liaison Officer with WHO, 8311 Kyburg Zh, Switzerland

*Secretariat*

Professor Paul B. Hammond, Department of Environmental Health, University of Cincinnati, The Kettering Laboratory, Cincinnati, Ohio, USA (*Temporary Adviser*)

Dr Y. Hasegawa, Medical Officer, Control of Environmental Pollution and Hazards, Division of Environmental Health, World Health Organization, Geneva, Switzerland

Dr J. E. Korneev, Scientist, Control of Environmental Pollution and Hazards, Division of Environmental Health, World Health Organization, Geneva, Switzerland

Dr V. Krichagin, Scientist, Promotion of Environmental Health, WHO Regional Office for Europe, Copenhagen, Denmark

Dr B. Marshall, Medical Officer, Occupational Health, Division of Environmental Health, World Health Organization, Geneva.

Professor L. A. Timofievskaja, Institute of Occupational Health, Moscow, USSR (*Temporary Adviser*)

Dr V. B. Vouk, Chief, Control of Environmental Pollution and Hazards, Division of Environmental Health, World Health Organization, Geneva (*Secretary*)

| | |
|---|---|
| ALA | $\delta$-aminolevulinic acid |
| ALA-U | $\delta$-aminolevulinic acid in urine |
| ALAD | porphobilinogen synthase (EC 4.2.1.24), $\delta$-aminolevulinate dehydratase, $\delta$-aminolevulinic acid dehydratase |
| ALAS | $\delta$-aminolevulinate synthase (EC 2.3.1.37), aminolevulinic acid synthetase |
| CP | coproporphyrins |
| CP-U | coproporphyrin in urine |
| CPG | coproporphyrinogen III |
| EDTA | ethylenediaminetetraacetic acid |
| FEP | free erythrocyte porphyrins |
| Hb | haemoglobin |
| $LD_{50}$ | median lethal dose |
| PP | protoporphyrin IX |
| PBG | porphobilinogen |
| Pb-B | lead in blood |
| Pb-U | lead in urine |
| RBC | red blood cells |
| SGOT | aspartate aminotransferase (EC 2.6.1.1), serum glutamic oxaloacetic transaminase |

A WHO Task Group on Environmental Health Criteria for Lead met in Geneva from 29 April to 5 May 1975. Dr B. H. Dieterich, Director, Division of Environmental Health, opened the meeting on behalf of the Director-General. The Task Group reviewed and revised the second draft criteria document and made an evaluation of the health risks from exposure to lead and its compounds.

The first and second drafts were prepared by Professor Paul B. Hammond of the Department of Environmental Health, The Kettering Laboratory, University of Cincinnati, Ohio, USA. The comments on which the second draft was based were received from the national focal points for the WHO Environmental Health Criteria Programme in Bulgaria, Czechoslovakia, Federal Republic of Germany, Greece, Japan, The Netherlands, New Zealand, Poland, Sweden, USA, and the USSR, and from the United Nations Educational, Scientific and Cultural Organization (UNESCO), Paris, from the United Nations Industrial Development Organization (UNIDO), Vienna, from the Centro Panamericano de Ingenieria Sanitaria y Ciencias del Ambiente (CEPIS) at Lima, Peru, and from the Health Protection Directorate of the Commission of the European Communities (CEC), Luxembourg. Comments were also received, at the request of the Secretariat from: Professor R. Goyer and Professor H. Warren, Canada; Professor J. Teisinger, Czechoslovakia; Dr S. Hernberg, Finland; Dr K. Cramer and Dr B. Haeger-Aronsen, Sweden; Dr D. Barltrop, Professor B. Clayton, Professor R. Lane, and Professor P. J. Lawther, United Kingdom; Dr J. J. Chisholm, Professor H. L. Margulis, and Dr G. Ter Haar, United States of America; and Dr D. Djurić and Professor K. Kostial, Yugoslavia.

Valuable comments were received on the third draft, resulting from the task group, from: Mr Joseph E. Faggan, Director of Petroleum Chemicals Research, Ethyl Corporation, Ferndale, Michigan, USA, and from Mr R. L. Stubbs, Director-General, Lead Development Association, London and Chairman, Statistical Committee, International Lead and Zinc Study Group.

The collaboration of these national institutions, international organizations, WHO collaborating centres, and individual experts is gratefully acknowledged. Without their assistance this document would not have been completed. The Secretariat wishes to thank, in particular, Professor Hammond for his continued help in all phases of the preparation of the document, and Dr H. Nordman of the Institute of Occupational Health,

Helsinki, who assisted the Secretariat in the final scientific editing of the document.

This document is based primarily on original publications listed in the reference section. However, several recent publications broadly reviewing health aspects of lead and its compounds have also been used. These include publications by Kehoe (1961), NAS-NRC (1972), NRC-Canada (1973), Goyer & Rhyne (1973), WHO Working Group (1973), Inter-Department Working Group on Heavy Metals (1974), SCEP (1974), Nordberg, ed. (1976). In addition, the document draws on comprehensive and useful data from the proceedings of several symposia and meetings, e.g. the "International Symposium on Environmental Aspects of Lead", Amsterdam, 1972, arranged by the Commission of the European Communities and the US Environmental Protection Agency; the "International Symposium on Recent Advances in the Assessment of the Health Effects of Environmental Pollution", Paris, 1974, jointly organized by the Commission of the European Communities, US Environmental Protection Agency, and the World Health Organization; the University of Missouri's Annual Conferences on Trace Substances in Environmental Health, Columbia, Missouri, 1967–1975; and the "International Symposium on Environmental Lead Research", Dubrovnik, 1975, organized by the Institute for Medical Research and Occupational Health, under the auspices of the Yugoslav Academy of Sciences and Arts.

Details of the WHO Environmental Health Criteria Programme, including some of the terms frequently used in the documents, may be found in the introduction to the publication "Environmental Health Criteria 1—Mercury", published by the World Health Organization, Geneva, in 1976.

# 1. SUMMARY AND RECOMMENDATIONS FOR FURTHER RESEARCH

## 1.1 Summary

### 1.1.1 Analytical problems

The procurement of environmental and biological samples requires careful consideration of the special problems relating to the particular material to be analysed. In air sampling, it is most important to ensure that the sampler is placed at the breathing zone of the population group under study. For all sampling procedures and particularly for blood, external contamination is a major problem.

The most successful analytical method in recent years has been atomic absorption spectroscopy. It has proved to be versatile and sufficiently sensitive for most purposes, but reliable results, particularly for biological specimens such as blood, can be obtained only after considerable experience has been acquired.

Determinations of haem intermediates and of porphobilinogen synthase (EC 4.2.1.24) (ALAD)[a, b] activity in blood are important methods for estimating the biological consequences of overexposure to lead. There is a great need for standardization of both these methods and of ways of expressing the results.

## 1.1.2 Sources and pathways of exposure

The major sources of lead in the environment that are of significance for the health of man, arise from the industrial and other technological uses of lead. The major dispersive non-recoverable use of lead is in the manufacture and application of alkyllead fuel additives. Because of current legislative actions with respect to the maximum permissible concentration of lead in gasoline, the consumption of lead for the production of alkyllead additives decreased from 1973 to 1975 and a further decline for the latter half of the 1970s may occur as more cars equipped with catalysts which require lead-free gasoline will come into use.

From a mass balance point of view, the transport and distribution of lead from stationary or mobile sources is mainly *via* air. Although large amounts are probably also discharged into soil and water, lead tends to localize near the points of such discharge. Lead that is discharged into the air over areas of high traffic density falls out mainly within the immediate metropolitan zone. The fraction that remains airborne (about 20%, based on very limited data) is widely dispersed. Residence time for these small particles is of the order of days and is influenced by rainfall. In spite of widespread dispersion, with consequent dilution, there is evidence of lead accumulation at points extremely remote from human activity, e.g. in glacial strata in Greenland.

The biota acquires lead both by surface deposition and by secondary transfer from soil to plants and from plants to animals. However, the impact of man-made lead pollution on the lead content of plants and

---

[a] In the first instance, enzymes are named according to the 1972 recommendations of the Commission on Enzyme Nomenclature but throughout the rest of the document the more familiar names or abbreviations are used.

[b] Formerly known as $\delta$-aminolevulinate dehydratase or $\delta$-aminolevulinic acid dehydratase.

animals is not perceptible except in localized areas of intense air pollution, e.g. around smelters and in the immediate vicinity of roads with heavy traffic.

The concentration of lead in air varies from 2–4 $\mu g/m^3$ in large cities with dense automobile traffic to less than 0.2 $\mu g/m^3$ in most suburban areas and still less in rural areas. The concentration of lead in drinking water is generally less than 10 $\mu g$/litre, but in some areas where the water is soft (low in calcium and magnesium) and where, at the same time, lead pipes and lead-lined water storage tanks are used, the concentration may reach 2000–3000 $\mu g$/litre. At this concentration (and even at concentrations of several hundred $\mu g$/litre) a perceptible rise in the body burden of lead occurs, which is reflected in elevated values of lead in the blood (Pb-B).

The contribution of food to man's exposure to lead is highly variable. Some recent studies in the USA have estimated the daily oral intake in food and beverages to be about 100 $\mu g$ whereas earlier studies and some recent European studies indicated the intake to be in the range of 200–500 $\mu g$/day. However, a recent Swedish study reported volumes of the order of 20 $\mu g$/day. No specific category of food has been identified as being especially high in lead content other than wine and foods that are stored in lead-soldered cans or lead-glazed pottery. Processed milk contains considerably more lead than fresh cow's milk which has a similar concentration to human milk. The reported lead concentrations range from less than 5 $\mu g$/litre to 12 $\mu g$/litre. If this information is correct, milk could be a significant source of lead for infants.

Various miscellaneous sources of lead have been identified as being highly hazardous. These include lead-glazed ceramics used for beverage storage, illicitly-distilled whisky, and discarded automobile battery casings when used for fuel.

In certain countries, gross overexposure of some infants and young children has been recorded. The major sources are lead-based paint in old houses and in the soil surrounding these homes, and the soil surrounding lead smelters. Lead in street dust due to atmospheric fallout, and miscellaneous lead-containing objects chewed or eaten by children are other possible sources of exposure, but their relative importance is not clear.

The highest exposure occurs in workers who come into contact with lead during mining, smelting, and various manufacturing processes where lead is used. The major pathway of exposure is inhalation. The concentration of air lead in the working environment of smelters and storage battery factories often exceeds 1000 $\mu g/m^3$. For other industries, data are either not available or indicate a lower level of exposure.

Extensive surveys have been made on blood concentrations in both adults and young children. Such data are useful indicators of overall exposure to lead.

### 1.1.3 Metabolism

A number of studies have been made which indicate that 35% of the lead inhaled by man is deposited in the lungs. The relative importance of the mucociliary escalator mechanism and of direct absorption from pulmonary deposition is poorly understood and the contribution of airborne lead to total daily intake cannot be estimated from metabolic data. But when sustained Pb-B is used as a measure of lead absorption, it can be assumed from human data that continuous exposure to 1 µg of lead per m$^3$ of air would contribute lead levels of about 1.0–2.0 µg/100 ml of blood.

About 10% of lead taken in from food and beverages is absorbed. However, using data from several sources, the dietary contribution to Pb-B can only be roughly estimated as 6–18 µg of lead per 100 ml of blood per 100 µg of dietary lead intake.

From both animal and human studies, the general features of lead distribution and excretion are fairly clearly defined. The body burden of lead can be subdivided into a large, slow-turnover compartment and a smaller more rapidly-exchanging compartment. Anatomically, the larger compartment is mainly located in bones. The amount of lead in this compartment increases throughout life. The smaller compartment consists of the soft tissues and includes the blood. Lead levels in soft tissues and in blood continue to increase up to early adulthood and then change little. Elimination of lead from the body is mainly by way of the urine (about 76%) and the gastrointestinal tract (about 16%). The other 8% is excreted by miscellaneous routes (sweat, exfoliation of the skin, loss of hair) about which little is known.

Alkyllead compounds (tetraethyllead and tetramethyllead) are de-alkylated both to trialkyl derivatives and to inorganic lead. Details of alkyllead metabolism have been learned from animal studies and have not been defined in man.

### 1.1.4 Experimental studies on the effects of lead

The extensive animal studies that have been conducted concerning the biological effects of lead indicate that, with rare exceptions, the toxic phenomena that have been observed in man have also been successfully reproduced in animals. Although animal studies have provided a more profound understanding of the effects of lead than could be learned from studies of man himself, they have not been of much use in the elucidation of dose–effect and dose–response relationships in man.

Major differences that have been noted are as follows: (1) benign and malignant tumour induction has occurred in rats and mice exposed to lead acetate and in rats exposed to lead subacetate and lead phosphate but carcinogenic effects have not been seen in man; (2) clear-cut reductions in fertility have been observed in experimental animals but not in man, although data have been reported which suggest that this might be so; (3) hyperactivity and other behavioural disturbances have been observed in rats, mice, and sheep without prior encephalopathy. This is especially important because of current suspicions that widespread, slight brain damage occurs in young children with relatively low exposure not preceded by encephalopathy. Evidence also exists for compensatory increases in ALAD in animals with continuing exposure to lead whereas all human studies to date have been negative in this respect.

### 1.1.5 Clinical and epidemiological studies on the effects of lead: Evaluation of health risk to man from exposure to lead

Studies of the effects of lead on man may be divided into two general types. The first type is the retrospective study of the causes of mortality in lead-exposed populations in contrast with those in matched control groups. Several studies showed that at high exposure levels (Pb − B > 80 µg/100 ml[a]), a slightly higher number of deaths occurred due to cerebrovascular disease and chronic nephritis. In one study, where the mortality rate due to cancer was observed, no statistically significant differences were found between the industrially exposed workers and the control group.

The second type of study concerns morbidity rates due to the effects of lead on specific organs and systems. In some cases, it has been possible to estimate the level of the exchangeable body burden (expressed as Pb-B) at which a given intensity of effect (dose-response relationship) has been observed in certain sections of a selected group. For other effects it has only been possible to specify the Pb-B level at which no effect was observed in reasonably large groups of people (no-detected-effect level).

The haematopoietic system shows effects at lower Pb-B levels than any other system. The effects are, in order of sensitivity: inhibition of erythrocyte ALAD, elevation of erythrocyte protoporphyrin IX (FEP), rise in urinary $\delta$-aminolevulinic acid (ALA) and coproporphyrin (CP) ex-

---

[a] In this document, the concentrations of lead in blood are expressed in µg/100 ml although in some original papers the values are given in µg/100 g. For practical purposes, the difference of about 5% can be neglected.

cretion, inhibition of erythrocyte sodium-potassium adenosine triphosphatase (EC 3.6.1.3) (Na-K-ATP'ase), and fall in haemoglobin level. A fall in haemoglobin level is clearly an indication of adverse effects. The no-detected-effect level for this effect is a Pb-B concentration equivalent to 50 μg/100 ml in adults and 40 μg/100 ml in children.

The effects of inorganic lead on the central nervous system have been under intensive investigation in recent years, particularly with regard to subtle effects on behaviour, mainly in children, but also to some extent in adults. Substantial doubts remain as to the validity of some of the studies because the relationship between the exposure to lead at the time the damage occurs and at the time the effects are first observed is not known. Nevertheless, a no-detected-effect level has been specified that is lower than for classical lead encephalopathy. The no-detected-effect level is estimated to be at Pb-B values of about 60–70 μg/100 ml for adults and of about 50–60 μg/100 ml for children.

The renal effects of lead are of two general types. The first is tubular, characterized by the Fanconi triad of aminoaciduria, hyperphosphaturia, and glycosuria. It occurs with relatively short-term exposure and is reversible. The second type of renal effect is characterized anatomically by sclerotic changes and interstitial fibrosis. Functionally, filtration capacity is reduced. These changes are of a progressive nature and may lead to renal failure. It is probable that exposures leading to this type of nephropathy are rarely encountered even in industry today. A no-detected-effect level cannot be specified.

The problem of the toxic effects of alkyllead is almost entirely restricted to workers who are occupationally exposed. There is very little information concerning dose–effect and dose–response relationships and even the frequency of occurrence of toxic effects and their relation to specific work activities is not well documented.

## 1.2 Recommendations for Further Research

### 1.2.1 Analytical methods

One of the major needs is for the standardization of analytical methods, particularly with regard to the haem intermediates, ALAD, and erythrocyte Na-K-ATP'ase. At the present time, it is often impossible to compare studies conducted in one laboratory with those of another. This is particularly true for enzymatic methods that give different results depending on pH, oxygen tension, and the presence or absence of other factors, e.g. other metals that can influence the action of lead. It is of equal importance that a standard mode of expressing results be introduced in order to achieve valid interlaboratory comparisons. Thus, measurements involving urine should be expressed per unit of creatinine excreted per unit time; this would probably take body mass into consideration.

In view of the highly variable results that have been obtained in the interlaboratory comparisons conducted to date, more cooperative efforts should be undertaken and maintained on a continuous basis. It is recommended that all published data include interlaboratory comparison results for the methods used. International standard specimens of the commonly investigated biological media with reliably determined concentrations of lead should be developed and made available to investigators.

Finally, standardized methods of statistical treatment of analytical data should be adopted and adhered to.

### 1.2.2  Sources of lead intake

It is apparent that the estimations of lead in the diet of man vary greatly. Future studies should include specifications concerning the characteristics of the individuals for whom lead consumption data are being reported, including sex, age, weight, and physical activity. Since the ultimate purpose of food studies is to evaluate the contribution made to the total dose, it is important that future reports also include the observed Pb-B levels and, preferably, other indices, such as $\delta$-aminolevulinic acid in urine (ALA-U), PP and ALAD in erythrocytes. Food studies should also include estimates of the lead concentration of various components of the total diet. Only with such studies will it be possible to arrive at decisions regarding the control of lead in foods.

More precise information is available concerning the contribution of airborne lead to Pb-B and although this seems to be a minor contributor to Pb-B for the general population compared with diet, additional studies are needed both in occupational situations, and for the general population. The studies should be of a relatively long-term nature and should be done, as far as possible, with personal air samplers maintained in operation continuously throughout the day during the period of study.

There is a great need to study the sources of lead affecting infants and young children including the contributions of food, milk and other beverages, and air, and also miscellaneous sources, e.g. paint, soil, and dust.

### 1.2.3  Epidemiological studies

Prospective studies are needed of the health effects of both inorganic and organolead compounds, with particular reference to a more thorough estimation of the nature of the lead exposure, Pb-B levels, and measurable effects. It would seem particularly useful to make further studies on occupational groups, beginning at the time of their entry into the high lead environment.

### 1.2.4 Interactions of lead with other environmental factors

In both epidemiological studies and in experimental studies on animals, not enough emphasis has been placed on the environmental variables that can affect man's response to lead. The list of such variables is long and is documented in this report. Particular attention should be paid to the influence of other metals, air pollutants, and the nutritional status of the subjects, since these factors have been identified as interacting with lead either in regard to its deposition in the body or in regard to its biological effects in target organs.

### 1.2.5 Significance of biological effects

Numerous abnormalities have been identified, the toxic significance of which is obscure, e.g. elevated free erythrocyte PP and marginal erythrocyte ALAD inhibition. There is an urgent need to study the significance of these findings in relation to human health.

## 2. PROPERTIES AND ANALYTICAL METHODS

### 2.1 Physical and Chemical Properties of Lead and its Compounds

Lead (atomic number, 82; atomic weight, 207.19; specific gravity, 11.34) is a bluish or silvery grey soft metal. The melting point is 327.5°C and the boiling point at atmospheric pressure 1740°C. It has four naturally occurring isotopes (208, 206, 207, and 204 in order of abundance), but the isotopic ratios for various mineral sources are sometimes substantially different. This property has been used to carry out non-radioactive-tracer environmental and metabolic studies.

Although lead has four electrons in its valence shell, only two ionize readily. The usual oxidation state of lead in inorganic compounds is therefore + 2 rather than + 4. The inorganic salts of lead (II), lead sulfide, and the oxides of lead are generally poorly soluble. Exceptions are the nitrate, the chlorate and, to a much lesser degree, the chloride (Table 1). Some of the salts formed with organic acids, e.g. lead oxalate, are also insoluble.

Table 1. Some physical and chemical data on lead and selected lead compounds[a]

| Name | Synonym and formula | Molecular weight | Melting point (°C) | Boiling point (°C) | Solubility in cold water (g/litre) | Soluble in |
|---|---|---|---|---|---|---|
| lead | $Pb$ | 207.19 | 327.502 | 1740 | insoluble | $HNO_3$; hot concentrated $H_2SO_4$ |
| acetate | $Pb(C_2H_3O_2)_2$ | 325.28 | 280 | — | 443 | hot water; glycerine; alcohol (slightly) |
| azide | $Pb(N_3)_2$ | 291.23 | | explodes 350 | 0.23 | acetic acid; hot water (0.9 g/litre) |
| carbonate | cerrusite $PbCO_3$ | 267.20 | 315 (decomposes) | | 0.0011 | acid; alkali; decomposes in hot water |
| chlorate | $Pb(ClO_3)_2$ | 374.09 | 230 (decomposes) | | very soluble | alcohol |
| chloride | cotunite $PbCl_2$ | 278.10 | 501 | 950 | 9.9 | $NH_4$ salts; slightly in dilute HCl and in $NH_3$; hot water (33.4 g/litre) |
| chromate | crocoite, chrome yellow $PbCrO_4$ | 328.18 | 844 | decomposes | 0.000058 | alcohol; alkali |
| nitrate | $Pb(NO_3)_2$ | 331.20 | 470 (decomposes) | | 376.5 | alcohol; alkali; $NH_3$; hot water (1270 g/litre) |
| ortophosphate | $Pb_3(PO_4)_2$ | 811.51 | 1014 | | 0.00014 | alkali; $HNO_3$ |
| oxalate | $PbC_2O_4$ | 295.21 | 300 (decomposes) | | 0.0016 | $HNO_3$ |
| oxide: di- | plattnerite $PbO_2$ | 239.19 | 290 (decomposes) | | insoluble | dilute HCl; acetic acid (slightly) |
| mono- | litharge $PbO$ | 223.19 | 888 | | 0.017 | $HNO_3$; alkali; $NH_4Cl$ |
| red | minium $Pb_3O_4$ | 685.57 | 500 (decomposes) | | insoluble | HCl; acetic acid |
| sesqui- | $Pb_2O_3$ | 462.38 | 370 (decomposes) | | insoluble | decomposes in acid and hot water |
| stearate | $Pb(C_{18}H_{35}O_2)_2$ | 774.15 | 115.7 | | 0.5 | hot water (0.6 g/litre); ether (0.05 g/litre) |
| sulfate | anglesite $PbSO_4$ | 303.25 | 1170 | | 0.0425 | $NH_4$ salts; concentrated $H_2SO_4$ (slightly) |
| sulfide | galena $PbS$ | 239.25 | 1114 | | 0.00086 | acid |
| tetraethyllead | $Pb(C_2H_5)_4$ | 323.44 | -136.80 | 200 decomposes; 91 | insoluble | benzene; petroleum; alcohol; ether |
| tetramethyllead | $Pb(CH_3)_4$ | 267.3 | -27.5 | 110 | insoluble | benzene; petroleum; alcohol; ether |

[a] Adapted from Weast, R. C., ed. Handbook of Chemistry and Physics, 55th edition, Cleveland, Ohio, Chemical Rubber Company, 1974.

Under appropriate conditions of synthesis, stable compounds are formed in which lead is directly bound to a carbon atom. Tetraethyllead and tetramethyllead are well-known organolead compounds. They are of great importance owing to their extensive use as fuel additives. Both are colourless liquids. Their volatility is lower than for most gasoline components. The boiling point of tetramethyllead is 110°C and that of tetraethyllead is 200°C. By contrast, the boiling point range for gasoline hydrocarbons is 20–200°C. Thus evaporation of gasoline tends to concentrate tetraethyllead and tetramethyllead in the liquid residue.

Both tetramethyllead and tetraethyllead decompose at, or somewhat below, the boiling point. Analysis of automobile exhaust gases shows that the ratio of tetramethyllead to tetraethyllead increases as the engine warms up, indicating that tetramethyllead is more thermostable than tetraethyllead (Laveskog, 1971). These compounds are also decomposed by ultraviolet light and trace chemicals in air such as halogens, acids, or oxidizing agents (Snyder, 1967).

## 2.2 Analytical Procedures

### 2.2.1 Sampling

Particular attention should be paid to the cleanliness of the instruments and the purity of chemicals to prevent the appearance of artifacts due to the secondary contamination by lead, especially in the sampling of foods and biological media.

In air sampling, high-volume samplers are preferable for accuracy (when it is necessary), but the low-volume technique is also useful for obtaining extensive data. As in all sampling for suspended particulate matter, the accuracy of volume meters should be checked periodically. The size of the pores of filters for collecting lead-containing particles should be small, possibly less than 0.2 µm for glass-fibre filters (Lee & Goransen, 1972). Liquid scrubbers containing iodine monochloride and solid scrubbers with activated carbon, cristobalite, or iodine crystals have been used for sampling organic lead compounds in air, in the range of about 1 µg/m³ or less (Snyder, 1967; ASTM, 1970; Laveskog, 1971; Coleville & Hickman, 1973; Purdue et al., 1973) up to 10 µg/m³ (Harrison et al., 1974).

Depending on the purpose of sampling, care should be taken to select the appropriate site for sampling devices and to achieve the best possible sampling conditions by:

— estimating the required amount of particulates before deciding on the sample volume and the sampling procedure;

— placing the sampling devices in the appropriate position (e.g. breathing air level, level of inlet tubes of house ventilators, window level in the case of a traffic-laden town street, at a reasonable distance from the highway in uninhabited zones, etc),
— taking the samples at appropriate rates and volumes (e.g. daily breathing volumes, daily ventilating capacities of installations) and for a sufficient time to make possible the estimation of the average concentration (e.g. during a work shift, or a 24-hour or longer period for general population exposure);
— taking into account the use of appropriate areas (cattle grazing, recreational zones, children's playgrounds etc.)

In addition, whenever possible, a procedure should be used that makes it possible to evaluate particle-size distribution and the physico-chemical properties of the lead compounds involved, including the shape of the particles and the state of their aggregation.

Stationary samplers can provide general indices of the exposure of individuals within a certain area. For estimating exposure through inhalation, personal samplers are highly desirable (Azar et al., 1973).

Techniques for sampling water are less complex than for air. The major question is whether or not the water should be filtered before analysis since it is known that lead occurs in water both in the particulate fraction and in solution. For most purposes at least, it is reasonable to sample water without any fractionation of the material collected.

However, in some cases it may be necessary to determine the biological availability for absorption of the various forms of lead that occur in water, and in soil. The latter is a dust source and may be a food contamination source as well.

The preparation of soil and soil dust samples for lead analysis usually involves drying (at 100°C), homogenization by grinding, and sieving (Thornton & Webb, 1975; Bolter et al., 1975).

For the study of lead in foods, two general methods have been used. These are the duplicate portions technique and the equivalent composite technique (theoretical diet). These two general techniques and others have been reviewed recently with reference to their advantages and disadvantages (Pekkarinen, 1970). The duplicate portions technique involves the collection for analysis of duplicates of the meals actually consumed by the individual. When carried out over a long enough period, the technique has the advantage of defining variability in consumption. Kehoe (1961) used this method for the daily determination of lead consumption over long periods. Considerable variation in lead consumption was found in individuals even when consumption was averaged for four- or eight-week collection periods. The disadvantages of the method are the expense and the exacting nature of the method of collecting samples; these factors tend to limit the numbers of individuals included in such studies.

The equivalent composite technique consists of formulating the ingredients of meals typical for subpopulations and analysing them. The advantages are economy and ease of collection. This approach may or may not include the cooking process. The disadvantage is uncertainty as to how typical or representative the formulation is. Even when the cooking process is included, there may be significant differences in the manner of preparation for the study in comparison with that carried out under actual home conditions.

The main problem in the sampling of body fluids and tissues for lead analysis is potential secondary contamination with lead. Special precautions must be taken to ensure that all blood-collecting and blood-storage materials are as free from lead as possible. All glass equipment involved in blood collection and storage should be made of lead-free silicate glass, rinsed first in mineral acid, then with copious amounts of glass-distilled or deionized water. Polypropylene syringes have been recommended (NAS-NRC, 1972). Needles should be of stainless steel with polypropylene hubs. Blood is often drawn directly from the needle into vacuum tubes. It is wise to confirm periodically the absence of significant amounts of lead in the anticoagulant used in the blood container, although this has not been reported as a problem.

New analytical techniques make it possible to determine lead concentrations in microlitre quantities of blood. The trend towards the procurement of micro-samples of blood by skin prick increases the hazard of secondary contamination of the blood. Only one systematic investigation on the significance of this problem has been reported. Mitchell et al. (1974) describe a procedure whereby sample contamination appears to be avoided. This is achieved by spraying collodion over the cleansed skin before lancing. The correlation between the concentration of lead in micro-samples and in macro-samples obtained by venipuncture was fairly good ($r = 0.92$). The same general precautions must be taken in the collection of urine samples as in the collection of blood samples.

Ceramic surfaces are analysed to determine the quantity of lead likely to be leached by different foods and beverages. In all cases acetic acid solutions are used but the concentrations vary from 1 to 4%. The temperature of the tests ranges from 20 to 100°C and the duration from 30 minutes to more than 24 hours (Laurs, 1976; Merwin, 1976).

### 2.2.2 Analytical methods for lead

The analytical methods currently in use for the estimation of lead content are of two general types, destructive and non-destructive. In the former, the sample is first oxidized to destroy all organic matter. The ash is then usually dissolved in an aqueous medium, either for further preparative steps or for direct instrumental analysis. Non-destructive methods are of more recent origin and are still too complicated for routine studies. They include X-ray fluorescence analysis and fast neutron activation. In selecting methods, consideration must be given to the cost of the equipment and the time involved in performing the analyses.

The oldest and best known of the general methods currently in wide use are those based on the formation of the red complex that lead forms with dithizone (diphenylthiocarbazone). Numerous specific procedures have been developed based on the spectrophotometric determination of lead dithizonate. A typical example is the "US Public Health Service" method commonly used for the determination of lead in biological materials (NAS-NRC, 1972). The method has evolved over many years. A study of its reliability was reported by Keenan et al. (1963). An interlaboratory comparison was made of analyses of blood and urine with and without the addition of lead. Ten laboratories participated in the study. For blood, the concentration of lead calculated in the principal laboratory was 20 µg/100 ml. The average reported by the participating laboratories was 26 µg/100 ml with a standard deviation of $\pm 0.82$ µg/100 ml. For samples of blood to which lead was added, the average result was right on the mark, 70 µg/100 ml $\pm 0.78$. For "spiked" urine, determined by the primary laboratory to contain 750 µg/litre, the average reported result was 679 $\pm 5.5$ µg/litre.

Perhaps no method of instrumental analysis for lead has enjoyed such a rapid acceptance in recent years as atomic absorption spectroscopy. In conventional atomic absorption spectroscopy, the source of heat is a flame into which the sample solution is aspirated. More recently, various procedures have been developed whereby the receptacle containing the sample is heated electrically. This type of modified procedure is termed flameless atomic absorption spectroscopy. The main advantage of this approach is that sample size is reduced from the millilitre to the microlitre range with no commensurate loss of sensitivity. Another advantage is that the heated receptacle can be used for ashing the sample immediately prior to the spectrophotometric analysis. Numerous reports have appeared describing various kinds of flameless instrumentation and their application in the analysis of the lead content of blood and other materials (Cernik, 1974; Delves, 1970; Ediger & Coleman, 1973; Matousek & Stevens, 1971; Kubasik et al., 1972; Hwang et al., 1971; Sansoni et al., 1973; Schramel,

1973; Schramel, 1974). It has been reported that the analytical capabilities of this method for determining lead in whole blood are comparable with that of the conventional flame atomic absorption method (Kubasik et al., 1972; Hicks et al., 1973).

Electroanalytical methods have also been found useful for lead determinations. These include polarography and, more recently, anodic stripping voltametry. The polarographic method was developed specifically for lead by Teisinger (1935). The low sensitivity of the method as applied to lead in blood and urine required working close to the detection limits. This is obviously a disadvantage when determining the normal levels of lead in blood and urine. Various modifications of the original method have been used for the evaluation of industrial exposures (Weber, 1947; Baker, 1950; Brezina & Zuman, 1958). This method found wide application until more effective masking procedures were developed to increase the specificity of the dithizone method. Anodic stripping voltametry is gaining in popularity for lead analysis. Results have been compared using a dithizone method, an atomic absorption method, and anodic stripping voltametry (Matson, 1971). Generally, there was good agreement between all three methods in the estimation of the lead contents of blood and urine. In another study, anodic stripping voltametry was compared with atomic absorption spectroscopy and polarography for the analysis of lead in blood and urine (Horiuchi et al., 1968). The authors concluded that there were no significant differences between the results obtained by the various methods. Anodic stripping voltametry has also been compared with conventional and flameless atomic absorption spectroscopy and with potentiometric determination using ion-specific electrodes to estimate the lead content of water (Kempf & Sonnenborn, 1973).

Two non-destructive methods for lead analysis have been under investigation in recent years. These are neutron activation and X-ray fluorescence. The first of these is not likely to find wide application for lead analysis in the near future because of the cost and the need for access to a fast neutron source. Its advantage is that the concentration of many elements can be determined simultaneously.

X-ray fluorescence is also theoretically capable of detecting, non-destructively, all elements in a substance. A major obstacle to the wide application of this method is the profound matrix effect of the substances being analysed. Another problem is the backscatter from the exciting source. These design problems and approaches to their solution have been discussed recently by Kneip & Laurer (1972). Lead analysis by means of X-ray fluorescence with proton excitation has been successfully used with biological samples (Möller et al., 1974). It has also been used as the standard method for the determination of lead on filters from air sampling equipment by the Warren Springs Laboratory in the United Kingdom. In the USA, the most extensive application of X-ray fluorescence for lead analysis has been for estimating the concentration and amount of lead on the walls of houses. For this purpose, several portable units have been designed and are being used in surveys of dwellings for hazardous concentrations of lead. Since the instruments in question scan surfaces, instrument response is in terms of lead detected per unit area and not per unit weight or volume of paint film. This creates difficulties, since the thickness of the total paint film varies depending on how many times a surface has been painted. Ordinances should perhaps be revised to specify tolerances based on surface area. The accuracy of these instruments is severely limited. These factors have been studied using one of the commercially available instruments (Spurgeon, 1973). In another report from the US National Bureau of Standards (Rasberry, 1973), four commercial instruments were tested as received from the manufacturer. It was found that all the instruments had a detection limit below 1 mg/cm$^2$, but that between 1 and 6.6 mg/cm$^2$, errors as large as 30–50 % occurred. It is difficult to evaluate the adequacy of such instruments since it is not at all clear where the cut-off is between hazardous and non-hazardous amounts of lead per unit area of paint film. Thus, if the cut-off were known to be at or above 1 mg/cm$^2$, the instruments would clearly be useful.

The accuracy and precision of various methods for the lead analysis of biological materials have been appraised in a number of interlaboratory comparison programmes both at the national (Keppler et al., 1970; Donovan et al., 1971) and international levels (Berlin et al., 1973). In general, these published studies have indicated that the accuracy of the measurements is unsatisfactory, with less than half of the laboratories performing adequately. More recently, in a programme involving sixty-six European laboratories, it was observed that even when only the laboratories that measured lead in blood and urine with a precision of greater than 10 % were selected, the interlaboratory variability still remained high. It is possible that the performance could be improved by rapid distribution of the sample and by improved sample preparation techniques, e.g. by subjecting blood samples to ultrasonic irradiation prior to despatch to participating laboratories.

The paper punch disc microtechnique (Cernik & Sayers, 1971; Cernik, 1974) was used in a population survey of blood lead content performed in Western Ireland (Grimes et al., 1975). Over 400 duplicate samples were analysed double-blind by one laboratory. The assay showed a satisfactory agreement with the results obtained by other laboratories using various techniques.

Comparisons have also been reported of the agreement between results obtained by the same investigator using different analytical methods. Yeager et al. (1971) compared the results obtained using a standard dithizone procedure and flame atomic absorption spectroscopy. The results from common digests of the same material were compared. The materials included blood, urine, tissue, faeces, food, and bone. Since the two methods are based on entirely different analytical principles, a straight line with a slope equal to 1 and an intercept equal to 0, obtained when the results of atomic absorption spectroscopy analyses were plotted against the results of the dithizone method, suggested that the two methods were equally accurate.

These studies show that blood sample preparation is important to ensure sufficient homogeneity for microanalytical techniques.

### 2.2.3 Methods for the measurement of some biochemical effects of lead

The classic method for the urinary $\delta$-aminolevulinic acid (ALA) determination was developed by Mauzerall & Granick (1956). The major procedural difficulty was separation from interfering substances. A number of modifications and simplifications have been made by several authors (Davis & Andelman, 1967; Grabecki et al., 1967; Williams & Few, 1967; Sun et al., 1969; Tomokumi & Ogata, 1972).

The original Mauzerall & Granick method does not discriminate between ALA and aminoacetone, a fact that these authors were careful to point out. This is probably not very important when ALA excretion is greatly increased due to lead exposure, but for marginal elevations, it may be a serious problem. In healthy humans on a normal diet, the urinary excretion of ALA and that of aminoacetone are nearly equal (Marver et al., 1966). These authors and also Urata & Granick (1963) separated ALA from aminoacetone by chromatography.

One interlaboratory comparison study of ALA methods has been reported (Berlin et al., 1973). The methods used by the laboratories were those of Mauzerall & Granick (1956), Davis & Andelman (1967) and of Grabecki et al. (1967). The results using the Grabecki method were significantly higher than those using the Mauzerall & Granick method. Results with the Davis & Andelman method gave a mean value intermediate between the other two. The coefficients of variation were quite high: 33%, Grabecki; 28%, Mauzerall & Granick; and 49%, Davis & Andelman. It should also be noted that in the case of the Grabecki method, the colorimetric reaction was influenced by various interfering substances in the individual urine samples. This source of error was not considered in the interlaboratory comparison (Mappes, 1972).

Comparisons have also been reported between these different techniques by Roels et al. (1974) who evaluated the critical factors in the urine preparation which affected the different methods. The ionic strength and pH of the urine can affect the results of some of the methods.

In the methods used for the determination of ALAD activity, the amount of porphobilinogen (PBG) formed per unit time by a standard amount of enzyme source is measured. Limited data indicate that ALAD in blood is stable for several hours, even at room temperature (Hernberg et al., 1970); however, storage at lower temperatures improves the stability. The major variables reported to influence the activity of the enzyme are pH (Nikkanen et al., 1972), oxygen tension (Gibson et al., 1955), the nature of the anticoagulant (Collier, 1971), and the presence or absence of activators (Bonsignore et al., 1965; Collier, 1971; Granick et al., 1973; Hapke & Prigge, 1973). Measurement of ALAD activity in erythrocytes is a relatively simple procedure that can be conducted without sophisticated equipment. This makes it attractive as a measure of the haematological effects of exposure to lead. A number of investigators have shown it to be fairly specific for lead.

In its simplest and most frequently used form, the method of Bonsignore et al. (1965) requires the incubation of a mixture of blood, ALA, and water under aerobic conditions at 38°C. However, many investigators have modified the procedure and results from different laboratories are not necessarily comparable. In a recent interlaboratory comparison (Berlin et al., 1973), nine participants used various modifications of the Bonsignore method. Thus, it was only possible to compare the activity ratios between different blood samples. For two blood samples this ratio showed a coefficient of variation of only 13%.

Recently a "European standardized method" has been developed, tested in a collaborative study, and agreed upon by nineteen laboratories. The results of these tests compare very favourably with blood lead determinations. The interlaboratory coefficient of variation for ALAD was 10% (Berlin et al., 1974).

Porphyrins exhibit intense fluorescence when excited by light at approximately 400 nm (Soret band). They may be quantitatively determined either by measurement of light absorption in the Soret band region or by the measurement of fluorescence (Sassa et al., 1973; Chisolm, 1974).

A number of methods have been reported for the measurement of protoporphyrin IX. Some of these methods discriminate between different porphyrins, measuring specifically the concentration of protoporphyrin IX in erythrocytes (Schwartz & Wikoff, 1952; Wranne, 1960; Schlegel et al., 1972; Granick et al., 1972; Sassa et al., 1973). Other methods measure the total concentration of free erythrocyte porphyrins including copro- and uro-porphyrins (Kammholtz et al., 1972; Piomelli, 1973; Schiele et al., 1974b). It is, however, scarcely necessary to make a distinction between the two kinds of procedure as over 90 % of the free erythrocyte porphyrins are made up of protoporphyrin IX (Baloh, 1974). A particular advantage of the more recently developed procedures for the measurement of FEP is that they can be performed on microcapillary samples of blood (Kammholz, 1972; Granick et al., 1972; Sassa et al., 1973; Piomelli, 1973; Schiele et al., 1974a). The Piomelli procedure utilizes two successive extractions into ethylacetate–acetic acid with subsequent transfer of porphyrins into hydrochloric acid. The procedure of Granick et al. (1972) is simpler. Ethylacetate–acetic acid and hydrochloric acid are successively added to the sample of blood. In both procedures the ethylacetate serves to remove and retain interfering impurities in the blood. The data obtained by these two methods are not strictly comparable.

All the methods described measure protoporphyrin in the free base form. Lamola & Yamane (1974) have recently demonstrated that the protoporphyrin IX associated with iron deficiency and lead intoxication is present as a zinc chelate. This is not so in the case of erythropoietic porphyria. On the basis of these observations they developed a fluorimetric method for zinc chelate (Lamola et al., 1975). The major advantage of this method is its simplicity and rapidity. Microlitre samples are analysed fluorimetrically, after dilution, without any extraction steps.

The measurement of coproporphyrins in urine is generally done by extraction of the porphyrins into either ethylacetate–acetic acid (Sano & Rimington, 1963) or diethyl ether (Askevold, 1951) followed by transfer into hydrochloric acid. Absorbance is then measured at 401 nm with the corrections recommended by Rimington & Sveinsson (1950). The method is apparently specific, since uroporphyrins, the most likely source of interference, are not extracted into the organic phase under these conditions (Rimington & Sveinsson, 1950). An alternative method has been reported whereby the fluorescence of the hydrochloric acid extract is measured after adsorption on to magnesium hydroxide (Djurić, 1964). Certain precautions are necessary if urine is to be analysed for coproporphyrins. Coproporphyrins are unstable in acid urine and, furthermore they are light-sensitive (Schwartz et al., 1951). They may be stored safely in the dark at 4°C if the pH is maintained between 6.5 and 8.5.

# 3. SOURCES OF LEAD IN THE ENVIRONMENT

## 3.1 Natural Occurrence

### 3.1.1 Rocks

Lead occurs naturally in the earth's crust in the concentration of about 13 mg/kg. As with all elements, there are some areas with much higher concentrations including the lead ore deposits scattered throughout the world.

The most important sources of lead are igneous and metamorphic rocks, with lead concentrations in the range of 10–20 mg/kg (Wedepohl, 1956, Vinogradov, 1956, 1962; Turekian & Wedepohl, 1961). The concentration of lead in sedimentary rocks is of the same order of magnitude. The lead content of carbonaceous shales from the United States of America and Europe ranges from 10 mg/kg to 70 mg/kg (Wedepohl, 1971; Davidson & Lakin, 1962). The lead contents of shale and sandstone are similar but that of phosphate rocks is higher, and may exceed 100 mg/kg (Sheldon et al., 1953). Unconsolidated sediments in bodies of freshwater and in shallow marine areas have a similar lead content to shales. Deep marine sediments have quite a high lead content by comparison, commonly containing 100–200 mg/kg (Riley & Skirrow, 1965).

The lead content of coal is relatively low. However, when expressed on an ash-weight basis, the concentration is generally higher than that of igneous, metamorphic, and sedimentary rocks, but not more than ten-fold (Abernethy et al., 1969).

### 3.1.2 Soils

Surface soils are in direct contact with the contemporary environment; thus, special care must be taken to distinguish between soils that acquire lead only from natural sources and soils that are polluted by man. Acidic soils generally have a lower lead content than alkaline soils. The nature of the organic matter in soil also has a considerable influence on its lead content. Some organic matter is rich in chelating components, and it binds lead, either promoting its movement out of the soil or fixing the metal, depending on the solubility properties of the complex. Although all of these factors no doubt play a role in determining the lead content of specific soils, the concentrations usually encountered in areas, remote from human activity, are similar to concentrations found in rocks, with an average range of 5–25 mg/kg (Swaine, 1955). More recent data from various parts of the world have confirmed this estimate.

### 3.1.3 Water

Analyses of groundwater have revealed lead concentrations varying from 1 to 60 µg/litre (Kehoe et al., 1933, 1944; Bagchi et al., 1940). Most data refer to water that has been filtered to remove particulate matter. Colloidal lead is only partially removed by filtration and to different degrees. Water that is pumped from the ground is usually not filtered prior to analysis. The content of colloidal material is probably insignificant in such samples owing to natural filtration which removes colloidal particles fairly effectively.

There have been a large number of investigations concerning the concentration of lead in natural surface waters. From the data available, Livingstone (1963) estimated that the global mean lead content in lakes and rivers is 1–10 µg/litre. Although this estimate includes man-made pollution, it probably still represents a fair approximation of natural conditions since water flowing through the ecosystems has a considerable self-cleaning capacity.

The concentration of lead in sea water has been found to be lower than in freshwaters. Tatsumoto & Patterson (1963) report 0.08–0.4 µg/litre in seawaters off the coast of California. In deep waters the concentration was even lower. According to Chow (1968) surface waters off Bermuda, which are free from continental influences, have lead concentrations averaging 0.07 µg/litre, while central Atlantic waters contain an average of 0.05 µg/litre. Although there seem to be somewhat higher lead concentrations in the surface waters of the Pacific and the Mediterranean, compared with the central Atlantic, the concentrations at depths below the 1000-m level are very similar, i.e. around 0.03–0.04 µg/litre (Chow, 1968).

### 3.1.4 Air

The atmospheric concentration of lead measured at points most remote from civilization is of the order of 0.0001–0.001 µg/m³ (Jernigan et al., 1971; Chow et al., 1969; Egorov et al., 1970; Murozumi et al., 1969). The sampling sites in these studies were mainly over remote areas of oceans and over Greenland. Patterson (1965) estimated from geochemical data that the concentration of lead in air of natural origin is about 0.0006 µg/m³. If that is a correct estimate, even the air over uninhabited, remote, continental areas may be contaminated by human activities. For example, Chow et al. (1972) reported that the concentration of lead in the air over remote, uninhabited mountains of southern California had a concentration of 0.008 µg/m³.

### 3.1.5 Plants

Lead occurs naturally in all plants, as well as in soil, air, and water. Extremely variable concentrations of lead in plants have been reported but nevertheless, certain generalizations have been made. Warren & Delavault (1962) have concluded that the normal concentration of lead in leaves and twigs of woody plants is 2.5 mg/kg on a dry weight basis. For vegetables and cereals they estimated normal concentrations to be 0.1–1.0 mg/kg dry weight. Mitchell (1963) found that the usual concentration of lead in pasture grasses was 1.0 mg/kg dry weight. These figures should be multiplied by a factor of 20 to convert concentration on a dry weight basis to an ash weight basis.

### 3.1.6 Environmental contamination from natural sources

The contribution of natural sources of lead to lead concentrations in the environment is small. As regards exposure of man, these sources are negligible. Through various breakdown processes, rocks yield lead which is transferred to the biosphere and the atmosphere and ultimately back to the earth's crust in the form of sedimentary rocks. Soluble lead has for thousands of years entered the oceans with river discharges, and the amount has been estimated by Patterson (1965) at some 17 000 tonnes per year. Sources contributing to airborne lead are silicate dusts, volcanic halogen aerosols, forest fires, sea salts aerosol, meteoric and meteoritic smoke, and lead derived from the decay of radon. The last mentioned source generates the lead isotope $^{210}Pb$ in trace amounts, the mean air residence time of which has been calculated to be about four weeks; the radioactive half-life is 22 years (Hill, 1960).

## 3.2 Production of Lead

### 3.2.1 Lead mining

Lead is produced from ores and recycled lead products. Lead occurs in a variety of minerals the most important of which are galena (PbS), cerrusite ($PbCO_3$) and anglesite ($PbSO_4$). Galena is by far the most important source of primary lead. It occurs mostly in deposits associated with other minerals, particularly those containing zinc. Mixed lead and zinc ores account for about 70 % of total primary lead supplies. Ores containing mainly lead account for about 20 % and the remaining 10 % is obtained as a by-product from other deposits, mainly zinc and copper-zinc deposits (Federal Institute for Minerals Research and German Institute for

Economic Research, 1972). The proportions of various metals may differ in the ores of different countries. Silver is the most important of the other metals frequently present in lead deposits but copper may also be present in concentrations high enough to be commercially important. Other minor constituents of lead ores are gold, bismuth, antimony, arsenic, cadmium, tin, gallium, thallium, indium, germanium, and tellurium. The lead content of ores is comparatively low, i.e. 3–8%, but even ores with lower lead contents may be commercially valuable.

The level of world mine production of lead concentrates from ores has increased in recent years. According to the International Lead and Zinc Study Group and the World Bureau of Metal Statistics, the world mine production of lead (lead content) was about 3.6 million tonnes in 1975, as compared with about 2.6 million tonnes in 1965. These figures include production estimates for socialist countries with a planned economy made by the World Bureau of Metal Statistics. The most important lead mining countries, producing over 100 000 tonnes each in 1975, were Australia (10% of the total world output), Bulgaria (3%), Canada (9.6%), China (3.8%), Mexico (4.5%), Peru (5.5%), United States of America (16%), USSR (14.5%), and Yugoslavia (3.5%). In addition, some other countries had a production of over 2% of the world total, e.g. Ireland, Japan, Democratic People's Republic of Korea, Morocco, Poland, Spain, and Sweden. There are about 40 countries producing only small amounts each, making together only some 12% of the world production. One estimate of proven lead reserves of the world is 93 million tonnes of lead metal content. (Federal Institute for Minerals Research and German Institute for Economic Research, 1972.)

### 3.2.2 Smelting and refining

Smelting and refining is classified as primary or secondary, the former producing refined lead from concentrates (primary lead); the latter recovering lead from scrap (secondary lead). The raw materials for secondary lead are process (new) scrap arising during manufacturing processes, and recycled (old) scrap which arises when lead-containing manufactured goods are discarded. Old material makes up the bulk of the scrap, the most important source being storage batteries, which account for 70–80% of the total supply of scrap.

Secondary lead accounts for about half the consumption in the United States of America and it has been estimated that about 35% of the total world lead supply comes from secondary sources (Federal Institute for Minerals Research and German Institute for Economic Research, 1972).

| Country | Lead ore production (metal content) | | | Metal production | | | Consumption (refined metal) | | |
|---|---|---|---|---|---|---|---|---|---|
| | 1973 | 1974 | 1975[b] | 1973 | 1974 | 1975[b] | 1973 | 1974 | 1975[b] |
| EUROPE | 1134 | 1134 | 1069 | 2054 | 2115 | 1871 | 2118 | 2125 | 1831 |
| Belgium | — | — | — | 98 | 95 | 103 | 52 | 64 | 54 |
| Bulgaria | 105 | 110 | 108 | 100 | 105 | 108 | 80 | 85 | 91 |
| Denmark | — | — | — | 13 | 15 | 13 | 19 | 23 | 20 |
| France | 25 | 24 | 22 | 186 | 178 | 150 | 214 | 199 | 188 |
| Germany, Federal Republic of | 40 | 35 | 37 | 300 | 319 | 260 | 290 | 260 | 210 |
| Ireland | 53 | 34 | 55 | — | — | — | 1 | 3 | 2 |
| Italy | 27 | 24 | 27 | 100 | 112 | 70 | 234 | 242 | 200 |
| Netherlands | — | — | — | 25 | 26 | 20 | 38 | 41 | 38 |
| Poland | 70 | 70 | 72 | 68 | 70 | 66 | 87 | 90 | 40 |
| Spain | 64 | 65 | 58 | 120 | 102 | 85 | 121 | 116 | 90 |
| Sweden | 74 | 73 | 69 | 42 | 41 | 37 | 34 | 36 | 32 |
| UK | — | — | — | 265 | 277 | 229 | 282 | 266 | 238 |
| USSR | 570 | 590 | 504 | 640 | 660 | 600 | 600 | 620 | 544 |
| Yugoslavia | 106 | 109 | 117 | 97 | 115 | 130 | 66 | 80 | 84 |
| AFRICA | 223 | 183 | 178 | 116 | 117 | 93 | 65 | 66 | 75 |
| Morocco | 90 | 86 | — | 1 | 1 | — | — | — | — |
| South Africa | 63 | 55 | 53 | 64 | 64 | 49 | 27 | 31 | 39 |
| AMERICA | 1430 | 1412 | 1379 | 1666 | 1677 | 1565 | 1718 | 1706 | 1339 |
| Canada | 388 | 314 | 348 | 187 | 127 | 172 | 69 | 63 | 55 |
| Mexico | 168 | 169 | 163 | 177 | 204 | 179 | 88 | 83 | 74 |
| Peru | 199 | 201 | 185 | 83 | 80 | 72 | 10 | 9 | 10 |
| USA | 570 | 616 | 575 | 1100 | 1128 | 1008 | 1423 | 1374 | 1027 |
| ASIA | 273 | 284 | 291 | 413 | 423 | 395 | 457 | 412 | 386 |
| Democratic Republic of Korea | 90 | 100 | 100 | 60 | 65 | 60 | 20 | 20 | 20 |
| Japan | 53 | 44 | 51 | 228 | 228 | 195 | 267 | 217 | 186 |
| People's Republic of China | 130 | 140 | 140 | 125 | 130 | 140 | 170 | 175 | 180 |
| OCEANIA | 396 | 360 | 384 | 221 | 225 | 191 | 82 | 79 | 75 |
| Australia | 396 | 360 | 384 | 221 | 225 | 191 | 74 | 72 | 68 |
| Other countries | 55 | 53 | 48 | 102 | 108 | 88 | 189 | 203 | 233 |
| TOTALS | 3617 | 3569 | 3497 | 4642 | 4723 | 4260 | 4883 | 4882 | 4154 |

[a] Sources: International Lead, Zinc Study Group, and World Bureau of Metal Statistics.
[b] Estimated.

Table 2 gives the production of lead ore, the total metal production, and the consumption of some industrialized countries.

### 3.2.3 Environmental pollution from production

Mining, smelting, and refining, as well as the manufacture of lead-containing compounds and goods, can give rise to lead emissions. According to a study of the industrial sources of air pollution by lead in the USA, Davis (1973) reported that 9 % of the total of 18 000 tonnes generated from such sources was attributable to the production of primary lead.

Smelters of lead ores are well known to create pollution problems in local areas. Their influence on the surrounding air and soil depends to a large extent on the height of the stack, the trapping devices in the stacks, the topography, and other local features. The emissions can cover a considerable area. The zone of air pollution for one large smelter in the USA extended to approximately 5 km from the smelter while soil contamination extended as far as 10 km (Landrigan et al., 1975b). The larger area of the zone of soil pollution compared to the zone of air pollution probably was due to the fact that current emission control devices are more effective than earlier ones used to be. The opposite situation was found around the Mežica mine and smelter in Yugoslavia (Djurić et al., 1971; Kerin, 1972, 1973). In this case, the zone of air pollution extended as far as 10 km from the smelter stack. Soil was grossly contaminated ($>200$ mg/kg) as far away as 7 km. There was also heavy pollution of water courses through effluents.

Secondary smelters producing lead from scrap are comparatively small, numerous, and frequently situated close to human settlements. Several studies showed that pollution in the surroundings of such smelters had been severe enough to produce an increase in the intake of lead by people living nearby (section 5.1.1).

### 3.3 Consumption and Uses of Lead and its Compounds

Figures for the consumption of lead are available for most industrialized countries. The estimated total world consumption of lead in 1975 was about 4.1 million tonnes (Table 2). The use of lead is greatly influenced by the growth of the automobile industry which in 1974 took about 56 % of total consumption. Table 3 is compiled from statistics of lead consumption for the Federal Republic of Germany, France, Italy, Japan, the United Kingdom, and the United States of America. There has been a notable increase in the consumption for batteries over the period 1969–1974.

### 3.3.1 Storage battery industry

The manufacture of electric storage batteries is responsible for the largest consumption of lead (Table 3). This industry uses both metallic lead in the form of a lead–antimony alloy, and lead oxides in about equal proportions. The metallic lead is in the grids and lugs, while the oxides, litharge (PbO), red lead ($Pb_3O_4$), and grey oxide ($PbO_2$), are used in the active material that is pasted on the plates. The demand for lead batteries decreased in 1974 and 1975 concommitantly with the decline in total consumption (Table 2) as a result of the economic recession in several of the major lead-producing countries. However, the fall in the demand for batteries has also been attributed to the longer life-time of batteries, (Stubbs, 1975) which in 1967 was considered to be about 29 months (US Bureau of Mines, 1969) but according to Stubbs is, at present, close to 4 years. The battery industry also constitutes the major source of lead for secondary lead production. It has been estimated that up to 80% of the lead in storage batteries is recovered at secondary smelters (Ziegfeld, 1964).

Table 3. Percentage of total lead consumption by different industries in six major industrial countries

| Industry | 1969 [a] | 1974 [b] |
|---|---|---|
| Batteries | 35.9 | 44 |
| Alkyllead | 12.0 | 12.0 |
| Cable sheathing | 10.9 | 9.2 |
| Chemical pigments | 10.9 | 12.0 |
| Alloys | 8.1 | 10.8 |
| Semi-manufacturers | 16.5 | 12.0 |

[a] Federal Institute for Minerals Research and German Institute for Economic Research, 1972.

[b] Based on data provided by Stubbs, R. L., Lead Development Association, London.

The lead battery is likely to retain its position as a convenient source of electricity in the forseeable future. The nickel–cadmium battery does offer some advantages but is about three times more expensive. Better battery design, improvements in the electrical systems in cars and lower mileages because of higher gasoline costs are factors that may retard the growth rate for lead consumption by the battery industry. New applications for batteries may, on the other hand, increase demand.

### 3.3.2 Alkyllead fuel additives

Alkyllead compounds have been in use as anti-knock additives in gasoline for almost 50 years. Use of these compounds (almost exclusively tetraethyllead and tetramethyllead) increased steadily up to 1973 (Table 4). In 1973, the world consumption of refined lead for the manufacture of lead additives was about 380 000 tonnes (International Lead and Zinc Study Group, 1976). The moderate decrease in consumption in 1974 was almost entirely attributable to a decrease of 22 000 tonnes in the use of lead for gasoline additives in the USA. A further decline in the consumption was estimated in the USA in 1975, amounting to some 50 000 tonnes (Table 4); thus, the consumption in 1975 declined by 30% in comparison with the 1973 consumption (Stubbs, 1975). In the USA, the manufacture of alkylleads is, after batteries, the largest lead consuming industry. By

| Country | 1972 | 1973 | 1974 | 1975[b] |
|---|---|---|---|---|
| USA | 253 | 249 | 227 | 175 |
| Europe: (total) | (87) | (89) | (89) | (91) |
| France | 13 | 14 | 14 | 14 |
| Germany, Federal Republic of | 9 | 9 | 10 | 9 |
| Italy | 15 | 12 | 10 | 10 |
| United Kingdom | 50 | 54 | 56 | 58 |
| Others | n.a.[b] | 40[b] | 40[b] | 35[b] |
| Total | 340 | 378 | 357 | 301 |

[a] From: International Lead and Zinc Study Group, 1976.

[b] Estimated data; n.a. = not available.

comparison, lead additives make up only 6 % of the European market for lead (International Lead and Zinc Study Group, 1973). The decrease in the use of lead for fuel additives is likely to continue in the latter half of the 1970s as more cars fitted with catalysts requiring lead-free gasoline will come into use. The regulations on the maximum permissible concentrations of lead in gasoline will further affect the consumption of lead in fuels. The US Environmental Protection Agency's reduction programme aiming at 0.13 g of lead per litre of gasoline by 1 January 1979 was ratified in March 1976 by the US Court of Appeals. The maximum permissible level in the Federal Republic of Germany has been 0.15 g of lead per litre since 1 January 1976, and in Japan has been, 0.31 g of lead per litre since July 1971. Some European countries introduced limits of 0.4 g of lead per litre (e.g. Austria, Norway, Sweden, Switzerland) but most European governments have deferred their decision because of the economic implications of lowering the lead content (International Lead and Zinc Study Group, 1976).

### 3.3.3 Cable industry

The relative importance of the cable industry as a lead consumer has declined considerably (Table 3), mainly owing to the introduction of plastic sheathing/insulation. However, the total amount of lead used is still notable (Table 5). The use of lead in cable production is comparatively greater in Europe and several developing countries than in the United States of America. Alloys used for cable sheathing contain small amounts of many other elements including cadmium, tellurium, copper, antimony, and arsenic.

### 3.3.4 Chemical industry

Although a wide range of lead pigments are still produced they are increasingly being substituted by other, less toxic, pigments. Red lead (minium) is used extensively in the painting of structural steel work and lead chromate is often used as a yellow pigment. The use of lead for pigment manufacture in 1974 is given in Table 5.

Lead arsenate was, at one time, an important insecticide but is now little used and current consumption figures are not available.

Table 5. Consumption of lead in cables and pigments in five industrial countries in 1974 (kilotonnes)[a]

| Country | Cable | Pigments |
|---|---|---|
| France | 40 | 32 |
| Germany, Federal Republic of | 52 | 80 |
| Italy | 50 | 47 |
| Japan | 21 | 50 |
| United Kingdom | 44 | 35 |
| Total | 205 | 244 |

[a] Data from International Lead Zinc Study Group statistics.

The use of lead for the manufacture of alkyllead additives was discussed in section 3.3.2. The petroleum industry also uses a small amount of litharge dissolved in sodium hydroxide solution to remove sulfur compounds in the refining of petroleum.

### 3.3.5 Miscellaneous

Industries producing semi-manufactured components account for an important proportion of the total consumption. The surface of lead oxidises readily and is then very resistant to corrosion. The building and construction industries use lead sheet for roofing and other flashings, wall cladding, and sound insulation. Lead also forms alloys readily and is used in solder, bearing metals, brasses, type metal, collapsible tubes, and for radiation shielding. The ammunition industry is another major consumer of lead. There are many minor uses of lead compounds but these account for only a very small proportion of total lead consumption.

### 3.3.6 Environmental pollution from consumption and uses of lead

The combustion of alkyllead additives in motor fuels accounts for the major part of all inorganic lead emissions. The consumption of lead for the manufacture of alkylleads was estimated at 380 000 tonnes in 1973 and 300 000 tonnes in 1975 (section 3.3.2). Of this amount, over 70 % is like to enter the environment immediately after combustion, the rest being trapped in the crank case oil and in the exhaust system of the vehicles (Davis, 1973; Huntzicker et al., 1975). Moreover, part of the lead retained in the lubricating oil will enter the environment through different pathways (section 3.4). The degree of pollution from the combustion of alkyllead naturally differs from country to country, depending on the car density. The importance of alkyllead combustion is exceptionally high in the USA, where 20% of the total lead consumed is for the manufacture of alkyllead compounds, the corresponding values in 1969 being only 5% for France and 11% for Italy and the United Kingdom. The estimated total world emissions from this source were, according to the figures mentioned above, at least 266 000 tonnes in 1973 and 210 000 tonnes in 1975.

In the study by Davis (1973) on lead emissions into the air from industrial sources in the USA, 11% (1900 tonnes) was attributed to the processing of alkyllead additives. The manufacture of storage batteries emitted smaller amounts (480 tonnes) and emissions were still smaller in the production of lead oxide, lead pigments, type metal, solder, etc. The amounts of effluent from these industries were not studied. The dispersion of lead through the exhausts of workrooms should also be considered. These emissions although not very large may still contribute significantly to the pollution of the surrounding areas. The possibility of contamination of the home environment through working clothes should be borne in mind.

The magnitude of the pollution arising from the vast number of lead containing items that are subjected to weathering or are decomposed in the course of time is difficult to appraise. According to one estimate, about 50% of paint is removed from surfaces protected by lead pigments in a period of about seven years before re-painting (Patterson, 1965). Heavy contamination of the dust and soil around houses painted with lead paints has been consistently reported (Ter Haar & Aranow, 1974).

Only an unknown, but probably small fraction of the lead used in metallic form for the production of sheeting, cable, printing metal, etc. is ever released into the environment. Contamination of domestic water supplies, foods, and beverages resulting from the use of lead pipes, PVC pipes, glazed ceramics, and from cans with lead containing solders may under certain conditions be hazardous to man's health (sections 5.1.2 and 5.1.4).

The lead content in tobacco has been attributed to lead residues present in the soils of tobacco fields as a result of the former use of lead arsenate as an insecticide (section 5.1.4).


### 3.4  Waste Disposal

A substantial part of lead wastes are remelted in secondary smelters (see section 3.2.2).

Municipal incinerators have recently been investigated for lead emissions. An unknown proportion of the non-recycled, lead containing, consumer products, e.g. collapsible tubes, bottle caps, cable scrap, battery casings, and products painted with lead pigments, are incinerated. Depending on the type of furnace and on purification devices, these emissions may be considerable (Davies, 1973; Mattsson & Jaakkola, 1974).

Waste lubricating oil has been contaminated through the combustion of lead alkyls. Over 50 % of the oil is dumped or used as road oil. In 1970, the total amount of waste oil generated in the USA was about 2400 million litres. Waste crankcase oil contains about 1 % lead. Thus, the estimated amount of lead discharged into the environment from this source in the USA was nearly twice the amount originating from, for instance, the production of primary lead (Davis, 1973).

The extent of environmental pollution by lead arising from the incineration of sewage and sludge is not known.

### 3.5  Miscellaneous Sources of Environmental Pollution

When studying all industrial sources emitting lead into air, Davis (1973) reported that out of a total of 18 000 tonnes, copper smelting accounted for 8 % and the production of steel and iron another 8 %. Smaller amounts were generated in the production of primary zinc and also in the production of cement.

Coal contains small amounts of lead with a wide range of concentrations in different coals. Concentrations found by Abernethy et al. (1969) in coal from various districts in the USA ranged from 0.6 to 33.1 mg/kg. According to Patterson (1965) about 5 % of the ash leaving boilers as stable fly-ash aerosols is made up of small particles of a few micrometres. This silicate matter contains about 100 mg of lead per kg. Large quantities of coal are burnt to produce steam in power stations, steel works, and in manufacturing industries.

Small amounts of lead are generated from burning oil, which also has a very broad range of lead concentrations. The average concentrations in oil appear to be below 0.5 mg/kg (Davis, 1973). The possible future use of sewage sludge as fertilizer is discussed in section 4.

## 4. ENVIRONMENTAL TRANSPORT AND DISTRIBUTION

From a mass-balance point of view, the transport and distribution of lead from stationary or mobile sources into other environmental media is mainly through the atmosphere. Large discharges may also occur directly into natural waters and on to the land but, in such cases, lead tends to localize near the points of discharge owing to the very low solubility of the compounds that are formed upon contact with soil and water. The mass transfer of lead from air to other media is as yet poorly defined and the various mechanisms involved in the removal of lead from air are not fully understood. Although some data indicate that an important proportion of the lead may be removed through sedimentation (Atkins, 1969) the most efficient clearing mechanism is probably rain (Ter Haar et al., 1967). In a study of the concentration of lead in rainfall at 32 stations in the United States of America the average was 34 µg/litre (Lazrus et al., 1970). Most of these data were collected in areas with a high population density. Over rural areas of the USA the concentration was found to be approximately 18 µg/litre (Ter Haar et al., 1967).

Lead is rapidly removed from water when it passes through soil and bottom sediments. This is due to the high capacity of organic matters to bind the lead firmly. Because of this clearing mechanism, lead concentrations in both natural waters and water supplies are generally low (section 5.1.2).

Table 6. Distribution of lead from motor vehicles in the Los Angeles basin[a]

| Environmental area | Fractional fallout |
|---|---|
| Retained in car | 0.25 |
| Near fallout | 0.40 |
| Far fallout | 0.08 |
| Airborne | 0.24 |
| Unaccounted for | 0.03 |

[a] Adapted from Huntzicker et al., 1975.

An attempt was made to account for the lead emitted by automobiles in the Los Angeles Basin (Huntzicker et al., 1975) which is an area of exceptionally dense motor traffic. Limited environmental monitoring data tended to confirm the approximate correctness of the calculations. The transport pattern was classified as "near fallout", "far fallout" and "airborne". "Near fallout" was defined as the deposition in the immediate vicinity of roadways. "Far fallout" was defined as the fallout away from roadways, but within the basin, and "airborne" designated small particles carried away from the basin and ultimately deposited elsewhere. The data are shown in Table 6 and indicate that most of the emission was deposited within the basin. The fallout figures are calculated from the estimated behaviour of the airborne particles based on particle size distribution. If these results are approximately valid for other metropolitan areas, soil and water pollution from automobile emission fallout is predominantly limited to the immediate metropolitan area. The particles carried away from the area by air transport are probably widely dispersed and diluted since the atmospheric retention time of small particles is probably fairly long. It has been estimated that the residence time of airborne particles ranges from 6 days to 2 weeks in the lower troposphere and from 2 to 4 weeks in the upper troposphere (SCEP, 1970). Residence time will vary with a number of factors such as wind currents and rainfall. Yamamoto et al. (1968) demonstrated that atmospheric turbidity varied inversely with rainfall, owing to the washout effect of rain.

In spite of the great dilution of airborne lead that occurs during transport from centres of human activity, there is evidence indicating that a long-term global accumulation of lead has occurred. This long-term accumulation has been studied in glacial ice and snow deposits. Studies in Greenland showed that ice formed in about 1750 had lead concentrations 25 times greater than ice estimated to have been formed in about 800 B.C. From 1750, the concentration increased steadily to about 1940. From 1940 to the present day, the rate of increase has risen even more sharply. The most recent ice layers examined (about 1968) had a concentration 400 times greater than the natural background. Similar studies in the region of the Antarctic have also shown a rise, but it has not been so dramatic (Murozumi et al., 1969). Jaworowski (1968) conducted studies of Polish glaciers similar to those conducted in Greenland. He observed an approximate 16-fold increase in the lead concentration over the past 100 years. Chronological increases in the lead content of Swedish mosses have also been reported from 1860 to 1968 (Rühling & Tyler, 1968). These increases, about 4-fold in the past hundred years, were thought to reflect first the increase in coal combustion and later the introduction of leaded gasoline.

The transfer of air lead to the biota may be direct or indirect. For plants, the fallout contribution may be direct via the above ground parts, or it may be indirect by way of the soil. The pattern and degree of lead accumulation appears to be substantially influenced by the state of growth. Mitchell & Reith (1966) found that the lead content of certain plants increased 10-fold or more from the period of active growth to the time when growth ceased in the late fall. Some trees apparently have the capacity to accumulate high concentrations of lead. Kennedy (1960) reported that the tips of larches, firs, and white pines contained 100 mg of lead per kg dry weight, when grown in the lead mining areas of Idaho where the soil lead concentration was 20 000 mg/kg. The total concentration of lead in soil does not correlate well with the concentration in the plant but a correlation does exist when adjustment is made for the degree to which the soil lead can be brought into an aqueous solution of ammonium lactate and acetic acid (Kerin et al., 1972).

Thus, there is no doubt that plants acquire lead from the soil and air, but interspecies differences are prominent (Dedolph et al., 1970). It does not seem likely, however, that lead deposited on the leaves of plants transfers readily to other parts. Thus, Ter Haar (1970) showed in greenhouse studies that atmospheric lead at 1.45 µg/m³ did not influence the lead content of tomatoes, beans, carrots, potatoes, wheat, and cabbage heads, but did have an effect on the lead content of lettuce and bean leaves.

Transfer of lead from plants to animals is not well-defined. However, the concentration of lead in meat and eggs is quite similar, on a wet weight basis, to the concentration found in vegetables and grains (Schroeder et al., 1961). There is no evidence of biological accumulation proceeding from plants to animals.

Much remains to be learned about the environmental transport and distribution of lead. The potential pathways of lead from air to man are indicated in Fig. 1. Special attention should be given to the potential

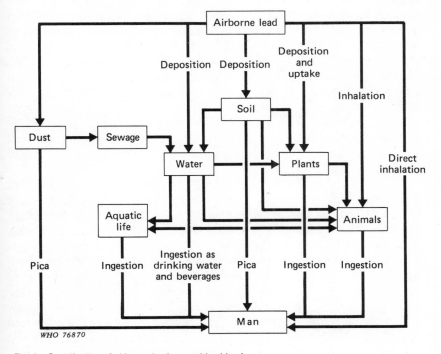

WHO 76870

Fig. 1  Contribution of airborne lead to total lead intake.

transfer of the fallout lead in cities that is washed into the sewage systems. Sewage sludge is currently being considered for use as fertilizer. Most cities have dual sewage system, i.e., storm and sanitary sewers, and it was shown in the report of the US Environmental Protection Agency Office of Research and Monitoring (1972) that storm runoff is far from being clean and probably warrants being treated in many instances. However, lead is not currently viewed as a hazard in this case because sludges have a high phosphate content which tends to minimize the bio-availability of the lead for plants (Chaney, 1973).

Little information exists with regard to the biotransformation of lead by microorganisms in the environment. However, Wong and his collaborators (1975) have reported that microorganisms in lake sediments can transform certain inorganic and organic lead compounds into volatile tetramethyllead. The authors were not able to explain completely the pathways of this transformation. A possible mechanism for the conversion of trimethyllead acetate into tetramethyllead in anaerobic systems was presented by Jarvie et al. (1975), who proposed that this takes place through the formation of an intermediate sulfide which decomposes into tetramethyllead. There is need for further research along these lines.

# 5. ENVIRONMENTAL LEVELS AND EXPOSURES

In the preceding chapter the general pattern of the environmental transport and distribution of lead was described. This chapter is more specifically concerned with the different circumstances under which people are exposed to lead to a degree that may be hazardous to their health.

## 5.1 Exposure of the General Population

The general population is exposed to lead by ingestion of food, and water, and by inhalation. In addition, children are exposed by eating non-food items, and those working in the lead industries suffer exposure over and above their exposure as members of the general population. These categories of exposure will be considered separately.

### 5.1.1 Air

The highest concentrations of lead in ambient air are found in dense population centres. The larger the city, the higher the ambient air lead concentration. As one moves away from the centre of the city, the concentration falls progressively. For urban stations, an average concentration of 1.1 μg/m³ has been reported; for non-urban stations (near the city) the average was 0.21 μg/m³; for stations somewhat farther removed it was 0.10 μg/m³, and for remote areas, 0.02 μg/m³ (McMullen et al., 1970). Air over streets with heavy traffic contained more lead than air over streets with light traffic, and considerably more than the ambient air over rural areas.

There is a clear pattern in this picture, the non-urban sites showing less than 0.5 μg/m³, while the urban sites have values ranging from 1 to 5–10 μg/m³. The highest levels have been recorded on highways during rush hours, 14–25 μg/m³ (WHO Expert Committee, 1969).

The results of continuous monitoring for 1971–72, in 27 European cities, by 43 uniform sampling stations are summarized in Table 7 (Commission of European Communities, 1973).

Table 7. Air lead concentrations in some cities of the European Community (1971–72)[a]

| Location | Continuous measurements | | Traffic-hour measurements | |
|---|---|---|---|---|
| Non-urban | monthly averages<br>daily maxima | <0.5 μg/m³<br><1 μg/m³ | —<br>— | |
| Small cities<br>residential areas | monthly averages<br>daily maxima | <1 μg/m³<br><2 μg/m³ | — | |
| traffic areas | — | | monthly averages<br>individual measurements | <3 μg/m³<br><8 μg/m³ |
| Metropolitan areas<br>residential areas | monthly averages<br>daily averages<br>up to | <2 μg/m³<br><br>8 μg/m³ | individual measurements | <4 μg/m³ |
| traffic areas | monthly averages<br>up to<br>daily values up to | <br>6.5 μg/m³<br>10 μg/m³ | monthly averages<br><br>single measurements<br>up to | <10 μg/m³<br><br><br>20 μg/m³ |

[a] Data from the Commission of the European Communities (1973).

The ambient air lead levels at 15 national sampling stations in Japan in 1973 were 0.30 μg/m³ for the average 24-hour value, 2.72 μg/m³ for the maximum 24-hour value, and 0.01 μg/m³ for the minimum (Environment Agency, Japan, 1975).

People who live in close proximity to dense automobile traffic are exposed to appreciably higher concentrations than others. In Los Angeles, California, where general ambient air levels are unusually high, the monthly mean concentration near traffic was as high as 6.4 μg/m³ (US Department of Health, Education and Welfare, 1965). This is in contrast to the general ambient air level of 2–4 μg/m³ reported for that city (Tepper & Levin, 1972). There is also a diurnal pattern whereby the concentration rises and falls in approximate proportion to the vehicular traffic activity

(US Department of Health, Education and Welfare, 1965; Lahmann, 1969; Heller & Kettner, 1969; Chovin et al., 1973). Most studies report that a seasonal variation also occurs (Tepper & Levin, 1972; Georgii & Jost, 1971).

Nearly all air lead measurements in communities have been made outdoors. Only a small number of indoor concentration studies are available (e.g. Fugaš et al., 1973; Yocom et al., 1971; Daines et al., 1972). Indoor levels vary from slightly lower than, to about $\frac{1}{3}$ of, comparable outdoor levels. Higher indoor levels are found only in lead industry environments. In the absence of specific data, reference should be made to the much more voluminous literature available on the penetration of undifferentiated particulate pollution into buildings. This was reviewed by Benson (1972). In general, very small particles enter buildings readily, and exist there at levels similar to those outside. Larger particles, near stationary sources and very close to roadways, penetrate buildings less readily.

Studies of air lead concentrations over a number of years or even a decade, at the same or similar locations, have produced quite variable results. Occasionally air lead levels have declined as in Cincinnati, Ohio (US Department of Health, Education and Welfare, 1965). This was attributed to greatly decreased coal consumption. The US National Air Surveillance Network, which has most of its stations in city centres, has shown little change in large cities and variable behaviour in smaller cities (NAS-NRC, 1972). Tepper & Levin (1972) and Chow & Earl (1970) have shown considerable increases in air lead levels at a number of stations in large cities. In 1967, Ott et al. (1970) developed a predictive model of increasing automotive pollution based on carbon monoxide emission patterns. Since air lead comes largely from vehicular sources, this report should be considered when changes in air lead with time are evaluated.

The respiratory uptake of lead from air depends on total lead concentration, particle size distribution, particle shape, chemical composition, physicochemical properties, and respiratory volume (section 6.1.1).

The particle size distribution of lead in ambient air has been studied by a number of investigators. As regards pulmonary deposition and absorption, the mass median equivalent diameter[a] rather than the microscopic particle size is considered appropriate. Robinson & Ludwig (1967) reported a mass median equivalent diameter of 0.25 μm, with 25% of the particles smaller than 0.16 μm and 25% larger than 0.43 μm. These data

---

[a] Mass median equivalent diameter = equivalent diameter above and below which the weights of all larger and smaller particles are equal.

were representative of a variety of areas in Los Angeles, the San Francisco Bay area, Cincinnati, Chicago, and Philadelphia. There was little variation from one city to another. Other studies conducted in the United States of America gave similar results (Mueller, 1970; Robinson et al., 1963). More recently Lee et al. (1972) have reported mass median equivalent diameters of 0.42–0.69 μm for six United States cities. Jost et al. (1973) reported that 50 % of particles had mass median equivalent diameters of less than 0.4 μm and 20 % of more than 0.5 μm.

Not much is known about the chemical form in which general ambient air lead occurs. Ter Haar & Bayard (1971) studied the composition of airborne lead particulates with an electron microprobe analyser. They studied particulates collected directly from the exhaust pipe of a car and also from air at various distances from a busy highway. Their results (Table 8) indicate that car exhaust lead is initially composed of halides that are converted to oxides, sulfates, and carbonates with aging.

Alkyllead vapours occur in ambient air because some of the alkyllead in gasoline escapes combustion. Purdue et al. (1973) have recently reported on the organic lead concentration in an underground parking garage and in the general ambient air of six major cities in the USA. In the parking garage, the total air lead level was 11.7 μg/m³, of which 16.7 % was organic lead. In the six major cities, the organic lead concentration was about 10 % of total lead. There is some uncertainty as to the accuracy of the organic lead data since the concentrations found approached the detection limit of the method. In another study in Los Angeles, using a different method for trapping organic lead, approximately 2 % of the lead was found to be organic (Snyder, 1967). Differences found in the concentration of organic lead relative to particulate lead can perhaps be explained in part by differences in proximity to the emitting source. Laveskog (1971) made repeated studies over several months on the presence of organic lead in the air at a number of locations in Stockholm. The levels were uniformly low (under 10 % of the total lead) except for 2 brief periods. These occurred near a gasoline station, and were attributed to the evaporation of spilled fuel. Colwill & Hickman (1973) verified this concept in similar studies near gasoline handling installations.

The air in the vicinity of lead smelters may be appreciably polluted and thus can affect the general population. A detailed study has recently been made of the environmental impact of a large ore smelter near El Paso, Texas (Landrigan et al., 1975b). The annual mean concentration in 1971 was approximately 80 μg/m³ in the immediate vicinity of the smelter and fell off rapidly, attaining a near-background level of 1 μg/m³ about 5 km away. Approximately 42 % of the particle mass had an aerodynamic diameter of less than 2 μm. In a similar study conducted near a smelter and a mine in

Table 8. Composition of airborne lead particles by electron microprobe analyser.[a]

| Sample | $PbCl_2$ | $PbBr_2$ | $PbBrCl$ | $Pb(OH)$ $Cl$ | $Pb(OH)$ $Br$ | $(PbO)_2$ $PbCl_2$ | $(PbO)_2$ $PbBr_2$ |
|---|---|---|---|---|---|---|---|
| Exhaust pipe |  |  |  |  |  |  |  |
| Zero time | 10.4 | 5.5 | 32.0 | 7.7 | 2.2 | 5.2 | 1.1 |
| 18 h | 8.3 | 0.5 | 12.0 | 7.2 | 0.1 | 5.6 | 0.1 |
| Eight Mile Road |  |  |  |  |  |  |  |
| Near road | 11.2 | 4.0 | 4.4 | 4.0 | 2.0 | 2.8 | 0.7 |
| 400 yards | 10.5 | 0.7 | 0.6 | 8.8 | 1.1 | 5.6 | 0.3 |
| Rural site | 5.4 | 0.1 | 1.6 | 4.0 | — | 1.5 | — |

| Sample | $(PbO)_2$ $PbBrCl$ | $PbCO_3$ | $Pb_3$ $(PO_4)_2$ | $PbO_x$ | $(PbO)_2$ $PbCO_3$ | $PbO$ $PbSO_4$ | $PbSO_4$ |
|---|---|---|---|---|---|---|---|
| Exhaust pipe |  |  |  |  |  |  |  |
| Zero time | 31.4 | 1.2 | — | 2.2 | 1.0 | — | 0.1 |
| 18 h | 1.6 | 13.8 | — | 21.2 | 29.6 | 0.1 | — |
| Eight Mile Road |  |  |  |  |  |  |  |
| Near road | 2.0 | 15.6 | 0.2 | 12.0 | 37.9 | 1.0 | 2.2 |
| 400 yards | 0.6 | 14.6 | 0.3 | 25.0 | 21.3 | 4.6 | 6.0 |
| Rural site | 1.0 | 30.2 | — | 20.5 | 27.5 | 5.0 | 3.2 |

[a] From: Ter Haar & Bayard (1971).

the Meža Valley, Yugoslavia, the air lead concentration, 10 km away, ranged from 1.3 to 24.0 µg/m³ (Djurić et al., 1971). Five air sampling stations were located at various distances within 10 km of the smelter stack. About 45 % of the particles had a diameter equal to or less than 0.3 µm and an additional 25 % were in the range of 0.31–0.8 µm. Although not specified, it is probable that these particle sizes were expressed as aerodynamic diameters, since the authors refer to this range as being of optimum size for absorption. The extensive pollution in some directions was probably due to the topography. The Meža Valley is only a few hundred metres wide and depending on wind direction, the lead particles can be conveyed long distances.

Roberts et al. (1974) reported the lead levels in air, dust, soil, and in the blood of children living near two secondary smelters in Toronto. Monthly mean air lead levels (1.0–5.3 µg/m³) were about twice those encountered in other parts of the city, but were subject to greater daily variation. Lead levels in dustfall of 200–1500 mg/m²/month, and soil levels of 16 000–40 000 mg/kg of soil near the smelter declined to background levels within 300–400 m of the stacks. From 13 to 30 % of the children living within 300 m of the stacks had blood lead levels of over 40 µg/100 ml.

Pollution of the surrounding country by a secondary lead smelter has been reported to have affected the lead absorption of adults living 1–4 km from the main emitter (Nordman et al., 1973). Air lead concentrations were not reported. The dustfall lead concentrations ranged from

10 mg/m$^2$/month (4 km from the chimney) to 200 mg/m$^2$/month (200 m from the chimney). There was a correlation between blood lead concentration and degree of erythrocyte ALAD inhibition on the one hand and the proximity of habitation to the smelter on the other. A correlation between Pb-B values and monthly dustfall lead was also demonstrated.

### 5.1.2 Water

Man's exposure to lead through water is generally low in comparison with exposure through air and food (WHO Working Group, 1973). The lead concentration in the water supplies of most of the 100 largest American cities, as determined in 1962, ranged from a trace to 62 μg/litre (Durfor & Becker, 1964). Since 1962, continuous monitoring of American water supplies has indicated that the US Public Health Service prescribed limit of 50 μg/litre has not been exceeded (NAS-NRC, 1972). In another study, only 41 out of 2595 samples of tap water contained more than 50 μg/litre, and 25 % contained no measurable amount of lead (McCabe, 1970).

Under some circumstances, the concentration of lead in drinking water can become extremely high. Gajdos & Gajdos-Török (1973) described two cases of severe clinical lead poisoning attributable to a municipal water supply that contained lead levels of 2.6 mg/litre. In another case, in rural Scotland, four people developed clinical lead poisoning and others showed biochemical evidence of grossly elevated lead exposure (Goldberg, 1974). The concentration of lead in the domestic water supply was 2–3 mg/litre. In this case, the reason for the extreme contamination was that the water was stored in lead tanks. In another study, conducted in Glasgow, Scotland, it was shown that lead pipes in the plumbing of homes can also result in high concentrations of lead in soft water (water low in calcium and magnesium) (Beattie et al., 1972a). Homes with both lead-lined water storage tanks and lead pipes had the highest concentration. The plumbosolvency of water standing in lead pipes is influenced significantly by several factors. The solvency increases about four-fold with increasing acidity over the pH range from 6 to 4. Increases of a somewhat lesser degree were also noted with increasing alkalinity over the range of pH from 8 to 10 (Moore, 1973). The same author also pointed out the increasing plumbosolvency of water with increasing temperature and with decreasing calcium concentration. Quite recently it was shown that lead concentrations in tap water were highly dependent on the volume of water flushed through the system before sampling. The concentrations were also considerably lower when a 95/5 (tin/lead) solder had been used in the copper piping instead of the 50/50 or 60/40 solders (Wong & Berrang, 1976).

When water was left standing overnight in plastic pipes, some degree of leaching of lead into the water was observed (Heusgem & DeGraeve, 1973).

49

The source of lead in this case was probably lead stearate which is used as a stabilizer in the manufacture of polyvinyl plastics. The problem of plastic pipes has been discussed recently (Schaller et al., 1968; Packham, 1971). Packham did not feel that there was a hazard associated with the use of such material in domestic water supply systems. But more study is needed, particularly of situations in which water stands in the pipes for prolonged periods.

Lead levels in surface and ground waters were recently reviewed by a WHO Working Group (1973). Natural surface waters have been reported to contain usually less than 0.1 mg/litre (Kopp & Kronen, 1965). In unpolluted areas the concentrations are of the order of 1 µg/litre or less (Žukovickaja et al., 1966). Some rivers in France were recently analyzed by Servant (1973) who found that, in the Midi-Pyrenees region, the mean concentration of dissolved lead varied from 6.7 to 10.4 µg/litre.

### 5.1.3 Food

The contribution of food to man's exposure to lead has been under study for many years, beginning with the study of Kehoe et al. (1933) who found lead in every item of food in both industrial and primitive societies. The concentration of lead in various items of food is best described as highly variable. In fact, there seems to be about as much variation within specific items of food as between different categories of foods. For example, Schroeder et al. (1961) found that the range was 0–1.5 mg/kg for condiments, 0.2–2.5 mg/kg for fish and seafood, 0–0.37 mg/kg for meat and eggs, 0–1.39 mg/kg for grains, and 0–1.3 mg/kg for vegetables.

Estimates of actual consumption of lead in food and beverages have been made using two general approaches. Some investigators have used the duplicate portions sampling method. Others have derived theoretical intakes based on nutritional tables and known concentrations of lead in the dietary components (composites technique, see section 2.2.1). The results of studies using the two methods are given in Table 9. In general, the composites technique appears to yield somewhat higher mean values for adults than the duplicate portion technique. There are considerable differences in the daily intakes reported from different countries. Whether these differences are real or due to factors associated with the methods used remains to be assessed. Inadequacies in sampling and in the analytical methods used may account for a considerable part of the differences; few of the studies cited present any evidence of interlaboratory quality control of the analytical assays. Most estimates do not specify the age, sex, or level of physical activity assumed in arriving at the estimates. These are very important determinants of dietary intake. Thus, the calorie requirement for a 25-year-old male in the United States of America is approximately 2900 calories, whereas it is only 1900 calories for women, age 35–55 (Altman & Dittmer, 1968). Horiuchi et al. (1956) were quite aware of the vast differences in food intake among different categories of adults and between adults and children. They made an effort to take these factors into account in developing their estimates of lead intake from dietary sources.

Daily faecal lead excretion can also be used as a means of estimating daily lead ingestion, since only approximately 10% of dietary lead is absorbed (Kehoe, 1961). This approach, which was used by Tepper & Levin (1972) (section 6.1.2.2), presents some uncertainty regarding actual absorption and neglects contributions to faecal lead from gastrointestinal secretion, which cannot be estimated.

A factor that is usually ignored is the occurrence of lead in various foods at concentrations below the practical detection limits. Thus, Kolbye et al. (1974) arrived at different estimates based on the assumptions they made regarding whether or not lead was present in all items eaten. There was uncertainty as to how to cope with the problem of lead concentrations reported as "zero" or "trace" in certain samples. When "zeros" and "traces" were accepted as meaning absolutely no lead, the estimated daily lead intake was 57.4 µg for an 18-year-old man. This seemed unduly low, particularly in the light of the fact that the faecal lead excretion of normal American women is 90–150 µg/day (Tepper & Levin, 1972). Certain assumptions were therefore made regarding the "zeros" and "traces". If it was assumed only that "traces" really represented 0.09 mg/kg, the

Table 9. Dietary lead intake.

| Method | Age | Sex | Activity | Lead/day (µg) range | Lead/day (µg) average | No. of subjects | References |
|---|---|---|---|---|---|---|---|
| Duplicate portions | adult | male | sedentary | 120–350 | 218 | 9 | Kehoe, 1961 |
| | adult | male | sedentary | 74–216 | 113 | 17 | Coulston et al., 1972[h] |
| | 21–30 years | 4 male 1 female | — | 237–306 | 274 | 5 | Thompson, 1971 |
| | adult | — | — | 4.8–83.0 | 17.8 | — | Schütz et al., 1971 |
| | adult | male | medium—heavy | 119–360 | 231 | 35 | Nordman, 1975 |
| | adult | female | medium | 89–305 | 178 | 36 | Nordman, 1975 |
| | 3 months—8.5 years | | | — | 40–210[h] | 8 | Alexander et al., 1973 |
| Composites technique | adult | male | heavy work | — | 455 | — | Horiuchi et al., 1956 |
| | adult | male | medium | — | 299 | — | Horiuchi et al., 1956 |
| | 10 years | male | | — | 254 | — | Horiuchi et al., 1956 |
| | 10 months | male | | — | 126 | — | Horiuchi et al., 1956 |
| | 18 years | male | medium | — | 57–233[a] | — | Kolbye et al., 1974 |
| | adult | male | medium | — | 139 | — | NRC (Canada), 1973 |
| | adult | male | medium | — | 518 | — | Lehnert et al., 1969 |
| | adult | male | medium | — | 505 | — | Zurlo et al., 1970 |

[a] See page 52
[h] Ranging from 40 µg in a breast-fed infant to 210 µg in an 8½-year-old child, as calculated using ICRP.

calculated intake became 159 µg/day. When the additional assumption was made that "zero" had a finite value of 0.05 mg/kg, the calculated daily intake of an 18-year-old male became 233 µg. Another source of error in establishing how much is consumed relates to food preparation. Lead may be either added to the diet or removed in the course of preparation. Both Horiuchi et al. (1956) and the report from the British Ministry of Agriculture, Fisheries and Foods (1972) took special pains to explain how this problem was handled.

Published reports on lead levels in wine (Truhaut et al., 1964; Zurlo & Griffini, 1973) show that important variations occur from sample to sample. Considering ordinary wines there does not seem to be any significant difference between white, red, and rosé (Truhaut et al., 1964). Average lead concentrations of 130–190 µg/litre could be calculated (range 60–255 µg/litre), but even higher mean values (299 µg/litre) have recently been reported (Boudene et al., 1975). Wine therefore is likely to be a substantial source of lead for some people, and may account for part of the differences in Pb-B levels (section 5.4) and in daily dietary lead intake between various countries.

The concentration of lead in milk is a matter of special concern because milk is a major dietary constituent for infants. Human breast milk has been reported to contain 12 µg/litre (Murthy & Rhea, 1971) and < 5 µg/litre (Lamm & Rosen, 1974). Cow's milk has been reported to have a similar concentration, when taken directly from cows for analysis; 9 µg/litre (Hammond & Aronson, 1964). The concentration of lead in processed cow's milk is higher than in human milk or in milk obtained directly from cows. The types of processing vary considerably as does the degree of apparent lead contamination. Thus whole milk concentrations are only moderately elevated. Mitchell & Aldous (1974) reported an average of 40 µg/litre in whole bulk milk and Kehoe (1961) reported 20–40 µg/litre for local USA market milk. By contrast, evaporated milk and formulas have still higher concentrations. Mitchell & Aldous (1974) reported an average of 202 µg/litre for evaporated milk. Somewhat higher values were reported by Murthy & Rhea (1971) (330–870 µg/litre) and somewhat lower by Lamm & Rosen (1974) (110 ± 11 µg/litre).

A major contribution to the lead content of processed milk as well as of other food products appears to be lead solder used in the seams and caps of cans. It has been shown that foods preserved in such cans frequently have much higher concentrations of lead than do the same items packed in glass containers (Mitchell & Aldous, 1974).

Although plants do not take lead up from the soil readily, fruits and vegetables grown in areas exposed to smelter emissions may be appreciably contaminated. Kerin (1972) determined lead in the total diet of peasants near a smelter and found that the daily ingestion of lead with food was 670–2640 µg.

### 5.1.4 Miscellaneous

The intake of lead in food, air, and water is a major concern as regards the general population because of the pervasive nature of these exposures. Another frequent exposure source, smoking, probably makes a small contribution to the lead burden (section 5.4.1). However, surprisingly little information is available concerning the concentration of lead in smoking tobacco. Cogbill & Hobbs (1957) reported the concentration of lead in two separate brands of cigarettes and in a composite sample of five brands. Concentrations were 19, 80, and 39 mg/kg at 58 % relative humidity or 21, 84, and 41 µg per cigarette. The amount of lead transferred to mainstream smoke was 1.0, 3.3, and 1.9 µg per cigarette which represented 4.8, 3.9, and 6 % transfer. Arsenic/lead ratios found in the tobacco indicated that the source of lead was probably lead arsenate. At one time lead arsenate was used extensively as an insecticide in American tobacco fields but other pesticides rapidly replaced it shortly after World War II. Residues of lead arsenate have probably persisted in the fields and could contaminate plants externally. More recently, Szadkowski et al. (1969) reported 0.483 $\pm$ 0.267 µg of lead per cigarette in the total smoke for eight brands of cigarette. This represented 19 % of the total lead in the tobacco or 2.6 µg per cigarette. No distinction was made between mainstream and sidestream smoke[a]. Untabulated data from a study by Menden et al. (1972) indicated that only about 2 % of lead in non-filter types of cigarettes was transferred to the mainstream smoke. The average content of lead in commercial cigarettes was given as 10.40–12.15 µg/cigarette (Petering & Menden, private communication). Most of the lead was found in the ash; the lead content of the sidestream of individual cigarettes varied considerably with a maximum value of 16 %. Assuming an average lead content ranging from 2.5 to 12.2 (Lehnert et al., 1967; Szadkowski et al., 1969; Rabinowitz, 1974; Petering & Menden, private communication) and a 2 % transfer to the mainstream smoke (Menden et al., 1972) are fair estimates, and without taking into account the possible contribution from the sidestream smoke, a crude assessment of the direct inhalation intake of lead from smoking 20 cigarettes a day would be about 1–5 µg.

Certain other sources of exposure are important. These sources do not affect any major segment of the population but collectively they no doubt

---

[a] i.e. the smoke which drifts off the burning end of cigarette between puffs.

account for the majority of the cases of clinical lead poisoning in the general population.

The presence of high concentrations of lead in illicitly distilled whisky occurs commonly in the USA and causes poisoning in adults. The condensers used in homemade stills are often discarded automobile radiators. These contain substantial amounts of lead in the soldered joints. The concentration of lead in the final product frequently exceeds 10 mg/litre. The problem of lead poisoning from this source exists predominantly in the southeastern parts of the USA where illicit whisky production is most common.

Another source of poisoning is improperly glazed earthenware vessels. Improper glazing results in the leaching of lead into the vessel, particularly when the contents are acidic. Cases of poisoning, both fatal and non-fatal, have been recorded from the use of improperly glazed pottery. Klein et al. (1970) reported two cases (one fatal) in which apple juice stored in the incriminated vessel for 3 days contained 1300 mg/litre. In another case the ceramic mug responsible was used for drinking cola[a], pH 2.7 (Harris & Elsea, 1967). After two hours of standing in the mug, the cola contained lead levels of 6.8 mg/litre. It was estimated that this patient drank 3.2 mg of lead per night in this fashion for two years. Other cases have been reported from Yugoslavia (Beritić & Stahuljak, 1961) and from the United Kingdom (Whitehead & Prior, 1960). The problem involves the storage of acid materials in the vessels. In a test of the leaching of lead from commercial and handcrafted pottery, Klein et al. (1970) found that 4% acetic acid allowed to stand at room temperature in the vessels for 18 hours often acquired concentrations of lead in excess of 100 mg/litre. In fact, in more than half of the cases, the concentration of lead exceeded 7 mg/litre.

Another source of lead poisoning in the general population is the use of discarded storage battery casings for fuel. There is some uncertainty as to whether the cases of poisoning that have been recorded (Williams et al., 1933; Gillet, 1955) were due to inhalation of lead fumes or to hand-to-mouth transfer of fallout material. The prevalence of children in the number of recorded cases supports the argument for hand-to-mouth transfer.

Because of the wide variety of applications of lead, additional potential hazards are still being identified. For example, the use of lead wire core wicks in candles was only recently called to the attention of the USA authorities (Bridbord, unpublished results, 1973).

---

[a] A popular carbonated non-alcoholic beverage.

## 5.2 Exposure of Infants and Young Children

### 5.2.1 Soil, dust, and paint

The young child of pre-school age is exposed to special hazards from environmental sources of lead. This is because such children frequently exhibit the habit of licking, chewing, or actually eating foreign objects. Lead-based paints have long been considered the major source of excessive lead intake in young children. Thus, Sachs (1974) reported that 80 % of patients seen because of evidence of excessive lead absorption had a history of eating paint or plaster and in another 10 % X-ray examination revealed paint in the abdomen. The author also was of the opinion that if X-ray examinations had been repeated at each visit to the clinic, evidence of paint ingestion would have appeared in all patients. A similar view was expressed by Chisholm & Harrison (1956). In their series of 105 children whose homes were investigated, 102 of the homes contained at least one source of paint containing 5 % lead or more. Of even greater significance was the fact that the painted surfaces identified as sources were flaking.

Other investigators have attempted to assess the importance of paint as a source of excessive lead exposure. Griggs et al. (1964) found a positive correlation between the presence of elevated urinary lead or coproporphyrin and the presence of flaking paint in the homes. Nonetheless, in many instances the homes of children with abnormal urine had no flaking paint indoors. Unfortunately data were not given as to the number of children with abnormal urine and no evidence of flaking paint indoors or outdoors. Guinee (1973) reported that in an extensive survey of the homes of children having blood lead concentrations equal to or greater than 60 µg/100 ml, 75 % of the homes had at least one surface in which the paint contained more than 1 % lead. Furthermore, children with elevated blood lead concentrations were more likely to live in homes where the painted surfaces were cracked than children with low blood lead values.

All of these studies indicate that lead in painted surfaces in houses is almost certainly the major source of lead for infants and young children. Some other studies suggest that the issue is not that clear-cut. Greenfield et al. (1973) reported that, in one study, 18 out of 19 rural children with elevated blood lead concentrations lived in homes having at least one accessible painted surface containing 1 % or more of lead, whereas paint containing 1 % or more of lead could be found on accessible surfaces in only 60 % of the homes of inner city children with excessive lead exposure. The implication is that sources of lead other than paint were often responsible for the exposure of city children. Two equally rational interpretations are that an insufficient number of surfaces were tested in the children's homes or that children often spend time in several homes, some of which might not have been tested for lead-based paint.

Studies of sources of lead all too often ignore the fact that painted surfaces on the outside of houses are a potential source of lead or, for that matter, that the soil surrounding the houses may have accumulated substantial concentrations of lead from the weathering of outer walls. With regard to the latter, Fairey & Gray (1970) reported that the concentration of lead in the soil near homes where paediatric lead poisoning had occurred was over 1000 mg/kg in 27 out of 30 cases. By contrast, only 30 out of 170 soil samples taken from yards selected at random (and not associated with cases of lead poisoning) had concentrations of lead in excess of 1000 mg/kg. Bertinuson & Clark (1973) have reported extremely high soil lead values close to residences in the older section of Cincinnati. In one case, because the distance across the yard from the base of the house to a road with heavy traffic was sufficient, it was possible to assess the relative contributions of lead from car exhaust and lead from the weathering of the house. The gradient ranged from 12 000 mg/kg adjacent to the house, down to 400 mg/kg, about 10 m from the road. This suggested that weathering of painted surfaces of the house could have been the major source of soil lead in this instance. Although the high concentrations of lead in the soil in the vicinity of houses may be due to weathering of lead-based paint, it is possible that in many cases it is also due to the accumulation of combusted alkyllead from car exhaust. In this connexion, recently-reported data of Ter Haar & Aranow (1974) are particularly informative. They surveyed the profile of lead in soil, extending from the base of 36 urban residences out to the street gutters. Eighteen of the residences were of brick construction and 18 were of frame construction. In summary form, their data were as shown in Table 10. The data reflect the likelihood of the major contribution of

Table 10. Lead in dirt in Detroit (mg/kg dry dirt)[a]

| Location | Painted frame houses | | Brick houses | |
|---|---|---|---|---|
| | Mean | Range | Mean | Range |
| Within 0.6 m of house | | | | |
| front | 2349 | (126–17 590) | 351 | (78–1030) |
| back | 1586 | (162–4951) | 501 | (72–2350) |
| sides | 2257 | (140–7284) | 426 | (91–1160) |
| | 1846 | (104–7000) | 595 | (40–2290) |
| 3 m from house | | | | |
| front | 447 | (58–1530) | 156 | (39–316) |
| back | 425 | (149–1410) | 200 | (72–480) |
| Near sidewalk | 627 | (152–1958) | 324 | (86–1130) |
| curb | 572 | (320–1957) | 612 | (147–2420) |
| gutter | 966 | (415–1827) | 1213 | (304–3170) |

[a] Adapted from Ter Haar & Aronow (1974).

weathered lead-based paint to soil lead. But they also strongly suggest that vehicular sources make a significant and sometimes very substantial contribution to soil lead near the sidewalks.

Street dust has also been found to contain high concentrations of lead. Using recent data from 77 midwestern cities in the USA, it was calculated that the concentration of lead averaged 1636 mg/kg dust in residential areas, 2413 mg/kg in commercial areas, and 1512 mg/kg in industrial areas (Hunt et al., 1971).

In order for soil or street dust to be a significant source of lead for man, it is, of course, necessary that it be ingested and/or inhaled. Evidence regarding the likelihood that young children would ingest soil or street dust is extremely fragmentary. However, in a recent study of 58 children with increased lead burdens, it was found that 37 had a history of eating dirt and sand, compared with 34 eating plaster, 20 eating paint flakes, 15 chewing on furniture, 14 chewing window sills, and 7 eating wallpaper (Pueschel et al., 1972). Further inferential evidence as to the possible significance of soil and dust as a source of lead is to be found in the recent Smeltertown episode near El Paso, Texas, referred to earlier (Landrigan et al., 1975b). This town is the site of a large smelter which processes lead ores, among others. The young children in the town have high blood lead concentrations. In a sample of 14 children of 1–5 years of age, 78.6–100 % were found to have lead concentrations equal to or greater than 40 µg/100 ml blood. The concentration of lead in the surface soil of Smeltertown has a median value of 3700 mg/kg. One is tempted to conclude that the blood lead levels of these young children increased owing to ingestion of this soil. However, the picture is somewhat confounded by the fact that older children also showed a high incidence of elevated blood lead concentrations but to a lesser degree; and older children are not generally considered to exhibit pica. Smeltertown adults had normal blood lead concentrations. Intake of lead by inhalation would probably have affected adults as well as children. Thus, it is likely that lead intake by the children was by direct oral intake. The painted surfaces in the residences were seldom in a flaking condition and were not found to be more than two or three layers thick, in contrast to the multiple layers usually found in city slum areas where lead poisoning is prevalent. The information available therefore suggests that the sources of lead were soil and dust. Indeed, there was a highly significant correlation between the concentration of lead in the blood of the children and the concentration of lead in household dust.

The presence of high concentrations of lead in soil is not necessarily hazardous. Thus, children living on soils containing lead levels of up to 8000 mg/kg showed only minimal elevations in blood lead concentration (Barltrop et al., 1974). This was found to be so even among the children with pica for soil. Perhaps climatic differences are important. Smeltertown in Texas is extremely dry and dusty whereas the region studied by Barltrop and coworkers was in England, where the soil is presumably not as accessible to children owing to the relatively heavy cover of vegetation. The play behaviour of children also determines to a certain extent their exposure to lead (Einbrodt et al., 1974).

Since dust and dirt occur indoors as well as outdoors, some attention has been directed recently to the significance of indoor dust. Transfer of lead-bearing house dust to the hands of young children has recently been demonstrated (Sayre et al., 1974). The house dust of inner city old houses contained far more lead than the dust of newer, suburban houses. Furthermore, the hands of the children in inner city houses were heavily contaminated with lead, whereas the hands of suburban children were not. It is not at all certain that the source of lead in the house dust was fallout from car exhaust. New housing in the inner city had very little lead in dust. The inference is that the lead was probably from the painted surfaces, since the paint in old houses has high concentrations of lead whereas the paint in new houses in the same area generally has a low lead content. But even the presence of lead-containing dust on children's hands provides little information concerning hazard since the critical question is how much is actually transferred from the hands to the digestive tract.

### 5.2.2 Miscellaneous

Facial cosmetics have long been a source of lead poisoning in Oriental countries. Kato (1932) discussed the problems encountered in Japan. Face powders, pastes, and liquids were found to contain as much as 67 % lead. Exposure of children was considered to be by inhalation of powders, or ingestion of powders and other formulations. More recently, there have been several reports of infant poisoning from a mascara-like cosmetic used by Indian and Pakistani women (Warley et al., 1968; Alexander & Delves, 1972). This substance may contain as much as 88 % lead sulfide.

Another source of lead exposure for young children is coloured newsprint (Hankin et al., 1973). It has been found that the coloured inks used in magazine illustrations contain extremely high concentrations of lead. Coloured pages were found to have lead concentrations of 1140–3170 mg/kg.

Children and other family members may be exposed to lead contamination at home by work clothing being worn at home or brought home for cleaning, or by small pieces of metal which may be brought in (Inter-Departmental Working Group on Heavy Metals, 1974).

## 5.3 Occupational Exposures

It is among the workers who smelt, refine, and use lead in manufacturing items of commerce that the highest and most prolonged exposures are found. Lead poisoning among these people was common at one time. Today, workers, management, and physicians are generally aware of the danger of lead and know how to handle the problem; so, the incidence and severity of poisoning have decreased substantially in recent years. However, much still remains to be done to eliminate lead poisoning completely as an occupational disease. The major hazard today seems to be in small enterprises (Engel et al., 1971) and in some large industries where adequate industrial hygiene programmes do not exist or are difficult to implement, or where awareness of the existence of hazardous circumstances may be lacking.

A recent WHO study of occupational health problems in the Andean countries (El Batawi, unpublished results, 1974) showed that, in Chile for instance, among 580 workers exposed to lead, 21.9 % had an increased level of ALA in the urine. In Colombia, 3370 workers exposed to lead were examined, of whom 4.3 % were considered to be suffering from lead poisoning.

The major route of lead exposure in industry is by inhalation. The generation of lead-bearing dusts and fumes is inevitable. The workers' clothes may also be an important source of exposure. Even the lesser problem of oral intake of lead is really a consequence of the generation of airborne dusts which settle out from the air on to food, water, or other objects that are transferred to the mouth in one fashion or another. Thus, good housekeeping and, above all, good ventilation have a strong impact on exposure. An industrial process may be quite safe in one factory and quite hazardous in another solely because of differences in ventilation engineering or because of differences in housekeeping practices and worker education.

### 5.3.1 Lead mining, smelting, and refining

The lead mining hazards depend, to some extent, on the solubility of the lead from the ores. The lead sulfide (PbS) in galena is insoluble and absorption through the lung is slight. However, in the stomach, some lead sulfide may be converted to slightly soluble lead chloride which may then be absorbed in moderate quantities.

The process of lead smelting and refining probably has the greatest potential for hazardous exposure of all the lead industries. The most hazardous operations are those in which molten lead and lead alloys are brought to high temperatures, resulting in the vaporization of lead. This is because condensed lead vapour has, to a substantial degree, a small ($<5$ µm), respirable particle size range. Thus, although the total air lead concentration may be greater in the vicinity of ore proportioning bins than it is in the vicinity of a blast furnace in a primary smelter, the amount of particle mass in the respirable size range may be much greater near the latter.

As an example, we can consider the processes involved in the preparation of lead bullion in typical primary lead ore smelters in Salt Lake City, Utah. The various processes are essentially grinding and smelting. The main operations are: (1) ore proportioning; (2) nodulizing and sintering; (3) blast furnace; (4) drossing and reverberation. Air lead concentrations have been determined using personal monitors worn by workers at the various stations. These data are summarized in Table 11.

Table 11. Air lead concentrations in three primary lead smelters (µg/m$^3$)[a]

| Smelter | Year | Location[b] | Means[c] | Mean of means | Range (all values) |
|---|---|---|---|---|---|
| A | 1972–75 | (1) | 610, 1930, 2860 | 1800 | 250–3670 |
|   |   | (2) | 970, 470, 450 | 630 | 250–1380 |
|   |   | (3) | 860, 950, 320 | 710 | 200–1700 |
|   |   | (4) | 1220, 350, 950 | 840 | 260–1640 |
| B | 1973–74 | (1) | 1310, 2330, 4720 | 2790 | 370–5160 |
|   |   | (2) | 2740, 3460, 770 | 2320 | 310–7570 |
|   |   | (3) | 860, 140, 530 | 510 | 120–1560 |
|   |   | (4) | 1270, 540, 5730, 4050 | 2900 | 60–7220 |
| C | 1973–74 | (1) | — | — | — |
|   |   | (2) | 3850, 8740, 830 | 4470 | <10–31 200 |
|   |   | (3) | 1320, 230 | 780 | 90–1340 |
|   |   | (4) | 80 | 80 | |

[a] Data provided by M. Varner, American Smelting and Refining Co., Salt Lake City, Utah, U.S.A.
[b] Locations: (1) Ore proportioning; (2) nodulizing and sintering; (3) blast furnace; (4) drossing and reverberation.
[c] Determined with personal monitors on separate occasions. Each sampling period was 5–7 hours.

Similar data for primary lead smelters elsewhere are not available. However, it is evident that lead exposure in primary smelters may be extremely high. The hazard to the workers in the example cited would be extremely serious were it not for the fact that the use of respirators is mandatory in these particular smelters.

Comparable data are not available for exposures in secondary smelters. Secondary smelters are to be found in or near most large cities. They depend on the local supply of lead scrap in the form of discarded electric storage batteries, cable casings, pipes, and other materials for their supply of lead. The nature of the operation is similar to the one described for primary smelters, except that no ore-processing is involved. Tola (1974) has recently reported on hazards in secondary lead smelters in Finland. The work practices involved were not described. Thus, it was not indicated whether or not these workers wore respirators on the job. But whatever the work practices may have been, they were not adequate. Out of 20 smelters and founders, 16 had blood lead concentrations equal to or greater than 70 µg/100 ml.

Foundries in which molten lead is alloyed with other metals have also been sources of high atmospheric exposure. In one such operation the concentration of lead was 280–290 µg/m$^3$ (Berg & Zenz, 1967).

### 5.3.2 Electric storage battery manufacturing

The electric storage battery industry has been studied fairly carefully with reference to the nature and degree of lead exposure.

Within the manufacturing process, there are numerous specific operations that are hazardous by virtue of the resultant high air lead concentrations. Plate casting is a molten metal operation. The hazard here is from spillage of dross, resulting in dusty floors. Mixing of lead oxide paste runs parallel to grid casting. Here, as in subsequent operations, the major hazard is from lead oxide dust, particularly when loading the mixer with lead oxide powder. Ventilation is needed during loading and frequent clean-up is necessary to prevent the accumulation of dust. Pasting of the plates follows, either by hand or by machine. In either case the hazard is from dust which accumulates as the paste dries. The plates are then cured, oven dried and removed for the forming process. Although the plates must be welded into circuits, the temperature is not high enough to generate

Table 12. Air lead concentrations (µg/m$^3$) in electric storage battery manufacturing

| Operation | Elkins, 1950[a] | Tsuchiya & Harashima, 1965[a] | | Williams, et al., 1969[b] | | Engels & Kühnen, 1973[c,d] | |
|---|---|---|---|---|---|---|---|
| | mean | mean | range | mean | S.E. | mean | range |
| Oxide mixing | 730 | 2000 | 250–13 000 | — | — | 5400 | 180–21 600 |
| Plate casting | 260 | 500 | 200–620 | 50 | 3 | — | — |
| Pasting, hand | 750 | — | — | 150 | 29 | 710 | 100–2700 |
| Pasting, machine | — | — | — | 220 | 25 | 1100 | 80–13 500 |
| Forming | — | — | — | 130 | 13 | 220 | 30–2200 |
| Stacking and breaking | 500 | — | — | — | — | 880 | 110–4500 |

[a] Air sampling time not stated.
[b] Personal air samplers worn for full work shift for 2 weeks.
[c] Air sampling time 40–60 minutes.
[d] Approximations derived by collation of various sub-categories from authors' data.

significant concentrations of lead fumes. Once more, the main problem is lead oxide dust, although the amount of handling involved generally does not require ventilation. After another drying process, the plates are stacked to make elements, either by hand or machine. In both cases the process is dusty and ventilation is needed, but particularly with machine stacking. The stacks are then burned to weld together the positive and negative lugs. This is done in a ventilated burning box. Final assembly and finishing are low-hazard operations that do not require ventilation if conducted with care.

Reports have appeared concerning the air lead concentrations associated with the various phases of battery manufacture. The data summarized in Table 12, show that oxide mixing is probably the most hazardous occupation, followed by machine pasting, assuming that the same accumulative time is spent at each activity. This conclusion is borne out by the result of a recent study. The blood lead concentration was found to be most elevated and the erythrocyte ALAD activity was most depressed among men engaged in oxide mixing and pasting (Tola et al., 1971).

The data cited above for air lead concentrations in the lead smelting and refining industry and in the electric storage battery industry may not of course be wholly representative of these industries. But they are sufficiently alarming to suggest that respirators must be worn in most of these operations, as indeed they were in the case of the smelters from which the data were gathered.

### 5.3.5 Shipbreaking and welding

Any process in which lead-containing metals are heated with torches to high temperatures are potentially hazardous. This is due to the formation of lead fumes with a high fraction of the airborne mass existing in the respirable particle size range. As an example, steel structures are coated with lead-based paint prior to final assembly. Thus, Tabershaw et al. (1943) found the average air lead concentration in the breathing zone of welders of structural steel to be 1200 $\mu g/m^3$. Welding can also be a hazard on occasion, when the coating is so-called zinc silicate, since zinc silicate can contain substantial concentrations of lead. Welding of zinc silicate-coated steel can give rise to breathing zone concentrations of lead far in excess of 150 $\mu g/m^3$, the current threshold limit value in the USA (Pegues, 1960). Even the welding of galvanized steel creates concentrations of 400–500 $\mu g/m^3$. These high values were recorded under conditions of poor ventilation. With good ventilation, welding of zinc silicate-coated steel resulted in lead concentrations of 180 $\mu g/m^3$ near the welder's nose and 70 $\mu g/100$ ml in his blood.

The recovery of scrap metal from the dismantling of ships requires extensive cutting of steel plates with electric torches. These plates are heavily coated with lead-based paint. Consequently, the evolution of lead fumes and their inhalation by the shipbreakers commonly results in lead intoxication. Air samples collected near the breathing zone of shipbreakers show that lead concentrations of as much as 2700 $\mu g/m^3$ are attained, even in the open (Rieke, 1969).

### 5.3.4 Printing

The hazard in a printing establishment is probably in direct proportion to the dispersion of lead oxide dust, secondary to the remelt operation. An early study was reported by Brandt & Reichenbach (1943) in which melting pots were located in a variety of places where used type was discarded. These pots were maintained at temperatures ranging from 318°C to 477°C. The highest air lead concentration recorded was 570 $\mu g/m^3$, and the highest average concentration for any room was 200 $\mu g/m^3$. Although working methods and industrial hygienic conditions have probably changed considerably since this report was published, a marginal degree of hazard still prevails. Tsuchiya & Harashima (1965) reported a range of lead levels of 30–360 $\mu g/m^3$ at breathing level in several printing shops in Japan.

Biological monitoring of workers in the printing industry has been reported. It was found that four of those engaged in smelting had blood lead concentrations greater than 50 $\mu g/100$ ml (Hernberg et al., 1969). There was only one blood lead value greater than 70 $\mu g/100$ ml among the 28 workers studied.

### 5.3.5 Alkyllead manufacture

Tetraethyllead was first distributed as an additive to automobile fuel in 1923. Tetramethyllead was introduced in 1960. Today, the annual production of these two alkyllead compounds accounts for approximately 12% of total lead consumption by industry (see 3.3). Inevitably, workers engaged in the manufacture of these compounds are exposed to both inorganic and alkyllead. Some exposure also occurs at the petroleum refineries where tetraethyllead and tetramethyllead are blended into gasoline.

The process of tetraethyllead manufacture consists of reacting a sodium–lead alloy with ethyl chloride. The alloy is made by combining molten lead with elemental sodium. The alloy is then transported to the autoclaves in hoppers. After the autoclave has been charged, ethyl chloride is added over several hours. The reaction takes place at about 75°C for a further period of 30–60 minutes. Steam distillation is then applied to remove residual ethyl chloride. The lead sludge is recovered, purified by smelting and re-used. The process generally in use for the manufacture of tetramethyllead is basically the same as for tetraethyllead. The final step is blending with dyes and scavengers. The product is shipped either in drums or tanker lorries.

Although there is a potential hazard from skin absorption of tetraethyl- and tetramethyllead, this is guarded against by the use of protective clothing. In a recent study, a good correlation was found between the organic air lead concentration in a plant and the rate of lead excretion in the urine (Linch et al., 1970). The average concentration of organic lead was 0.179 mg/m$^3$ for the tetramethyllead operation and 0.120 mg/m$^3$ for the tetraethyllead operation. The somewhat higher level registered for tetramethyllead was probably because the reaction between the organic reagent and the lead alloy takes place at a somewhat higher temperature and pressure than that employed in tetraethyllead production. Categories of hazard have been established based on the frequency with which workers are removed from exposure because of excessive urinary lead excretion (Table 13).

No exposure data are available for the blenders who mix tetraethyllead and tetramethyllead with gasoline at the refineries, but some exposure is likely to occur. Even at the filling stations where gasoline is pumped into cars, the concentration of organic lead in the vicinity of the pumps is appreciably greater than in the ambient air. Organic lead concentrations of 0.2–1.5 µg/m$^3$ were found in the vicinity of pumps (Colwill & Hickman, 1973; Harrison et al., 1974), and the concentration of tetraalkyllead emitted from the exhaust pipe of cars varied from 50 to 1000 µg/m$^3$ when the engine was idling (Laveskog, 1971).

### 5.3.6 Other industrial exposures

The diversity and extent of the industrial applications of lead makes it impossible to consider all cases. Furthermore, in most instances the actual

Table 13. Degree of hazard from lead exposure in the alkyllead industry[a]

| High | Moderate | Low |
|---|---|---|
| 1. smelting furnaces | 1. drumming plant | 1. blending |
| 2. charging autoclaves | 2. steam distillation | 2. pressure vessel inspection |
| 3. unloading and movement of lead pigs | 3. alloying | |
| 4. lead recovery | 4. autoclave area | |
| 5. maintenance | | |

[a] Data provided by: M. R. Zavon, Medical Director, Ethyl Corporation, Ferndale, Michigan, USA.

exposure levels have not been assessed. Some technological applications of lead are too recent to have provided much industrial hygiene experience. For example, the use of lead stearate as a stabilizer in the manufacture of poly(vinylchloride) is emerging as a new hazard. In the 1971 Annual Report of the British Chief Inspector of Factories, the number of reported cases of lead poisoning in the plastics industry was second only to that in the lead smelting industry (HM Chief Inspector of Factories, 1973). Other individual cases have been reported in recent years (Scarlato et al., 1969; Maljković, 1971). Lead stearate is milled and mixed with the poly(vinylchloride) and the plasticizer, to the extent of about 1-3 %. It seems probable that the source of the problem is the dust that is generated in the mixing process. It appears too, that lead exposure occurs in the rubber tire industry (Sakurai et al., 1974), probably as a result of using lead dithiocarbamate as an accelerator in rubber manufacture.

Table 14. Relative hazard of lead poisoning in some occupations or operations[a]

| High hazard | Moderate or slight hazard |
| --- | --- |
| Primary and secondary lead smelting | Lead mining |
| Welding and cutting of lead-painted metal constructions | Plumbing |
| Welding of galvanized or zinc silicate coated sheets | Cable making |
| Shipbreaking | Wire patenting |
| Nonferrous foundries | Lead casting |
| Storage battery manufacture: pasting, assembling, welding of battery connectors | Type founding in printing shops |
| | Stereotype setting |
| Production of lead paints | Assembling of cars |
| Spray painting | Automobile repair |
| Mixing (by hand) of lead stabilizers into poly(vinyl chloride) | Shot making |
| | Welding (occasionally) |
| Mixing (by hand) of crystal glass mass | Lead glass blowing |
| Sanding or scraping of lead paint | Pottery/glass making |
| Burning of lead in enamelling workshops | |
| Repair of automobile radiators | |

[a] From: Hernberg, 1973.

Drawing from his own experiences and knowledge of the field, Hernberg (1973) has provided a classification of hazard for common industrial activities where lead is used (Table 14).

## 5.4 Blood Lead Concentrations of Various Populations

Under certain conditions, blood lead levels (Pb-B) are a useful indicator of exposure and are therefore discussed in this section dealing with environmental levels and exposures (see also section 6.1.1.2).

### 5.4.1 Adult populations

A great deal of data is available on the blood lead levels of adult populations. By far the major proportion of these studies have reported that Pb-B mean values for occupationally unexposed, rural, and urban, populations range from 10 to 25 µg/100 ml (Hofreuter et al., 1961; US Department of Health, Education and Welfare, 1965; Butt et al., 1964; Holmquist, 1966; Lehnert et al., 1970; Horiuchi, 1970; Tepper & Levin, 1972; McLaughlin et al., 1973; Tsuchiya et al., 1975). Studies relating to populations from northern Italy have consistently revealed somewhat higher mean values, ranging from 24 to 35 µg/100 ml (Zurlo et al., 1970; Secchi et al., 1971; Secchi & Alessio, 1974). Similar Pb-B levels were also reported from rural and urban population groups in France (Boudene et al., 1975). In contrast, relatively low values (8.5 µg/100 ml) have been reported for 50 women from southern Sweden (Haeger-Aronsen et al., 1971). These are consistent with recently reported values for the Finnish female general population, ranging from 7.9 (rural), to 9.7 (urban) µg/100 ml (Nordman, 1975).

As a rule, the Pb-B levels of urban populations, and of people heavily exposed to automobile exhausts, have been found to be higher than those of rural populations or of populations living in areas with less traffic (Hofreuter et al., 1961; US Department of Health, Education and Welfare, 1965; Lehnert et al., 1970; Tepper & Levin, 1972) (Table 15). In one recent study, Pb-B levels were determined among adults before and after the opening of a motorway interchange with a high traffic density. Pb-B levels were found to be considerably higher among men and women living in the immediate vicinity of the interchange after it was opened than before (Waldron, 1975). In the evaluation of the results of this study, allowance must be made for the facts that no control group was studied, the procedure of drawing blood samples was changed after opening the interchange, the sampling took place at different times of the year and no data were given pertaining to the control of the analytical method used (atomic absorption spectroscopy). Thus, the possibility of systematic errors cannot be ruled out. On the other hand, Stopps (1969) found that the Pb-B levels of people living in various places remote from civilization had group means of 12–23 µg/100 ml, values not significantly different from group means reported for people living in urban areas of highly industrialized countries. No information was given in the report concerning procedures or quality control of the analytical methods.

A distinct increase in the lead absorption has been recorded in people living in the vicinity of lead smelters (Secchi et al., 1971; Nordman et al., 1973; Martin et al., 1975; Graovac-Leposavić et al., 1973).

Men have higher Pb-B levels than women (NAS-NRC, 1972). This

Table 15. Summary of concentration of lead in blood of selected groups of males, USA[a]

| Mean (µg/100 ml) | No. of subjects | Identity of groups |
|---|---|---|
| 11 | 9 | Suburban nonsmokers, Philadelphia |
| 12 | 16 | Residents of rural California county |
| 13 | 10 | Commuter nonsmokers, Philadelphia |
| 15 | 14 | Suburban smokers, Philadelphia |
| 19 | 291 | Aircraft employees, Los Angeles |
| 19 | 88 | City employees, Pasadena |
| 21 | 33 | Commuter smokers, Philadelphia |
| 21 | 36 | City Health Dept. employees, Cincinnati |
| 21 | 155 | Policemen, Los Angeles |
| 22 | 11 | Live and work downtown, nonsmokers, Philadelphia |
| 23 | 140 | Post Office employees, Cincinnati |
| 24 | 30 | Policemen, nonsmokers, Philadelphia |
| 25 | 191 | Firemen, Cincinnati |
| 25 | 123 | All policemen, Cincinnati |
| 25 | 55 | Live and work downtown, smokers, Philadelphia |
| 26 | 83 | Police, smokers, Philadelphia |
| 27 | 86 | Refinery handlers of gasoline, Cincinnati (1956) |
| 28 | 130 | Service station attendants, Cincinnati (1956) |
| 30 | 40 | Traffic police, Cincinnati |
| 30 | 60 | Tunnel employees, Boston |
| 31 | 17 | Traffic police, Cincinnati (1956) |
| 31 | 14 | Drivers of cars, Cincinnati |
| 33 | 45 | Drivers of cars, Cincinnati (1956) |
| 34 | 48 | Parking lot attendants, Cincinnati (1956) |
| 38 | 152 | Garage mechanics, Cincinnati (1956) |

[a] From: US Department of Health, Education and Welfare, 1965.

difference does not appear to be totally attributable to the higher haematocrit values of men (Tepper & Levin, 1972; Nordman, 1975). At least part of the difference is likely to be accounted for by the higher food consumption of men.

No association has been established between Pb-B levels and age in adults (NAS-NRC, 1972; Tepper & Levin, 1972; Nordman, 1975).

The influence of cigarette smoking is not fully evaluated; some researchers have reported higher Pb-B levels for smokers than for non-smokers (Hofreuter et al., 1961; US Department of Health, Education and Welfare, 1965; Tepper & Levin, 1972), while others have been unable to confirm such an association (Lehnert et al., 1967; Jones et al., 1972; McLaughlin & Stopps, 1973; Nordman, 1975; Tsuchiya et al., 1975).

### 5.4.2 Children

European studies of Pb-B levels in children indicate that, in general, the values are similar to or possibly even lower than those in adults. Pb-B levels of 200 children aged 4–13 years in rural western Ireland have been reported to be below 13 µg/100 ml with 45% of the results below 10 µg/100 ml (Grimes et al., 1975). A group of 363 children aged from 8 days to 8 years was surveyed in the Nüremberg/Erlangen area. The children displayed a mean Pb-B level of $3.3 \pm 2.6$ µg/100 ml in the first year of life; the Pb-B level increased year by year and reached a mean of $11.5 \pm 4.9$ µg/100 ml at the age of 6–8 years (Haas et al., 1972a). However, most of the available data on Pb-B levels in children have been obtained as a result of case-finding programmes conducted in the USA. In one study, the average blood level of 230 children, aged 1–5 years, in two rural counties was found to be 22.8 $\pm 11.0$ µg/100 ml (Cohen et al., 1973). An upward correction was made for all haematocrit values below 40%; more than half of the children lived in older houses (more than 25 years old) one-quarter of which had flaking paint or holes in the plaster.

There has been great concern in the USA that a very large number of inner city children have abnormally elevated Pb-B levels. The concern is for children in the blood lead range of 40–80 µg/100 ml. Thus, Blanksma et al. (1969) reported that in 1967, and 1968, 8%, and 3.8%, respectively, of Chicago slum children had Pb-B concentrations in excess of 49 µg/100 ml. This study involved 68 744 children, the majority of whom were between 1 and 6 years of age. The problem is not limited to large cities. Fine et al. (1972) reported on a survey of 14 Illinois communities with populations ranging from 9641 (Robbins) to 126 963 (Peoria). Of a total of 6151 children, 18.6% had Pb-B levels higher than 39 µg/100 ml and 3.1% had levels higher than 59 µg/100 ml. Some of the communities were in the Chicago urban complex, but a considerable number were not. There did not appear to be any great difference in the percentage of children having an excessive concentration of lead among the Chicago urban communities as compared with the downstate and western Illinois communities. The findings are certainly not unique to Illinois. In a recent survey, 34% of 343 children in an impoverished area of Boston had Pb-B levels in excess of 39 µg/100 ml and 12% were over 49 µg/100 ml (Pueschel et al., 1972). Similar data have been gathered recently in New York City and elsewhere.

# 6. METABOLISM OF LEAD

## 6.1 Absorption[a]

The absorption of lead from environmental sources is not solely dependent on the amount of lead presented to the portals of entry per unit time. It is also dependent on the physical and chemical state in which the metal is presented and it is influenced by host factors such as age and physiological status. The amount of food eaten and the amount of air breathed, with the proportionate ingestion or inhalation of lead, are functions of metabolic activity. Men engaged in heavy work breathe more air and eat more food than sedentary individuals of the same weight, and children eat almost as much food and breathe almost as much air as middle-aged adults.

## 6.1.1 Absorption by inhalation

A large amount of information has accumulated regarding the factors that determine the degree of deposition and retention of inhaled aerosols in general (Task Group on Lung Dynamics, 1966). With appropriate knowledge of the aerodynamic characteristics of lead aerosols, it would be possible to make reasonable predictions from the lung model developed by the ICRP Task Group on Lung Dynamics, concerning the fractional deposition that would occur in the human airways. It would also be possible to predict the pattern of regional deposition in the airways. Unfortunately, the knowledge necessary for making accurate predictions is not available, particularly in the case of industrial exposure.

The ICRP lung model would predict that approximately 35% of the lead inhaled in general ambient air would be deposited in the airways, since the aerodynamic diameter[b] of the lead particles is approximately 0.1–1.0 μm (see section 5.1.1). The lung model would also predict that regional deposition would be predominantly in the alveolar bed and in the deeper regions of the tracheobronchial system. Furthermore, it would predict that fractional deposition of lead dusts generated in an industrial environment would be greater than it would be for lead in general ambient air; however, the deposition would be mainly in the nasopharynx rather than in the pulmonary bed or tracheobronchial region, owing to the larger particle size. Industrial lead fumes, such as those generated in the process of cutting metals with electric torches, would be of small particle size and would behave accordingly. But even the lead aerosols breathed by the general population are not well enough characterized to predict deposition. This is particularly true for the very small particles (<0.1 μm) which are largely deposited by diffusion (Lawther, 1972).

The adequacy of the ICRP lung deposition model is open to question, at least for small particles. The model predicts a total airway deposition of 40–50% for 0.5-μm particles, whereas a study in human volunteers indicated a deposition of only 6–16% depending on the rate and depth of respiration (Muir & Davies, 1967).

Predictions concerning the characteristics of airway clearance of lead aerosols using the ICRP lung model are even more difficult to make than

---

[a] In this document, absorption and uptake are used synonymously.

[b] Aerodynamic diameter = diameter of a unit density sphere with the same settling velocity as the particle in question (Task Group on Lung Dynamics, 1966).

predictions regarding deposition. The lung model would predict that the fate of lead deposited in the airways would vary greatly depending on its solubility characteristics and on the inherent toxicity of the particles to the clearance mechanism (lung macrophages and cilia). The chemical forms of lead in air are both numerous and variable, depending on the source and on residence time in the air (see section 5.1.1). In many types of industrial exposure, lead is probably mainly in the form of lead oxide.

### 6.1.1.1 *Human studies*

Actual studies on the fractional deposition of particles in the respiratory tract of man have not been extensive, especially in the case of lead. Kehoe (1961) studied the deposition of lead in human volunteers with an air lead level of 150 µg/m³. The source of lead was combusted tetraethyllead which produced lead (III) oxide ($Pb_2O_3$) in the air. Subjects breathed air containing particles with an average diameter of 0.05 µm viewed under the electron microscope, with 90 % ranging from 0.02–0.09 µm. A diameter of 0.05 µm for lead (III) oxide as seen under the electron microscope represents a mass median equivalent diameter of approximately 0.26 µm (NAS-NRC, 1972). Subjects also breathed air containing particles having an average diameter of 0.9 µm (mass median equivalent diameter = 2.9 µm); 36 % of the smaller particles, and 46 % of the larger particles, were deposited.

Nozaki (1966) also reported on lung deposition of inhaled lead in man. Lead fumes were generated in a high-frequency induction furnace and were inhaled at a concentration of 10 000 µg/m³. Particle size was closely controlled according to the method of Homma (1966). The results (see Table 16) were similar to those of Kehoe (1961) and were reasonably consistent with the ICRP lung deposition model (Task Group on Lung Dynamics, 1966).

These data suggest that an estimate of 30 ± 10 % deposition is reasonable for the usual general ambient air situation and that lead oxide deposition characteristics will vary considerably, depending on the particle size and on the depth and frequency of respiration.

However, one cannot predict the contribution of airborne lead to the body burden of lead on the basis of deposition studies alone. Regional deposition probably varies greatly from one exposure situation to another, that is, the industrial setting *versus* the ambient environment. Also, the nature of lung clearance is unknown and is difficult to study. Nevertheless, it is possible to determine short-term lung clearance by carrying out gamma ray lung scans following inhalation of ²¹²Pb. Such a study in man has been reported (Hursh & Mercer, 1970) but its relevance to the rate of clearance of the chemical and physical forms of lead usually inhaled by man is highly questionable. Such radioactive lead studies involve the adsorption of $^{212}Pb$ atoms on carrier aerosol particles. The desorption of lead atoms from aerosol nuclei under these artificial circumstances may be quite significant, and the estimated rate may be totally unlike the clearance rate for ambient air lead particles.

Kehoe (1961) has reported that when a subject breathed large-particle aerosols of lead (III) oxide (approximately 2.9 µm mass median equivalent diameter) for many weeks at 150 µg/m³ a very substantial increase in faecal excretion occurred, probably reflecting the fact that the particles were largely trapped in the nasopharynx and swallowed. When the same subject inhaled air with a lead concentration of 150 µg/m³, with the lead in small particles (approximately 0.26 µm mass median equivalent diameter), only a small rise in faecal lead excretion was observed.

During inhalation of particulate air pollutants, the lead dust comes into contact with lung cells, which are primarily responsible for phagocytosis. It must be remembered that alveolar macrophages are damaged *in vitro* by inorganic lead compounds (Beck et al., 1973), and that similar effects have been demonstrated *in vivo* in rats and guinea-pigs (section 6.1.1.4). It

Table 16. Deposition of lead fumes in the airways of human subjects[a]

| 10 respirations/min : 1350 cm³ tidal air | | 30 respirations/min : 450 cm³ tidal air | |
|---|---|---|---|
| Particle diameter[b] (µm) | % Deposition | Particle diameter[b] (µm) | % Deposition |
| 1.0 | 63.2 | 1.0 | 35.5 |
| 0.6 | 59.0 | 0.6 | 33.5 |
| 0.4 | 50.9 | 0.4 | 33.0 |
| 0.2 | 48.1 | 0.2 | 29.9 |
| 0.1 | 39.3 | 0.1 | 27.9 |
| 0.08 | 40.0 | 0.08 | 26.5 |
| 0.05 | 42.5 | 0.05 | 21.0 |

[a] Adapted from Nozaki, 1966.
[b] Mass median equivalent diameter.

seems possible, therefore, that the lung defence mechanisms are, to some extent, impaired in an environment with a high air lead concentration, and that the rate of absorption of inhaled particles under such circumstances is affected.

In summary, studies of airway deposition and clearance of lead in man have not, as yet, provided any clear indication of the daily absorption to be expected under realistic conditions. They have only emphasized the necessity to consider other kinds of data to obtain this information.

Since the concentration of lead in the blood is thought to reflect current and recent lead exposure, the degree of lead intake from air should be reflected in this factor.

### 6.1.1.2 *The relationship of air lead to blood lead in the general population*

The risk to man from lead in air has become a matter of considerable concern in recent years. Studies of lead deposition and retention in the airways of man have not been very enlightening. A more indirect but nonetheless useful approach to the problem starts from the assumption that the concentration of lead in the blood is proportional to the concurrent level of total uptake by way of the several portals of entry. It follows that each environmental source (mainly air, food and water) would contribute to the blood lead concentration in direct proportion to its contribution to the total daily lead uptake. Up to the present time, such a relationship has never been rigorously demonstrated. Goldsmith & Hexter (1967) developed a linear regression plot of log Pb-B *versus* log lead concentration in air. The air lead samples were not necessarily taken at the same time and place as the blood samples. Thus, the regression line was calculated on the basis of rather imprecise information. However, data from experimental human subjects breathing known high concentrations of lead oxide were found to fit the regression line rather well. A cogent criticism is the fact that the validity of the air lead data as applied to the specific blood lead data is very uncertain. The contribution of air lead to blood lead, as inferred from the Goldsmith-Hexter curve, is about 1.3 μg of lead per 100 ml of blood per 1 μg of lead per $m^3$ of air. Other epidemiological studies have been made of the relationship between air lead and blood lead. Azar et al. (1973) monitored the inhaled air of 150 individuals using personal air samplers continuously, 24 hours per day. The air lead exposure ranged from 2 μg/$m^3$ to 9 μg/$m^3$. There was a significant correlation between log air lead level and log blood lead level, when data from all the cities involved were pooled. The contribution of air lead to blood lead was found to be somewhat less (approximately 1.0 μg of lead per 100 ml of blood per 1 μg of lead per $m^3$ of air over the range of air lead concentrations studied), than was estimated from the Goldsmith-Hexter curve.

Another recent epidemiological investigation which examined the relationship between air lead and blood lead levels was the Seven Cities Study (Tepper & Levin, 1972). No significant correlation was found between air lead and blood lead levels over an air lead range of 0.17–3.39 μg/$m^3$. A major deficiency was the fact that the air data were obtained from fixed outdoor sampling stations in the 11 cities involved.

Two studies have been reported recently in which the relationship between blood lead and air lead levels was investigated in human volunteers. In one study, 14 male volunteers were exposed to a lead oxide aerosol for 23 hours per day at an average concentration of 10.9 μg/$m^3$ for up to 17 weeks. Blood lead concentrations and other parameters were measured before, during, and following the exposure period. A plateau of blood lead concentration was attained during the exposure, and a return to pre- or near pre-exposure levels was observed during the post-exposure period. The air contribution to the Pb-B levels was approximately 1.4 µg of lead per 100 ml blood per 1 µg of lead per m³ of air (Coulston et al., 1972b). In another study, male volunteers inhaled an air lead concentration of 3.2 µg/m³. The blood lead level increased from 18 µg to 25 µg/100 ml,

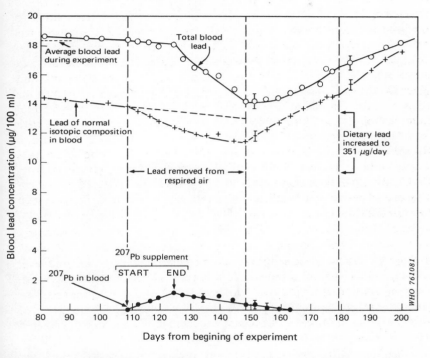

Fig. 2    Effect on blood lead of removal of lead from room air. Adapted from Rabinowitz (1974) with the addition of the dashed line.

that is approximately 2 µg of lead per 100 ml blood per 1 µg of lead per m³ (Coulston et al., 1972c). Rabinowitz (1974) reported a study of a single volunteer using stable lead isotope tracers in which the sudden removal of the normal lead in air by filtration resulted in a reduction of the blood lead concentration from approximately 14.5 µg/100 ml to approximately 11.3 µg/100 ml over a period of 40 days (Fig. 2). The average air lead levels were estimated taking into account measurements made indoors and outdoors, and the time spent in both locations. Prior to the experiment (day 109), the average air lead concentration was 1.6 µg/m$^3$ and during the experiment (day 109–148) it was 0.2 µg/m$^3$. In calculating the contribution of air lead to blood lead at day 148, allowance should be made for the fact that the concentration of lead of normal isotopic composition was decreasing prior to the removal of lead from the air. If this is taken into account the contribution of air lead to blood lead at day 148 would be about 1.7 µg of lead per 100 ml of blood per 1.4 µg of lead per m$^3$ of air, or 1.2 µg of lead per 100 ml of blood per 1 µg of lead per m$^3$ of air. It is unfortunate that it was not possible to follow the blood lead concentrations for a longer period of time after removal of lead from room air, since a new steady state had not been fully achieved. The study is also of limited value for application to the general population because only one individual was studied.

On the other hand, the Coulston study was deficient in that the form of air lead breathed (lead (III) oxide) may be deposited in, and cleared from, the airways in a significantly different manner from lead, as it actually occurs in general ambient air.

In conclusion it seems, that there is probably a perceptible effect of air lead on blood lead in the range of air lead concentrations applicable to the general population. The data available suggest that with blood lead levels in the range found in the general population, air lead levels may contribute from 1.0 to 2.0 µg of lead per 100 ml of blood per 1 µg/m$^3$ of air.

6.1.1.3  *The relationship of air lead to blood lead in occupational exposure*

There is very little precise information concerning the relationship between the concentration of lead in air (Pb-A) and Pb-B levels in subjects who are occupationally exposed. The air sampling technique used in the study of this relationship is of great importance. Personal monitors should be used since in most industrial situations the air lead concentrations to which individuals are exposed may be highly variable, depending on the particular tasks being performed and on the individual's work habits.

Only one study has been reported in which the subjects wore personal monitors and in which the estimated individual Pb-A could be related to Pb-B and some biochemical tests (Williams et al., 1969). In this study, workers in various departments of an electric storage battery factory wore personal samplers for the full work shift for two weeks. There were considerable variations in the measured concentrations of air lead both among departments and among individual personal samples. Relevant data are presented in Table 17.

Using the data reported by Williams et al. (1969) an attempt was made to estimate very crudely the potential contribution of Pb-A to Pb-B in

Table 17. Means and standard errors of measured lead in air and Pb-B levels in different departments of an electric storage battery factory[a]

| Department | No. | Pb in air ($\mu g/m^3$) | | Pb-B ($\mu g/100$ ml) | |
|---|---|---|---|---|---|
| | | mean | S.E. | mean | S.E. |
| Machine pasting | 6 | 218 | 25 | 74.2 | 4.7 |
| Hand pasting | 8 | 150 | 29 | 63.2 | 9.2 |
| Forming | 9 | 134 | 13 | 63.0 | 2.7 |
| Casting | 6 | 52 | 3 | — | — |
| Plastics dept. A | 5 | 12 | 0.8 | 27.2 | 1.4 |
| Plastics dept. B | 5 | 9 | 0.8 | 29.1 | 1.6 |

[a] Adapted from Williams et al., 1969.

subjects who were occupationally exposed to lead. Several arbitrary assumptions were made in this estimation:

(1) that the weekly time-weighted average concentration of lead in air ($c$) is a good measure of the effective inhalation exposure, irrespective of the probable differences in breathing rates during work hours. For a 40-hour working week $c = 0.24$ (Pb-A)$_o$ + 0.76 (Pb-A)$_a$, subscripts $o$ and $a$ referring to the occupational and ambient concentration of lead in air.

(2) that (Pb-A)$_a$ was 1 $\mu g/m^3$ and that it had contributed 1.4 $\mu g/100$ ml to the measured Pb-B values (see 6.1.1.2), and that for each further increase of Pb-A = 1 $\mu g/m^3$, the increase in Pb-B would be 1.4 $\mu g/100$ ml in the range of Pb-A values up to about 10 $\mu g/m^3$.

(3) that the contributions of the occupational inhalation exposure, non-occupational inhalation exposure, and exposures from other sources (such as food) to the Pb-B levels are additive.

A further oversimplification was that the probable differences in the chemical composition and physical characteristics of air-borne lead in occupational and non-occupational environments were completely neglected.

The contribution, (Pb-B)$_F$, of non-inhalation exposures such as food intake to the measured levels of lead in blood was assumed to be the same for all workers and constant over the two week period of observation. It was calculated from the data of Table 17 for the workers in plastics departments A and B used as control groups, by subtracting the estimated contribution of $c$ to blood lead from the measured Pb-B values, and taking the mean, i.e. (Pb-B)$_F = \frac{1}{2}[27.2 - (3.6 \times 1.4) + 29.1 - (2.9 \times 1.4)]$ $= 23.6$ $\mu g/100$ ml. (Pb-B)$_o$ was then obtained by subtracting 23.6 from measured Pb-B values for all other departments.

The results are shown in Table 18.

Table 18. Estimation of blood lead levels potentially derived from effective inhalation exposure $c$

| Department | Measured Pb-B μg/100 ml | $(Pb-B)_o$ μg/100 ml | $c$ μg/m³ | $(Pb-B)_o/c$ |
|---|---|---|---|---|
| Machine pasting | 74.2 | 50.6 | 53.1 | 0.96 |
| Hand pasting | 63.2 | 39.6 | 36.8 | 1.1 |
| Forming | 63.0 | 39.4 | 32.9 | 1.2 |
| Casting | — | — | 13.2 | — |
| Plastics A | 27.2 | 3.6 | 3.6 | 1.0 |
| Plastics B | 29.1 | 5.5 | 2.9 | 1.5 |

From these calculations it would appear that an increase of 1 μg/m³ in the weekly time-weighted average concentration of lead in air would correspond to an increase of approximately 1 μg/100 ml in Pb-B.

A similar but somewhat lower figure for the air lead contribution to Pb-B levels can be arrived at using data from a study, parts of which are reported in two different publications (Prpić-Majić et al., 1973; Fugaš et al., 1973). From their data, they calculated that the time-weighted average concentration of respirable lead particles for 52 workers in unspecified lead trades was 35 μg/m³. Their average Pb-B level was 44.3 μg/100 ml, while the Pb-B level of a control population living in an air environment of 0.2 μg/m³ was 22.4 μg/100 ml. Assuming the Pb-B levels due to non-air sources to be the same for the two groups, i.e. 22.1 μg/100 ml (total (22.4) minus the ambient air contribution to Pb-B ($0.2 \times 1.4 = 0.3$)), the air contribution to the Pb-B level for the industrially exposed group would be 44.3–22.1 or 22.2 μg/100 ml. Since the air lead concentration was 35 μg/m³, 1 μg of lead per m³ contributes 0.6 μg of lead per 100 ml of blood.

Another possible method of estimating the contribution of Pb-A to Pb-B in the occupationally exposed subjects is to find first a functional relationship that fits the Pb-B data from Table 17 and $c$. A power function $\ln y = \ln 18.9 + 0.34 \ln c$ gives a good fit in the range of $c = 10$ to $c = 50$ (correlation coefficient $r = 0.994$), and enables the estimation of the increase in Pb-B per unit increase in $c$. The results of these calculations are shown in Table 19. Although still a gross oversimplification, this method seems to give more realistic results because it reflects the fact that, at the higher Pb-A level, Pb-B does not increase linearly with Pb-A, and that therefore, the expected increase in Pb-B per unit increase in $c$ ($dy/dc$, Table 19) gets smaller and smaller as Pb-A levels grow.

### 6.1.1.4 Animal studies

Animal studies have been useful in the development of the ICRP lung deposition and clearance models, but they have not contributed much to resolution of the specific questions concerning the fate of inhaled lead in man. However, observations made on the effects of inhaled lead on lung

Table 19. Power curve fit[a] to the plot of Pb-B against the time-weighted average concentration of lead in air ($c$)

| Department | $c$ μg/m³ | Measured Pb-B μg/100 ml | $y$ μg/100 ml | dy/d$c$ |
|---|---|---|---|---|
| Machine pasting | 53.1 | 74.2 | 73.2 | 0.47 |
| Hand pasting | 36.8 | 63.2 | 64.6 | 0.60 |
| Forming | 32.9 | 63.0 | 62.2 | 0.64 |
| Casting | 13.2 | — | 45.6 | 1.09 |
| — | (10) | — | 41.3 | 1.41 |
| Plastics A | 3.6 | 27.2 | 29.3 | — |
| Plastics B | 2.9 | 29.1 | 27.2 | — |

[a] $y = 18.9c^{0.34}$ = Pb-B calculated.

macrophages are of special interest. A pronounced reduction in the number of lung macrophages has been demonstrated in rats and guinea-pigs owing to inhalation of lead (III) oxide at both 10 and 150 μg/m³ (Bingham et al., 1968). Maximum reduction occurred within approximately one week. This phenomenon has also been reported by others (Beck et al., 1973; Bruch et al., 1973, 1975). These observations suggest that, with high air lead concentrations at least, the lung clearance mechanism may not be functioning as effectively in diverting lead deposited in the lower airways to the gastrointestinal tract as the ICRP lung clearance model predicts. Thus, Pott & Brockhaus (1971) reported that large doses of lead bromide solution and of lead oxide suspension administered intratracheally to rats (1.5 mg of lead oxide per dose on 8 successive days) were retained by the body as completely as intravenous doses. However, at $\frac{1}{3}$ of this dose, retention was significantly less.

### 6.1.2 Absorption of lead from the gastrointestinal tract

6.1.2.1 *Human studies*

The uptake of lead from the gastrointestinal tract has been studied fairly extensively, but as with the uptake of lead from air, the evidence concerning a number of important points is somewhat uncertain. Long-term balance studies conducted by Kehoe (1961) showed that the daily excretion of lead into the urine was a little less than 10% of the intake from food and beverages. He surmised that this fraction represented the amount absorbed from the gastrointestinal tract. In estimates made on this basis, the amount of urinary lead that could have originated from the air is disregarded, as well as the fact that some of the lead absorbed from the gastrointestinal tract is re-excreted into the bowel.

Recent studies by Rabinowitz et al. (1974), using orally administered ²⁰⁴Pb, indicate that the absorption of lead incorporated into the diet is a

Table 20. Comparison of daily oral lead intake with Pb-B levels

| Study design | Oral intake (µg/day) | Pb-B[a] (µg/100 ml) | Pb-B per 100 µg oral Pb | Reference |
|---|---|---|---|---|
| Duplicate portion | 113 (men) | 20.7 | 18.3 | Coulston et al., 1972b |
| Faecal excretion | 119[b] (women) | 15.3 | 13.0 | Tepper & Levin, 1972 |
| Duplicate portion | 230 (men) | 12.3 | 5.4 | Nordman, 1975 |
| Duplicate portion | 180 (women) | 7.9 | 4.4 | Nordman, 1975 |
| Composites technique | 505 (men) | 34.6 | 6.8 | Zurlo & Griffini, 1973[c] |

[a] Contributions of air to Pb-B levels are not reported in most of these studies and could not be subtracted from total Pb-B levels.

[b] Calculated from daily faecal excretion of 108 µg of lead assuming gastrointestinal absorption of 10%.

[c] Pb-B levels from Secchi et al. (1971).

little less than 10 %, which is consistent with Kehoe's conclusions based on a different experimental approach.

Attention has been directed recently towards the absorption of lead from the gastrointestinal tract in infants and young children. In a study of eight normal children, from 3 months to 8.5 years of age, Alexander et al. (1973) found a high degree of lead absorption (53 %). There did not appear to be any significant reduction in fractional retention within the age range studied. This work is subject to criticism because of the large scatter of values and because the conclusions were based on 3-day balances, a period that is probably insufficient for reaching any reliable conclusions.

6.1.2.2 *The relationship of oral intake of lead to blood lead levels in man*

It would be of great interest to be able to relate oral intake of lead to blood lead levels. It is obvious that, as the intake of lead increases, blood lead levels will rise, but a quantitative expression of this relationship at any particular level of lead intake has not been determined. Table 20 compares daily oral lead intake (µg/day) and Pb-B levels (µg/100 ml) found in adult populations without known excessive exposure to lead, from several parts of the world.

From the data in Table 20 it is not possible to draw any reliable conclusions regarding the contribution of foods and beverages to Pb-B levels. The contribution is calculated to be greater in the two American studies than in the European ones. One of these two American studies (Tepper & Levin, 1972) was actually of faecal lead excretion, not of dietary lead. But even if this study were discounted, there remains a considerable discrepancy between the other American study (Coulston et al., 1972b) and the European studies, which cannot be explained.

Each of these studies involved a different number of subjects and involved different analytical techniques. It is also probable that there was also exposure from other environmental sources.

At levels of lead intake above 1000 µg per day, the rise in blood lead level does not appear to increase linearly with dose, but, in fact, may fit a logarithmic function.

From data published by Kehoe (1961) concerning balance studies on human volunteers, a single individual with a total daily lead intake of 600 µg had blood lead levels in the range of 30–35 µg/100 ml registered over several months, which is consistent with the relationships suggested in Table 20. However, individuals with larger daily additions of lead did not have proportionately higher blood lead levels. A single individual with oral lead intake of 3300 µg per day had a blood lead level in the 50–60 µg/100 ml range, again followed up for several months.

For children, the dietary contribution to blood lead is more difficult to estimate than for adults. Because of the higher absorption of lead, particularly in infants, the contribution of dietary lead to blood lead levels may be higher than for adults.

### 6.1.2.3   Animal studies

The effect of age on gastrointestinal absorption of lead has been studied in experimental animals. The absorption of lead from food has been investigated in many animal studies. Values between 5 and 10% are usual (Pott & Brockhaus, 1971; Schlipköter & Pott, 1973; Horiuchi, 1970).

Kostial et al. (1971) demonstrated that 5–7 day old rats absorb at least 55% of single oral tracer doses of [203]Pb. In an extension of these studies, Forbes & Reina (1972) observed that the gastrointestinal absorption of tracer doses of [212]Pb, [85]Sr and [59]Fe was high prior to weaning and decreased rapidly thereafter. In the case of lead, absorption which was 83% at 16 days, decreased gradually to 74% on the day of weaning (22 days) and rapidly thereafter to about 16% at 89 days. The addition of tracer doses of metals to the diet is, however, an artificial situation. Results might have been quite different had appreciable amounts of carrier lead been included. Nevertheless, these observations are consistent with those reported in young children.

Certain dietary factors have also been shown to influence the gastrointestinal absorption of lead. Kello & Kostial (1973) have shown that milk enhances lead absorption in 6-week-old-rats. Fasting enhances lead absorption, at least as determined by Garber & Wei (1974) in mice. Low dietary levels of calcium and of vitamin D enhance lead absorption (Sobel et al., 1938b; Six & Goyer, 1970). It has also been demonstrated that rats on an iron-deficient diet accumulate more lead in their bodies than do rats on an iron-sufficient diet (Six & Goyer, 1972). This seems particularly significant in the light of the fact that young children in socially and economically deficient homes have a high incidence of anaemia and excessively high blood lead concentrations.

The absorption of lead ingested in the form of paint has received attention because of the hazard of lead-based paint to young children. Recent data from experiments on rats indicate that lead chromate and lead naphthenate incorporated into dried paint films are substantially available for absorption, although to a somewhat lesser degree than lead naphthenate in oil or lead nitrate in aqueous solution (Gage & Litchfield, 1968, 1969).

## 6.2 Distribution and Retention

As with all substances entering the body, a single dose of lead distributes initially in accordance with the rate of delivery of blood to the various organs and systems. Redistribution then occurs to organs and systems in proportion to their respective affinities for lead. Under conditions of continuous intake over long periods of time, a near-steady state is achieved with respect to intercompartmental distribution.

Perturbations in the pattern of distribution occur when large, short-term peaks of lead intake are superimposed on this well-defined pattern of long-term distribution.

### 6.2.1 Human studies

The kinetics of lead distribution and accumulation in man have not been well defined in man directly. However, from autopsy data, the general pattern of lead metabolism is clearly discernible. Above all, it is clear that lead has a strong tendency to localize and accumulate in bone. The accumulation of lead in the human body begins in fetal life (Horiuchi et al., 1959; Barltrop, 1969). Lead is readily transferred across the placenta and the concentration of lead in the blood of newborn children is similar to that of their mothers, indicating mother–fetus equilibration processes (Haas et al., 1972b; Hower et al., 1975). The distribution of lead in fetal tissue is quite similar to the distribution in adults (Barltrop, 1969).

The total lead content of the body may reach more than 200 mg in men aged 60–70 years, but is lower for women. Barry & Mossman (1970) calculated that in non-occupationally exposed adults, 94–95 % of the total body lead (body burden) was in the bones. A similar estimate was made by Schroeder & Tipton (1968), by Horiuchi et al. (1959), and by Horiguchi & Utsunomiya (1973). These recent reports serve to reaffirm the long-recognized affinity of lead for bone. They also provide the additional observation that the concentration of lead in bones increases throughout most of life. This is in contrast to soft tissues. Most soft tissues do not show a significant age-related change in lead concentration after the second decade of life (Barry, 1975). This is also true of the concentration of lead in whole blood (US Department of Health, Education and Welfare, 1965; Horiuchi & Takada, 1954) and in blood serum (Butt et al., 1964). Thus, it appears that the skeleton is a repository for lead that reflects the long-term accumulative human exposure, whereas the body fluids and soft tissues equilibrate reasonably fast and therefore reflect current and recent exposure. Little is known as to whether the mobilization of lead lying inactive in the bones can occur so rapidly that signs of poisoning appear. There is need for more studies in this field.

The concentration of lead in the blood is of prime importance in the evaluation of lead exposure. It is relied upon as an aid to the diagnosis of poisoning and as an index of exposure to assess hazardous conditions both in occupationally-exposed people and in the general population. It has long been known that lead circulating in the blood is mainly found in the erythrocytes (Cantarow & Trumper, 1944). The concentration of lead in erythrocytes is about 16 times greater than in plasma (Butt et al., 1964). The nature of the association of lead with the erythrocyte is not clearly understood. Numerous studies have been reported concerning the *in vitro* addition of lead to erythrocytes suspended in plasma or saline solutions. But the validity of such studies is open to serious question. Thus, Clarkson & Kench (1958) found that lead added *in vitro* was readily removed by EDTA, whereas residual lead present in the cells prior to the addition of lead could not be removed. This suggests a difference in regard to: (1) the degree of binding, (2) the site of binding in or on the cell, or (3) the type of binding of the lead. Recent studies indicate that lead is mainly bound to human erythrocyte protein, notably to haemaglobin, rather than to stroma (Barltrop & Smith, 1971, 1972).

The rate of equilibration of lead in blood with sources of input and with other body compartments has been studied in man by Rabinowitz et al. (1973, 1974) using a stable lead isotope tracer ($^{204}$Pb). The data reported indicate that with a constant daily oral input of $^{204}$Pb, a virtually constant concentration of the tracer in the blood is achieved after approximately 110 days. Upon withdrawal of the tracer $^{204}$Pb from the diet, the $^{204}$Pb concentration in the blood disappears with a half-time of approximately 19 days. The kinetics of disappearance and accumulation suggest that first order rate processes of exchange are involved with regard to this relatively mobile compartment. Tola et al. (1973) also provided data which indicate that the concentration of lead in the blood rises fairly rapidly to a new steady state level when men are newly introduced into an occupational lead environment. The time required for the blood lead concentration to achieve a new plateau reflecting the new environment is about 60 days.

The body burden of lead increases from birth to old age (Schroeder & Tipton, 1968; Barry & Mossman, 1970; Barry, 1975). When data for various specific organs and systems are examined, it becomes evident that there are two general pools of lead within the total organism. The major one, in terms of total lead, consists of bone. This pool is clearly highly accumulative. As a consequence, lead in bone accumulates through most of the life span. Other organs and systems are much less accumulative and, to different degrees, tend to stabilize relatively early in adult life reflecting a greater turnover rate of lead compared with that in bone.

There is good reason to make a distinction between total body burden and exchangeable body burden since the organs and systems comprising the exchangeable body burden are the ones having the greater toxicological significance. It is also extremely important to note that lead in whole blood is a part of the exchangeable fraction of the body burden. Among adults in the general population there is no age-related difference in regard either to the concentration of lead in whole blood or in blood serum. Thus, in a general way, the Pb-B level reflects the concentration of lead in soft tissues, and long-term changes in Pb-B levels with changes in exposure levels are probably accompanied by corresponding long-term changes in the rest of the exchangeable pool.

Nuclear inclusion bodies containing lead have been found in man subjected to lead exposure (Cramer et al., 1974; Galle & Morel-Maroger, 1965; Richet et al., 1966) as well as in experimental animals (see section 7.1.3). Although most frequently reported to occur in the kidney, they have been found in other organs as well. There is a suggestion from limited data that inclusion bodies are associated with short-term lead exposure and not with long-term exposure (Cramer et al., 1974).

The concentration of lead in deciduous teeth has received special attention because they are readily available from young children and because they provide a long-term record of lead exposure, much as is the case with bone. Dentine in the area adjacent to the pulp is particularly useful in this respect because it is laid down from the time of eruption to the time the tooth is shed. It has been reported that the concentration of lead in dentine is considerably lower in suburban schoolchildren than it is in children in areas of high lead exposure (Needleman & Shapiro, 1974).

There has been some interest in the possible use of hair lead as an index of exposure. Unfortunately, there is no reliable information, as yet, to indicate just how hair analyses should be interpreted in relation to the frequency and degree of exposure.

## 6.2.2   Studies in animals

Animal studies have been particularly useful in defining more precisely the nature of the kinetics of lead distribution and removal from various tissues. Following administration of a single dose of lead to rats, the concentration of lead in soft tissues is relatively high and falls rapidly, mainly as a result of transfer into the bone (Hammond, 1971). The distribution characteristics of lead were found to be independent of the dose of lead over a wide range. The rate constants for the elimination of lead from various tissues in rats following a single dose of lead have been described by Castellino & Aloj (1964). The rate of elimination was much slower from bone than from other tissues. In studies on rats, Bolanowska et al. (1964) noted that the rate of elimination of a single dose of lead from the body by spontaneous excretion became slower with time, reflecting progressively decreasing mobility of the residual body burden. This is no doubt mainly due to the fact that as lead becomes progressively more deeply buried in the bone matrix, its exchangeability with other compartments and its availability for excretion decrease.

Rather striking age-related differences have been observed concerning the distribution and retention of lead in rats (Momčilović & Kostial, 1974). The rate of elimination of a single tracer dose of $^{203}$Pb from the whole body, blood, and kidney was faster in adults than in sucklings. In the case of the brain, there was actually a slight increase in the $^{203}$Pb content of the brain of the sucklings while the content was falling in other soft tissues. Numerous animal studies have also demonstrated placental transfer of lead to the fetus (see Carpenter, 1974, for relevant literature).

The intracellular distribution of lead has been studied in rat tissue, mainly by cell fractionation techniques (Castellino & Aloj, 1969; Barltrop et al., 1971). Lead has an affinity for membranes of the cell, particularly mitochondria. These organelles undergo functional and ultrastructural changes in organs showing lead effects, e.g. renal tubular cells (Goyer & Krall, 1969). Little lead is found in lysosomes (Barltrop et al., 1971) in contrast with the intracellular distribution of many other metals, e.g. mercury, copper, iron.

There are few studies indicating the concentration of lead in target organs that will produce effects. Formation of nuclear inclusion bodies is observed in rats with renal lead concentrations of about 10 mg/kg (wet weight) of kidney (Goyer et al., 1970a). Other effects of lead were found to occur at higher levels of organ concentration. Death in cattle is associated with lead levels of about 50 mg/kg of kidney cortex (wet weight) (Allcroft & Blaxter, 1950).

The concept of estimating the lowest level of metal accumulation that results in adverse effects in a target organ has not been well-explored in the case of lead. This is in contrast with cadmium where estimates have been made of the minimum concentrations of cadmium in the kidney cortex at which evidence of renal damage appears (Friberg et al., 1974).

### 6.3 Elimination of Lead

The elimination of lead from the body is thought to be mainly by way of the urine and the gastrointestinal tract. Little is known about the miscellaneous routes of excretion such as sweat, exfoliation of skin, and loss of hair.

### 6.3.1 Human studies

An approximation of the relative contributions of the various routes to lead excretion in man has been given by Rabinowitz et al. (1973). This study refers to only one non-occupationally exposed human subject. Excretions via the kidneys and the gastrointestinal tract were measured directly. Loss via other routes, e.g. hair, fingernails, and sweat, was estimated from data on the efflux of $^{204}$Pb from the blood compartment. Losses per day were as follows:

| | |
|---|---|
| urine | 38 µg (76%) |
| gastrointestinal secretions | 8 µg (16%) |
| hair, nails, sweat, other | 4 µg (8%) |

The figure of 38 µg for daily urinary excretion is consistent with the data of Teisinger & Srbova (1959). They reported an average daily urinary lead excretion of 31 µg.

The mechanism of urinary lead excretion in man is not well understood. However, the studies of Vostal (1966) provide strong evidence that the process of renal clearance of lead is essentially glomerular filtration. Extrapolation of a curve of glomerular filtration rate plotted against lead excretion rate resulted in zero lead excretion at zero filtration. The form of lead appearing in the urine has not been defined. One study suggests that the form in which lead appears in the urine depends on whether exposure to lead is normal or elevated. Thus, in lead workers with high urinary lead excretion, it has been found that only one-half to two-thirds of the urine lead can be precipitated with co-precipitating agents such as oxalate, phosphate, or carbonate. By contrast, virtually all the lead in the urine of people with normal lead exposure can be co-precipitated (Dinischiotu et al., 1960). This suggests that a stable lead chelate species arises with elevated exposure. Nuclear inclusion bodies or lead–protein complexes are found in the urine of children with acute lead poisoning (Landing & Nakai, 1959).

The rate of biliary excretion of lead in man is not known.

The biological half-time of lead is extremely difficult to estimate. The constantly decreasing availability of the major stores of lead in osseous tissue makes it virtually impossible to describe the rate of loss from the body in simple terms. It is at least clear that, in man, clearance of one-half of a body burden of lead would require a number of years.

### 6.3.2 Animal studies

Animal data on the routes of lead excretion suggest a considerable species variation. In rats (Castellino et al., 1966) and in sheep (Blaxter & Cowie, 1946) excretion by biliary and transmucosal routes is greater than urinary excretion. On the other hand, the ratio of urinary to gastrointestinal lead excretion in the baboon is 2:1 (Eisenbud & Wrenn, 1970). Vostal (1966) studied the mechanism of lead excretion in dogs. In mild chronic intoxication, excretion was by glomerular filtration, without evidence of any tubular secretion or reabsorption. With more severe poisoning, there was evidence of renal tubular reabsorption. Evidence was also presented for a tubular secretory mechanism in the chicken.

### 6.4 The Metabolism of Alkyllead Compounds

The characteristic toxic effects of tetraethyllead and tetramethyllead are not caused by the tetraalkyl compounds themselves, but rather by the trialkyl derivatives formed by dealkylation in the liver (Cremer, 1959; Cremer & Callaway, 1961). Tetraethyllead is initially converted mainly to triethyllead and partly to inorganic lead (Bolanowska, 1968). The triethyllead concentration in organs then falls only slowly. Even after several days, there is no significant reduction. The behaviour of tetramethyllead is quite simi'ar to the behaviour of tetraethyllead. Tetramethyllead is much less toxic probably because it is dealkylated to the trialkyl toxic form much more slowly than is the case with tetraethyllead (Cremer, 1965).

Since both these compounds have toxic and biochemical effects unlike those of inorganic lead, it is not to be expected that the biochemical tests used in assessing inorganic lead exposure would have the same significance as in exposure to organic lead. Indeed, in severe cases of tetraethyllead poisoning, urinary coproporphyrins and ALA excretion are usually not elevated, and free erythrocyte porphyrins are only moderately and inconstantly elevated (Gutniak et al., 1964; Beattie et al., 1972b). These biochemical tests are therefore of little use in short-term exposure situations. But, in long-term exposure situations, it is possible that some of them may be useful. Indeed, Robinson (1974) has shown that in workers industrially exposed to tetraethyllead, the urinary excretion of ALA is increased, but not to the same degree as in workers exposed to inorganic lead who have equivalent levels of total urinary lead excretion (organic plus inorganic). This suggests that some portion of total urinary lead is reflecting alkyllead exposure. Bolanowska et al. (1967) demonstrated that, in three fatal cases of tetraethyllead poisoning, the ratio of inorganic lead to triethyllead ranged from 67:1 to 18:1 in the urine. But this ratio did not reflect the ratio of inorganic to triethyllead in tissues at all accurately. In tissues, including the brain, the ratios were approximately 1:1.


# 7. EXPERIMENTAL STUDIES ON THE EFFECTS OF LEAD

The major part of published experimental work on animals describes or aims to explain pathological or pathophysiological changes caused by lead. It does not contribute much to the understanding of the relationship between the dose administered, its distribution in a period of time, and the biological effect. The doses used in most animal experiments have, as a rule, been far above the levels that can occur in environmental or occupational contact with lead, with the exception of accidental ingestion of soluble lead compounds.

## 7.1 Animal Studies

### 7.1.1 Haemopoietic system

Experimental studies on the effects of lead on blood and haemopoiesis have been carried out essentially to study pathogenic mechanisms. There are few studies dealing with the relationship between the lead dose and blood changes.

There is a great deal of evidence showing that lead inhibits several enzymes that participate in haem synthesis. Inhibition of these enzymes is invoked to explain the rises in haem intermediates that occur as a result of lead exposure. Thus, the rise in erythrocyte protoporphyrin is readily explained on the basis of the well known inhibitory effect of lead on the mitochondrial enzyme ferrochelatase (EC 4.99.1.1) (haem synthetase). This action was first proposed by Rimington (1938) as the probable explanation for the anaemia in lead poisoning. Numerous studies have since confirmed that lead is indeed a rather potent inhibitor of haem synthetase (Dresel & Falk, 1954; Goldberg et al., 1956; Klein, 1962).

Although specific inhibition of the enzyme haem synthetase is usually invoked to explain the accumulation of protoporphyrin, it is also possible that the availability of iron for coupling with protoporphyrin is inhibited by lead. It has been shown that lead interferes with the transfer of iron from transferrin to human reticulocytes (see section 8.2.1). Further support for the idea that lead interferes with the availability of iron is to be found in studies showing that lead causes accumulation of iron as "ferruginous micelles" in developing erythrocytes (Bessis & Jensen, 1965). Mitochondrial damage was evident in these studies, suggesting the possibility that globin synthesis may be compromised, along with haem synthesis.

The increased excretion of coproporphyrin III in urine is suggestive of an inhibition by lead of the enzyme coproporphyrinogen oxidase (EC 1.3.3.3), which converts coproporphyrinogen III to protoporphyrin IX (PP) (Goldberg, 1972). There is no supportive evidence showing a direct inhibitory effect on this enzyme. One would imagine that inhibition of coproporphyrinogen oxidase (EC 1.3.3.3) would result in decreased blood levels of protoporphyrin IX, however, the opposite is true. Perhaps the concurrent rise in erythrocyte protoporphyrin, ALA excretion, and excretion coproporphyrin III in urine can be explained on the basis of $\delta$-aminolevulinate synthase (EC 2.3.1.37) (ALAS) stimulation. Stimulation of ALAS activity by lead acetate *in vivo* has been demonstrated in the avian hepatocyte, probably due to impairment of haem synthesis (Strand et al., 1972). By contrast, Gajdos & Gajdos-Török (1969) found no change in the ALAS activity of bone marrow or liver in experimental lead intoxication of rabbits.

Animal studies have been reported concerning ALAD inhibition by lead in tissues concurrently with inhibition in the circulating erythrocytes. This has been shown in the blood, brain, and liver of suckling rats (Miller et al., 1970). After 30–40 days of exposure, erythrocyte ALAD underwent an 80–90 % reduction. The blood lead concentration in these rats is not given but can be estimated from data in the report. It is stated that a maximum of 3 ml of blood was obtained from each rat. It also appears that the blood specimens each contained about 4.5 µg of lead. Therefore the blood lead concentration must have been at least 150 µg of lead per 100 ml of blood, and was probably nearer to 200.

In other studies, in which long-term lead exposure of rats resulted in about 50 % inhibition of erythrocyte ALAD, there was no inhibition of brain or liver ALAD (Coulston et al., 1972a): this may be due to the fact that the exposure levels were lower in this study than in the others cited.

The question of the significance of lead exposure in relation to haemoglobin formation has been studied in dogs by Maxfield et al. (1972). These authors were mainly concerned with the question of whether the depression of ALAD activity in the peripheral blood was in any way associated with depressed formation of haemoglobin. Dogs were given lead over a period and the ALAD activity fell to a very low level. But the ability of the dogs to regenerate haemoglobin after removal of half of the circulating blood volume remained essentially normal. Although this indicates that the inhibition of ALAD activity in peripheral blood may not be significant, it should be pointed out that the lead exposure was not sufficiently high to cause any substantial rise in ALA excretion in the urine. ALA excretion was only approximately two–three times the baseline level.

There is evidence that the synthesis of globin is affected by lead in animals as well as in man. In an *in vitro* study, it was shown that the incorporation of $^{14}$C-glycine into globin in duck erythrocytes was reduced by 25% by a lead concentration of $5 \times 10^{-4}$ M (Kassenaar et al., 1957). The reduction of $^{14}$C-glycine incorporation into haem was considerably greater.

Relatively little is known about the effects of lead on the formation or activity of other haem-containing compounds in the body. There is some evidence, however, that lead can inhibit formation of cytochrome P-450, a haemoprotein intimately involved in the drug-metabolizing mixed function oxidase system of hepatic microsomes (Alvares et al., 1972). Long-term lead administration has also been shown to affect the activity of cytochrome $c$ oxidase (EC 1.9.3.1) (Makašev & Verbolovič, 1967; Verbolovič, 1965). The effects seen were of a mixed nature, involving first stimulation then depression of activity. A decrease was also observed in the myoglobin concentration of some muscle groups. It is not clear from these studies whether the effects were due to inhibition of haem synthesis or of protein synthesis.

Administration of lead to rats (over a period of 6 months) in doses of 2–4 g per rat, resulted in a change in cytochrome $c$ oxidase (EC 1.9.3.1) activity and in the amount of haemoglobin. The magnitude of the change increased with larger doses (Verbolovič, 1965). Dogs were given a solution of lead acetate over a 2-year period resulting in a reduction in the activity of cytochrome $c$ oxidase (EC 1.9.3.1) that was in proportion to the dose of lead administered (Makašev & Verbolovič, 1967).

By means of electron microscopy, Pernis et al. (1964) showed grossly swollen mitochondria in the erythrocytes of lead-poisoned guinea-pigs, diverse vacuolar formations, and aggregates of molecules of ferritin. Electron microscopy of erythrocytes of rabbits receiving an intravenous dose of 20 mg/kg of 2% lead acetate solution showed vacuolization of the cytoplasm and swelling of the plasma membrane. An intensive vacuolization in thrombocytes and a reduction in the quantity of organelles, particularly those containing serotonin, also took place. In addition, a swelling of mitochondria after the complete disruption of cristae was noted (Hačirov, 1972). An experiment on rats showed ultramicroscopic changes of mitochondria in the red bone marrow cells in the early stages of poisoning.

### 7.1.2 Nervous system

#### 7.1.2.1 *Inorganic lead*

In view of recent concern about subtle impairments of cerebral function at sub-encephalopathic levels of lead exposure, there has been a renewed interest in lead and its toxic effects. High doses of lead will produce encephalopathy; this has been reported in cats (Aub et al., 1926) and dogs (Staples, 1955).

The brains showed histopathological features similar to those described in human encephalopathy. Others have since reproduced this syndrome in the rat (Thomas et al., 1971; Michaelson, 1973; Clasen et al., 1974) and in the mouse (Rosenblum et al., 1968; Silbergeld & Goldberg, 1974). The effects may be explained on the basis of retardation of brain development (Michaelson, 1973; Krigman et al., 1974).

Paraplegia was reported in suckling rats by Pentschew & Garro (1966). The disease was produced by transfer of lead from the mother's milk until weaning, with subsequent post-weaning feeding of lead to the young.

Behavioural abnormalities such as excessive self-grooming and aggressiveness occur, even when the lead intake is reduced to a point where paraplegia no longer occurs (Michaelson & Sauerhoff, 1974). It was estimated that the minimum daily lead intake causing behavioural effects (Michaelson & Sauerhoff, 1974) rose from 0.08 mg/kg body weight at birth to 3 mg/kg at day 16 as a result of suckling. Post-weaning, this minimum intake rose from 50 mg/kg at day 20 to 60 mg/kg at day 28. From day 16 to day 20, intake was difficult to estimate since the infant rats were eating and suckling to different degrees. Other studies in rats (Snowdon, 1973) and sheep (Carson et al., 1974) indicate that offspring of mothers exposed to lead during pregnancy show learning defects. Older animals are refractory to this type of effect (Brown et al., 1971).

Future behavioural studies should probably be extended to include subhuman primates since it has been shown that the histological and clinical features of lead encephalopathy can be produced in both infant and adult baboons (Cohen et al., 1972; Hopkins & Dayan, 1974).

Studies of lead neuropathy in animals indicate that demyelination and axonal degeneration are more consistent findings than neuronal damage in the anterior horn cells or dorsal root ganglia of the spinal cord (Lampert & Schochet, 1968; Schlaepfer, 1969; Fullerton, 1966). This is consistent with findings in man. The slowing of nerve conduction found in man has also been produced experimentally in the guinea-pig (Fullerton, 1966).

It is known that lead interfers in some manner with synaptic transmission in the peripheral nervous system and that the effects can be reversed by calcium (Kostial & Vouk, 1957). But, in addition, an increased frequency of miniature end-plate potentials has been reported (Manalis & Cooper, 1973). Neuromuscular blockade has also been demonstrated in the rat phrenic nerve-hemidiaphragm preparation (Silbergeld et al., 1974). Again, as in the other studies, the effect was antagonized by calcium. The significance of these findings with regard to the central nervous system remains to be determined.

Studies at the biochemical level have been very limited. It has been shown, using the "Pentschew model", that incorporation of $^{14}$C-glucose carbon into dicarboxylic amino-acids of the brain is reduced (Patel et al., 1974a, 1974b). These results were interpreted to indicate delayed brain maturation.

Recent work in dogs (Stowe et al., 1973) has mapped the variation in lead concentrations in different parts of the brain of lead-poisoned dogs. The studies show a relationship between areas of the most marked histological change and high lead concentration. Male pups from the same litter were fed a purified diet, low in calcium and phosphorus, with and without 100 mg/kg of lead as lead acetate from the age of 6–18 weeks. The concentration of lead in the various brain segments is given in Table 21.

Table 21. Distribution of lead in the brains of control and lead-intoxicated dogs[a]

| Brain segment | Lead concentration (mg/kg of wet tissue) | |
| | Control | Lead intoxicated |
| --- | --- | --- |
| Cerebellum | 0.160 ± 0.052 | 0.587 ± 0.113 |
| Medulla | 0.155 ± 0.007 | 0.713 ± 0.112 |
| Frontal white | 0.053 ± 0.027 | 0.920 ± 0.156 |
| Thalamus | < 0.10 | 1.023 ± 0.142 |
| Occipital white | 0.020 ± 0.012 | 1.030 ± 0.115 |
| Caudate | 0.120 ± 0.083 | 1.613 ± 0.345 |
| Frontal grey | 0.033 ± 0.015 | 1.767 ± 0.254 |
| Occipital grey | 0.080 ± 0.071 | 2.357 ± 0.181 |

[a] Adapted from Stowe et al., 1973.

### 7.1.2.2 Alkyllead compounds

Unlike the case with inorganic lead, intoxication by tetraethyllead in juvenile or adult rats caused a characteristic encephalopathy, involving restlessness, ataxia, combativeness progressing to convulsions, coma, and death (Davis et al., 1963). In dogs there was extensive muscular tremor and twitching, which progressed to convulsions, coma, and death.

Biochemical studies of the respiration of brain slices incubated with inorganic lead compared with triethyllead (the active metabolite of tetraethyllead) have substantiated the fundamental difference in the action of alkyllead compounds on the brain (Cremer, 1959). The toxic moieties in tetraethyllead and tetramethyllead poisoning are the trialkyl metabolites and not the inorganic lead ion.

Essentially there is no qualitative difference between the toxic effects of tetramethyllead and tetraethyllead. However, there is a quantitative difference in that the inhalation $LC_{50}$ for tetramethyllead (8870 mg/m$^3$) is about 10 times higher than that for tetraethyllead (850 mg/m$^3$) (Cremer & Callaway, 1961). The intravenous $LD_{50}$ for tetraethyllead is about 10 mg of lead per kg of body weight in the rat (Cremer, 1959). This is in contrast to the intravenous $LD_{50}$ for inorganic lead, which is approximately 70 mg/kg in the rat (Fried et al., 1956). The precise manner in which the trialkyllead ion acts to cause altered brain function is not clearly known but the mechanism may involve inhibition of amine oxidase (flavin-containing) (EC 1.4.3.4) (monoamine oxidase) (Galzigna et al., 1964). In rabbits, administration of toxic doses of tetraethyllead results in a loss of copper, iron, and zinc in certain areas of the brain (Niklowitz & Yeager, 1973), suggesting that triethyllead may act by displacing certain essential trace metals from metalloenzymes in the brain.

Tetramethyllead injected into rats in overtly neurotoxic doses did not depress ability to learn a simple task (Bullock et al., 1966).

### 7.1.3 Renal system

Animal studies have contributed to an understanding of the order of appearance of the various manifestations of renal toxicity in lead exposure. The spectrum and train of events as related to the exposure time and to the dose of lead have recently been reviewed (Goyer & Rhyne, 1973). In the earliest stage of renal response to lead exposure, reversible tubular effects occur. These include the appearance of intranuclear inclusion bodies, which is probably a mechanism for sequestration of lead. These bodies have been isolated and found to be composed of a lead–protein complex. The protein is insoluble in physiological solutions and is rich in acidic amino acids. It has not been characterized further (Moore & Goyer, 1974). The intranuclear inclusion bodies appear to have a high and specific affinity for lead compared with that for calcium, iron, zinc, copper or cadmium and about 90 % of the lead in the kidney is associated with them (Goyer et al., 1970a; 1970b).

The appearance of these bodies is accompanied by amino aciduria, glycosuria, and hyperphosphaturia. Morphological and functional changes in tubular epithelial cells also occur at this stage, including impaired respiratory and phosphorylative ability.

After further lead administration, more severe changes occur in the renal tubular epithelium such as hyperplasia and cystic changes. There is a progressive increase in interstitial fibrous tissue and atrophy of tubular cells. These are irreversible changes that lead to a third stage of renal failure, manifested by azotaemia and hyperuricaemia. Sclerotic glomeruli appear, but the hypertension seen in some cases of chronic lead nephropathy in man has not been reproduced in experimental animals

The sequence described above in animals is probably generally valid for man.

### 7.1.4 Gastrointestinal tract

The effects of lead on the gastrointestinal tract have been studied in some detail in the guinea-pig (Mambeeva & Ahmiedova, 1967). Spastic contractions occurred from the stomach to the jejunum. Inhibitory effects were also noted, accounting for the frequent constipation seen to accompany lead colic.

### 7.1.5 Cardiovascular system

Experimental animal data on the question of hypertension are conflicting. Among rats given 70 mg of lead acetate per day orally, only a few survived 40 days and all were hypertensive (Griffith & Landauer, 1944). Hypertension has also been produced in the rabbit (Beckmann, 1925). Others have not seen hypertension with lead exposure in rats (Padilla et al., 1969) or dogs (Fouts & Page, 1942). From all the above animal studies it seems that hypertension can occur with heavy lead exposure.

There are conflicting reports regarding whether lead can cause atherosclerosis in experimental animals. Sroczynski et al. (1967) observed increased serum lipoprotein, cholesterol, and cholesterol deposits in the aortas of both rats and rabbits receiving large doses of lead. On the other hand, Přerovská (1973) did not produce atherosclerotic lesions in rabbits using similar doses of lead given over an even longer period of time.

Cardiac myopathy has also been shown experimentally in lead-intoxicated rabbits (Kosmider & Sroczynski, 1961). The mechanism for this effect is not known.

Kuz'minskaja (1964) and Mirončik & Timofeeva (1974) observed that rabbits receiving lead after a cholesterol load showed more intense sclerotic changes in the aorta and myocardium than rabbits on a normal diet without lead, or than rabbits given cholesterol alone.

Makašev & Krivdina (1972) observed a phased change in the permeability of blood vessels (first phase—increased permeability; second phase—decreased permeability) in rats, rabbits and dogs, that received a solution of lead acetate. A phased change in the content of catecholamines in the myocardium and in the blood vessels was observed in subacute lead poisoning in dogs (Mambeeva & Kobkova, 1969). This effect appears to be a link in the complex mechanism of the cardiovascular pathology of lead poisoning.

### 7.1.6 Respiratory system

Alveolar macrophages from guinea-pigs are damaged *in vitro* by inorganic lead compounds ($3 \mu g/1 \times 10^6$ cells) thus releasing a rapidly occurring lysis and a slowly developing, coarsely blistered vacuolization. More than 90 % of the cells are damaged within 20 hours (Beck et al., 1973).

Similar effects seem to occur in the organism, since in rats that had inhaled 10 µg lead/$m^3$ for 3–12 months, the number of macrophages that could be flushed from the lungs was reduced by 60 % (Bingham et al., 1968).

Electron microscope investigations of the lungs of rats that had been exposed for 14 days to concentrations of 100–200 µg of lead oxide/$m^3$ revealed toxic effects in the alveolar macrophages and the type I alveolar epithelial cells. The structures of the endoplasmatic reticulum and mitochondria were changed (Bruch et al., 1973).

In the lungs, the alveolar macrophages have the capacity to degrade noxious substances and are important for other defence reactions. The ability of alveolar macrophages of guinea-pigs that had inhaled concentrations of 70–170 µg of lead per $m^3$ of air for four days, to degrade benzo[a]pyrene was distinctly decreased, the benzopyrene 3-monooxygenase (1.14.14.2) activity being only about 10 % of the original value. The activity returned to normal after three days without any lead exposure (Bruch et al., 1975). The elimination of bacteria from the lungs was also reduced, when rats were exposed to 70 µg of lead per $m^3$ of air (Schlipköter et al., 1977).

### 7.1.7 Reproductive system

Animal studies support the contention that behavioural deficiencies can occur in infants and newborn as a result of intrauterine exposure to lead via their mothers (see section 7.1.2). Others have shown a reduction in the numbers and size of offspring (Dalldorf & Williams, 1945; Puhae et al., 1963). Data in rabbits (Cole & Bachhuber, 1914), guinea-pigs (Weller, 1915), and rats (Stowe & Goyer, 1971) indicate that paternally-transmitted effects can occur, including reductions in litter size, weights of offspring, and in survival rate. Several investigators have reported that oral administration of lead to animals even at doses in the microgram per kilogram range can cause changes in spermatogenesis (Egorova et al., 1966;

Golubovič et al., 1968), and an increase in testicular RNA and DNA content (Golubovič & Gnevkovskaja, 1967; Golubovič et al., 1968).

## 7.1.8 Endocrine organs

The effects of lead on thyroid function that have been reported in man have also been demonstrated experimentally in rats (Zel'tser, 1962; Sandstead, 1967).

## 7.1.9 Carcinogenicity

### 7.1.9.1 *Inorganic lead compounds*

The carcinogenic risk to man of lead salts and the relevant studies in animals have recently been discussed in an IARC publication (IARC, 1972).

The induction of benign and malignant renal neoplasms has been observed in both Swiss mice and rats fed on diets containing 100 or 1000 mg of basic lead acetate ($Pb(C_2H_3O_2)_2.2Pb(OH)_2$) per kg of diet (Van Esch & Kroes, 1969; Van Esch et al., 1962; Mao & Molner, 1967; Azar et al., 1973). Similar results were observed in rats fed 1000 mg of lead acetate ($Pb(C_2H_3O_2)_2.3H_2O$) per kg of diet (Boyland et al., 1962). In addition to renal neoplasms, tumours of the testes, the adrenal, thyroid, pituitary, and prostrate glands and of the brain have been reported in rats fed lead acetate or basic lead acetate, but the results await confirmation (Zawirska & Medras, 1968; Oyasu et al., 1970). Rats given intraperitoneal or subcutaneous injections of lead phosphate also developed renal tumours. Total doses of 120–680 mg of lead were effective (Zollinger, 1953; Roe et al., 1965). No kidney tumours were reported in hamsters fed 100 or 500 mg of basic lead acetate per kg of diet for up to 2 years (van Esch & Kroes, 1969).

In Syrian golden hamsters given a combination of lead oxide and benzo[a]pyrene intratracheally once weekly for 10 weeks, lung adenomas occurred in 11/26 animals within 60 weeks. One adenocarcinoma of the lung was also observed. Such tumours did not occur in animals given the same dose of lead oxide or benzo[a]pyrene alone (Kobayashi & Okamoto, 1974).

### 7.1.9.2 *Alkyllead compounds*

Epstein & Mantel (1968) reported that subcutaneous injection of 0.6 mg of tetraethyllead (given as 4 equally divided doses) to Swiss mice between birth and 21 days of age produced malignant lymphomas in 1/26 males and 5/41 females, compared with 1/39 and 0/48 controls. In treated females, the

94

tumours were observed between 36 and 51 weeks after the first injection. The significance of this finding in female mice is difficult to assess since this tumour occurs frequently and with variable prevalence in untreated mice of this strain.

### 7.1.10 Mutagenicity

Chromosomes from leukocyte cultures from mice fed 1 % lead acetate in the diet showed an increased number of gap-break type aberrations (Muro & Goyer, 1969). These changes involved single chromatids, suggesting that injury followed DNA replication.

### 7.1.11 Teratogenicity

There have not been any adequate animal studies to provide evidence to support the suggestion that lead may have a teratogenic effect.

## 7.2 Acquisition of Tolerance to Lead

Although human studies suggest that there is no acquired tolerance in regard to haem-synthesis mechanisms, there may be for other toxic effects. In this regard, it is interesting to note that the blood lead level at which cattle develop severe encephalopathy from eating paint is often less than 80 µg/100 ml (Hammond et al., 1956). However, in cattle receiving 5–6 mg of lead per kg per day orally, the concentration of lead in the blood exceeded 100 µg/100 ml within 2–4 months and remained at about that level for as long as four years with continuous administration, without any apparent harm to the animals (Allcroft, 1951). In these studies, haemoglobin did not fall until a terminal illness developed. Hapke (1974) found that in cattle and sheep the sensitivity to acutely toxic amounts of lead was reduced by a pretreatment with lead for 5 months. Goyer et al. (1972) have suggested from their studies on rats that the intranuclear inclusion bodies that develop during lead exposure serve as a protective mechanism by binding lead in the kidney, making it less toxic. But in the recent study of Cramer et al. (1974) (see section 7.1.3) it was shown that renal intranuclear inclusion bodies are present only in workers exposed to lead for a relatively short period of time. Thus, if inclusion bodies serve some protective function, it is only during a limited period of exposure. The formation of the cadmium-binding protein, metallothionein, which appears to have a protective role in cadmium exposure, is induced by a number of metals but not by lead (Webb, 1972).

## 7.3 Factors Influencing Lead Toxicity

### 7.3.1 Age and sex

It has recently been reported that the intraperitoneal lethal dose of lead in rats is significantly lower for adult male rats than for adult female rats (Kostial et al., 1974). In the same study, it was observed that the lethal dose in mg/kg body weight for 3-week-old rats was about the same as for adult females.

### 7.3.2 Seasonal variations

The same seasonal pattern of high incidence of poisoning has been reported in dogs belonging to urban families as has been reported in children (Zook et al., 1969). It has also been shown experimentally in rats and mice (Baetjer, 1959; Baetjer & Horiguchi, 1963) and in rabbits (Blackman, 1937; Horiuchi et al., 1964) that susceptibility is greater at high ambient temperatures than at normal temperatures.

### 7.3.3 Nutrition

Experimental studies have shown that nutritional factors may influence the absorption of lead from the gastro-intestinal tract and thus alter susceptibility to the toxic effects of lead (Goyer & Mahaffey, 1972). Low phosphorous and calcium in the diet (Sobel et al., 1938b; Six & Goyer, 1970), high vitamin D (Sobel et al., 1938a), and low iron (Mahaffey, 1974) all enhance lead absorption. The amount and the composition of dietary protein may also influence lead toxicity. Low protein diets appear to increase the susceptibility to lead intoxication as compared to high protein diets (Baernstein & Grand, 1942; Goyer & Mahaffey, 1972).

The significance of these findings for the susceptibility of people to lead poisoning has not been established. However, many children, even in developed countries like the USA, have sub-optimal dietary intakes of calcium, iron, and other nutrients (US Department of Health, Education and Welfare, cited by Mahaffey, 1974). This may have a bearing on the problem of increased lead absorption frequently found in children in poor, urban areas.

### 7.3.4 Intercurrent disease, alcohol, and other metals

High lead exposure increases the susceptibility of mice to *Salmonella typhimurium* infection (Hemphill et al., 1971). Lead administration also increases the susceptibility of rats (Filkins & Buchanan, 1973; Selye et al.,

96

1966; Erve & Schumer, 1972), mice (Clercq de & Merigan, 1969), and baboons (Hoffman et al., 1974) to endotoxin shock, but such studies have been performed using extremely large intravenous doses of lead simultaneously with the endotoxin.

Administration of ethanol (10% *ad libitum* in drinking water) had no effect on the toxicity of lead to rats as measured by urinary ALA excretion, renal weight, or lead concentration in the kidneys, liver, or bones (Mahoffey, 1974).

Very little is known about metal interactions and how they might affect the toxicity of lead, except at the nutritional level (see section 7.3.3). Beyond that, a synergistic effect has been noted between lead and cadmium with regard to experimental teratogenesis (Ferm, 1969). It was also found that zinc, given in the diet with lead, protected horses against the toxic effects of lead. Probably, this effect was not due to inhibition of lead absorption. Zinc supplementation actually caused an increase in the lead content of liver and kidney, but a decrease in the lead content of brain and bone (Willoughby et al., 1972). It might be inferred that zinc displaced lead from lead-inhibited enzymes that are zinc-dependent, such as ALAD (Cheh & Neilands, 1973). A dose-dependent effect of zinc, antagonistic to the depression of ALAD by lead, has recently been shown *in vivo* and *in vitro* as well as an *in vitro* antagonism of zinc on the cytotoxic effect of lead on macrophages (Schlipköter et al., 1975; Ruiter de et al., 1977).

### 7.4 Human Studies

Planned experimental studies on the effects of lead in man are sparse. Kehoe (1961), in his famous experimental studies, in which human volunteers were exposed to a known amount of lead over various periods of time, confined himself to studying the lead balance only, and did not report on the effects of lead.

Three subjects ingested 1 and 3 mg of lead daily, in the form of lead (II) nitrate, for 33 weeks. The ALA-U, CP-U, and erythrocyte protoporphyrin IX were measured regularly while Pb-B and Pb-U measurements were performed at irregular intervals (Schlegel et al., 1973). Exposure from food and ambient air was not controlled during the experiment. A rise in FEP was obtained with both doses and a rise in ALA-U and CP-U only with the 3 mg dose. Evaluation of the results obtained in this study is difficult, partly because of the small number of subjects studied and partly because the results were rather erratic.

Coulston et al. (1972b; 1972c) conducted two exposure chamber experiments on male volunteers (see section 6.1.1.2). The volunteers were exposed to air lead concentrations with an average of 10.9 and 3.2 $\mu g/m^3$

for up to 17 weeks. In the 10.9 µg/m³ exposure study, 24 volunteers participated, 6 of whom served as controls. In order to control dietary lead exposure, total diet for one full day was collected at intervals of eight days; the results indicated an average lead intake of about 110 µg/day only. The variables measured were the Pb-B, ALAD, ALA-U, and CP-U. Blood lead levels increased in all of the exposed men and appeared to stabilize after about 12 weeks of exposure. The mean Pb-B level at that time was about double the pre-exposure mean, i.e., an increase from 19 to 37 µg/100 ml. A concomitant increase of the urinary excretion of lead was reported; the faecal excretion remained unchanged however. The rise in blood lead levels was followed by a decrease in ALAD activity, which after 5 weeks of exposure was about 50 % of the pre-exposure level. No change in ALA-U and CP-U was reported. Five months after the termination of the exposure, all but one of the participants had Pb-B values similar to those before exposure. The ALAD activity returned to normal almost immediately after cessation of exposure. No changes in the haemoglobin level were noted during the experiment. In the 3.2 µg/m³ experiment a rise in the Pb-B level from 20 to 26 µg/100 ml was obtained, followed by a slight decrease in ALAD activity, which after five weeks of exposure was about 85 % of the pre-exposure level. Other changes were not reported.

In a recent experimental study, a greater susceptibility to inorganic lead was demonstrated in females (Stuik, 1974; Stuik & Zielhuis, 1975). The volunteers were healthy male and female students aged 18–26 years. Groups of 5 males and 5 females received 20 µg of lead per kg per day orally for 21 days. Lead was administered as lead acetate in glycerol.

The control blood lead levels remained fairly constant at approximately 17 µg/100 ml during the experiment. The exposed male subjects showed an increase from 20.6 µg/100 ml to 40.0 µg/100 ml at the end of the second week of exposure (40.9 µg/100 ml in the third week). The blood lead in females rose from 12.7 µg/100 ml to 30.4 µg/100 ml, the highest level being reached in the first part of the third week.

The protoporphyrin IX content of the erythrocytes showed no change in either the control or the exposed male group. However, in the female group, it showed a rise beginning in the third week and rising to 48.0 µg/100 ml erythrocytes. The findings were confirmed in a second experiment.

It is suggested that the increase of the erythrocyte protoporphyrin IX was a result of interference in the use of iron in the formation of haemoglobin. The synergism of lead exposure and iron deficiency might be suggested as being responsible for the increased response of FEP in females but this will have to be tested further in experimental and epidemiological work.

# 8. EFFECTS OF LEAD ON MAN—
# EPIDEMIOLOGICAL AND CLINICAL STUDIES

Two types of study characterizing the effects of lead on man have been reported:

— retrospective studies of the causes of mortality and morbidity in lead-exposed populations compared with unexposed populations, and

— studies of the effects of lead on specific organs and systems.

The findings from these two types of study will be considered separately. In both cases, the main objective will be to establish, as far as possible, the dose of, or exposure to lead which is associated with specified effects, and the frequency of such effects.

From the toxicological point of view, "the dose should be defined as the amount or concentration of a given chemical at the site of effect, i.e. where its presence leads to a given effect" (Nordberg ed., 1976). The application of this definition is difficult because the dose as defined above can rarely be measured directly and has to be estimated in various ways. In experiments, it is estimated from the amount injected or ingested or from dermal and other topical applications (using appropriate absorption factors and body distribution factors). In inhalation experiments it is estimated from the concentration as measured in air, the time of exposure, and the relevant deposition, retention, and absorption factors (if available). The same considerations apply for dose estimation from occupational exposure where, in addition to inhalation, the possible dermal exposure, ingestion during work-time, and exposure which workers are subject to as members of the general population, should be taken into account. The dose for the general population is estimated from inhalation of air, ingestion of food, water, and other beverages, and various other contacts, including drugs and consumer products, smoking, and in children, ingestion of soil, settled dust, and paint chips. A more direct way of estimating the dose is from measurements in body tissues and fluids such as blood, urine, faeces, sweat, or hair. Other organs, tissues, cells, and subcellular elements can be used for this purpose in animal experiments or in autopsy or biopsy material.

Although the biological effects of lead on man have been characterized in some detail, the precise doses of lead responsible for the effects are rarely, if ever, known. With all its acknowledged shortcomings, the Pb-B level is the vital link between exposure and an effect. In section 6, an effort was made to define, as far as possible, the relationship between the lead in air and in the diet and Pb-B levels. The main objective of this section is to establish the relationship between Pb-B levels and biological effects. Only in this way is it possible to estimate the possible biological consequences of specific levels of lead in environmental media.

Some biological effects of lead bear a close relationship to concurrent Pb-B levels, others do not. Thus, the degree of ALAD inhibition in peripheral blood rises and falls more or less concurrently with the Pb-B level, while some renal effects of lead are the consequence of an exposure to lead that may have occurred at a point remote in time and which is not reflected in the Pb-B level at the time the effect is first manifested clinically. The fidelity with which the Pb-B level reflects lead concentrations in target organs is subject to serious problems of analytical error as described in section 3.

Beyond these considerations, there is the additional problem of variation in the inherent susceptibility of individuals, and the influence of co-existent variables that may modify this susceptibility, such as nutritional status, age, and presence or absence of diseases such as alcoholism. For all the above reasons, the Pb-B level cannot be used as a reliable indication of dose or exposure in dealing with individual patients. They should be used only in assessing population group exposures at which effects may occur in a certain proportion of individuals.

Other tests for assessing dose have been proposed, e.g. lead excretion in response to chelating agents. Regardless of potential merits and special applications, most information relating health effects to dose has been obtained using Pb-B levels as an estimate or index of dose.

## 8.1 Retrospective Studies of Lead-exposed Populations

### 8.1.1 Epidemiology of lead poisoning in industry

In many countries there has been a considerable improvement over the past forty years with respect to hygienic conditions in the lead-using industries. The exposure of workers to lead was considerably higher before 1930 than after. In the United Kingdom, the number of reported cases of poisoning fell dramatically in the decade 1920–30 (Lane, 1964). Against this background, it is useful to consider the studies of Dingwall-Fordyce & Lane (1963). They found a higher than expected incidence of death due to cerebrovascular disease among men with past high lead exposure. The men studied retired from work between 1926 and 1960. All those studied had at least 25 years of service. Men in the heavy exposure category had an average urine lead concentration of 100–250 μg/litre[a] over the last 20 years

---

[a] 100 μg/litre corresponds to a Pb-B level of approximately 60 μg/100 ml and 250 μg/litre corresponds to a Pb-B level of approximately 120 μg/100 ml (Williams et al., 1969).

of employment. Men in the moderate exposure group had urine lead concentrations in the normal range. The third group had no exposure. As can be seen from Table 22, in the heavy exposure group deaths from cerebrovascular diseases (cerebral haemorrhage, thrombosis, and arteriosclerosis) were much higher than normal.

Table 22. Deaths from cerebrovascular disease in retired and employed workers from a lead industry[a]

| Status | Year of death | Grade of exposure | | | | | |
|--------|---------------|-------------------|--------------------|-----------------------|--------------------|-----------------------|--------------------|
| | | None | | Medium | | Heavy | |
| | | Expected incidence | Observed incidence | Expected incidence | Observed incidence | Expected incidence | Observed incidence |
| Retired | 1926–50 | 0.7 | 0 | 0.2 | 3 | 0.8 | 5 |
| | 1951–61 | 7.2 | 6 | 3.2 | 3 | 8.5 | 19 |
| | 1926–61 | 7.9 | 6 | 3.4 | 6 | 9.3 | 24[b] |
| Employed | 1946–61 | 3.2 | 3 | 3.1 | 3 | 5.6 | 9 |

[a] Adapted from Dingwall-Fordyce & Lane, 1963.
[b] $P < 0.001$.

The data also suggest that in this group the excessive death rate was most pronounced among men who retired prior to 1951 when exposure conditions were probably considerably worse than they were later. In the same study, it was found that the death rate from malignant neoplasms was not above the expected rate in any exposure grade. Unfortunately, the incidence of death due to chronic nephritis was not reported. A very similar survey was reported by Malcolm (1971) in which the subjects studied had, with few exceptions, been exposed to lead at moderate levels (average Pb-B level—65 µg/100 ml). There was no statistically significant excess mortality in any of the following disease categories: heart disease, chest disease, cerebrovascular accidents, cancer, renal disease, and "miscellaneous".

A recent American study is in general agreement with the conclusions of the British investigators concerning longevity and causes of death in the lead industries as they have operated over the last 25–30 years (Tabershaw & Cooper, 1974; Cooper & Gaffrey, 1975). The subjects were 1356 workers employed in the lead battery and smelter industries from 1946 to 1970. Both blood lead levels and urinary lead excretion were quite high. For example, 78.7% of 47 smelter workers had Pb-B levels of 80 µg/100 ml or more, from 1946 to 1961. The figure was still 13.5% after 1965 (489 total workers sampled). The percentage of battery workers with Pb-B levels above this was somewhat lower. But for all the various categories of duration of employment and type of work, 81.5–95.7% of the Pb-B levels were equal to or greater than 40 µg/100 ml. About 50% of the workers were employed for more than 10 years. The total mortality in this group was approximately the same as in the general population. The authors concluded that there was no evidence that work associated with lead increased the risk of death due to the major categories of cardiovascular and renal diseases. However, when chronic renal disease (chronic nephritis or other renal sclerosis) was segregated as a separate cause of death, there did appear to be a significant excess number of deaths. Thus, among smelter workers, the ratio of observed deaths to expected deaths was 7:2.8 and among the battery workers the ratio was 14:8.6. A similar association was found for a category of death classified as "other hypertensive disease": 7:1.9 among smelter workers and 13:6.3 among battery workers. For the two disease categories this adds up to 21 excess deaths out of 1267 for whom cause of death was listed. The authors emphasize that many of the workers in the study group were probably exposed to air lead concentrations considerably in excesss of 0.15 mg/m$^3$.

Table 23. Urine and blood lead content of persons in the Wenatchee study according to severity of exposure[a]

| Group | Urine lead content | | | Blood lead content | | |
|-------|--------------------|---|---|--------------------|---|---|
| | No. analyses | Average, µg/litre | S.D. µg/litre | No. analyses | Average, µg/100 ml | S.D. µg/100 ml |
| Low exposure | | | | | | |
| men | 146 | 35 | 21 | 148 | 26 | 11 |
| women | 123 | 28 | 19 | 124 | 26 | 10 |
| Intermediate exposure | | | | | | |
| men | 102 | 43 | 30 | 108 | 30 | 11 |
| women | 25 | 27 | 15 | 27 | 22 | 10 |
| High exposure | | | | | | |
| men | 386 | 88 | 60 | 329 | 44 | 16 |
| women | 61 | 46 | 25 | 58 | 34 | 13 |
| Children under 15 years | | | | | | |
| boys | 81 | 53 | 39 | 17 | 37 | 15 |
| girls | 65 | 54 | 40 | 14 | 36 | 10 |

[a] From Neal et al. (1941).

Although most epidemiological studies on occupational exposure have been carried out on industrial populations, one extensive study on orchard workers in the Wenatchee area of the state of Washington, has been reported (Neal et al., 1941). This study was somewhat complicated by the fact that exposure was to lead arsenate. In view of the known toxicity of arsenic, studies were included on the combined toxicities of lead and arsenic in animals. No synergism was found in these animal studies (Fairhall & Miller, 1941). The blood lead concentrations of the orchard workers and their families are summarized in Table 23.

This study may have been crude in comparison to some more recent ones, but it had the rather unique merit of examining health effects not only in men, but also in women and children. Furthermore, the exposure levels, as reflected in the urine and blood data of Table 23, were only slightly higher than the approximate upper limit for people living in highly polluted cities today. The study was concerned with weight, blood pressure, diseases of the cardiovascular system, skin disorders, eye irritation, chronic nervous diseases, blood dyscrasias, kidney diseases, neoplastic diseases, and fertility. There was no evidence, based on data available at the time, that the health profile of these people was any different from that of the general population.

In 1968, a follow-up study was undertaken of the people who had participated in the original study (Nelson et al., 1973). Over 97 % of the original participants were successfully traced. There had been 452 deaths among the 1231 original participants. A life table method of analysis of the standard mortality ratio was used. The overall mortality was less than the average for the state of Washington. The standard mortality ratios of exposed groups were not consistent with the exposure gradient. The mortality pattern for increasing duration of exposure was not consistent either.

### 8.1.2  Epidemiology of lead poisoning in the general adult population

Adequate studies of the relationship between lead exposure and health status in the general adult population have not been carried out. The limitations that apply to the epidemiological studies of occupational groups are magnified when applied to the general population. The range of exposure levels is smaller between sub-groups of the general adult population and their socioeconomic, physiological, and health profiles are probably more diverse.

### 8.1.3  Epidemiology of lead poisoning in infants and young children

There has been only one study reported of general mortality and disease-specific morbidity rate in children exposed to lead. The Wenatchee study referred to in section 8.1.1 included 146 children under the age of 15. As with the adults in this study, no abnormal pattern of disease incidence was noted. These children had moderately high lead exposure (see Table 23).

### 8.2  Clinical and Epidemiological Studies of the Effects of Lead on Specific Organs and Systems

In the following discussion of the effects of lead on various organs and systems, consideration will be given to dose–effect and dose–response relationships. The word "dose" as used here will refer to Pb-B levels, as described in the introductory remarks of this chapter.

The diversity of the effects of lead on haemoglobin formation and the complexity of the process itself make it difficult to determine which inhibitory effect is most sensitive and what is their relative importance at different levels of exposure (or dose).

Dose–effect refers to the relationship between dose and the intensity of a specified effect in an individual, e.g. Pb-B level *versus* percentage inhibition of blood ALAD.

Dose–response refers to the relationship between the dose and the proportion of a population showing a defined effect, specified as to the level of intensity, e.g. the proportion of a population showing more than 50 % inhibition of blood ALAD at a Pb-B of 20 µg/100 ml.

Some effects of lead are not graded, for example, the effects on the kidney and the central nervous system are usually reported in all-or-none terms, i.e. a certain proportion of individuals in a population are reported to have shown the effect at a given range of Pb-B concentrations. With many effects of lead it is difficult to specify a dose–response or a dose–effect relationship because the available data are inadequate.

### 8.2.1 Haemopoietic system

The evidence for disturbances in haem synthesis is clearly shown in man by the appearance of abnormal concentrations of haem precursors in blood and urine. The levels of lead exposure at which these various manifestations of disturbed haem synthesis first appear have been studied extensively in man. The sequence of reactions affected by lead, and the consequences thereof, are shown in Fig. 5.

Lead interferes with the biosynthesis of haem at several enzymatic steps, with the use of iron, and with globin synthesis in erythrocytes. Inhibition of ALAD and haem synthetase is well documented, and accumulation of the substrates of these enzymes (ALA and PP) is characteristic of human lead poisoning. Inhibition of ALAS is based on experimental evidence only. Whether there is enzymatic inhibition or whether other factors affect the conversion of coproporphyrinogen III (CPG) to protoporphyrin IX (PP) is not clear; nevertheless, increased urinary excretion of coproporphyrin III is prominent in human lead poisoning. Minor increases in porphobilinogen (PBG) and uroporphyrins in urine are occasionally reported in severe lead poisoning. Although the *in vivo* mechanisms are not clear, nonhaem iron (ferritin and iron micelles) accumulates in red blood cells with damaged mitochondria and other fragments not found in normal mature erythrocytes. Serum iron may be increased in persons with lead

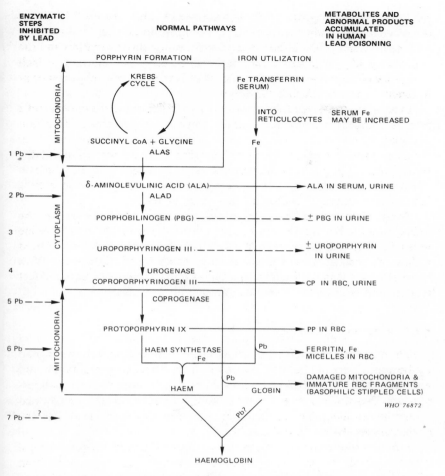

Fig. 3   Lead interference with the biosynthesis of haem (NAS-NRC 1972).

poisoning, but without iron-deficiency states. Globin synthesis in red blood cells is apparently impaired, although the mechanisms responsible for reduced globin synthesis remain unknown.

The evidence available suggests that mild anaemia with a small reduction in blood haemoglobin may occur at, or slightly above, dose levels that are associated with minimal increases in urinary excretion of ALA, (Tola et al., 1973).

Increased urinary excretion of ALA is accompanied by an elevation of the concentration in plasma in adults (Cramer et al., 1974) and in children (Chisolm, 1968a). This could indicate either an increased rate of ALA formation or a decrease in the rate of use of ALA. In view of the well-known inhibition of the enzyme ALAD, most authorities favour the view that elevated plasma levels reflect decreased use of ALA. The alternative possibility is that ALA formation is increased, presumably by increased formation or activity of the enzyme ALA-synthetase (ALAS). This may in fact be a significant factor. Berk et al. (1970) studied the rate of haem labelling in one case of lead poisoning with anaemia. They observed an increase in the rate of $^{14}$C-glycine incorporation into the "early labelled peak" of stercobilin, and into haemin, indicating an increased rate of haem synthesis in response to an anaemia due to increased erythrocyte destruction. Coproporphyrin (CP) and ALA excretion were both elevated. This indicates that haem biosynthesis may be increased in lead poisoning in spite of increased excretion of haem precursors.

It is also possible that the rate-limiting step in the pathogenesis of lead-induced anaemia may involve globin synthesis rather than haem synthesis. White & Harvey (1972) reported that the incorporation of $^{3}$H-leucine into $\alpha$- and $\beta$-chain globins of reticulocytes was differentially affected in a pair of 3-year-old twins with clinical lead poisoning accompanied by anaemia. The radioactivity associated with the different globin chains shifted systematically as the blood haemoglobin values of the children returned towards normal.

The major effects of lead on haemopoiesis that are readily measured in man, are on the rate of excretion of ALA or CP in the urine, on the concentration of PP in the blood, and on the degree of inhibition of ALAD in the blood. None have been evaluated in relation to the fidelity with which they reflect the actual amount of lead absorbed per unit time, but they have been evaluated extensively with reference to their correlation with the concentration of lead in the blood. The literature since 1955, concerned with these interrelationships, has been reviewed recently by Zeilhuis (1971).

### 8.2.1.1 $\delta$-aminolevulinic acid dehydratase (ALAD)

The effect of lead that most closely correlates with the concentration of lead in the blood is the inhibition of erythrocyte ALAD activity. Within the range of lead exposure encountered in the general population, the higher the concentration of lead in the blood, the lower the activity of the enzyme. Above this range, enzyme inhibition is almost complete and changes little with increasing dose. The relationship between Pb-B levels and ALAD activity was first reported by Makao et al. (1968) in a group of twelve men industrially exposed to lead. Later, Hernberg et al. (1970) reported on a much larger population of adults having a wide range of lead exposures. Granick et al. (1973) suggest the interesting possibility of correcting for individual variations in total ALAD by calculating the ratio of activity with, *versus* activity without, enzyme reactivation using dithiothreitol as a reactivator. This calculation presumably expresses the inhibitory activity of lead for the particular sample. The normalization procedure improved the correlation between ALAD and blood lead. They found that the average no-effect Pb-B level for inhibitory effects in children, using this correction procedure, was about 15 µg/100 ml. Tola (1973) reached a similar conclusion from his study of 1370 workers. His observations suggested that the average threshold was at a Pb-B level of 10–20 µg/100 ml. However, a recent study on the Finnish general population puts the existence of a no-effect level into some doubt. In their study, Nordman & Hernberg (1975) obtained a statistically significant correlation between ALAD activity and Pb-B values not exceeding 10 µg/100 ml (Pb-B mean value 8.4 µg/100 ml).

Based on data concerning male workers and children, Zielhuis (1975) calculated a dose–response relationship for over 40% and over 70% inhibition of ALAD (see Table 24).

Table 24. Percentage of adults and children with more than 40% and 70% inhibition of the mean ALAD activity found in control subjects with Pb-B <14 µg/100 ml[a]

| Pb-B level (µg/100 ml) | adults | | | children | | |
|---|---|---|---|---|---|---|
| | No. | >40% | >70% | No. | >40% | >70% |
| 14 | — | — | — | 9 | 11 | 0 |
| 15–24 | 30 | 13 | 3 | 37 | 73 | 8 |
| 25–34 | 26 | 62 | 12 | 24 | 88 | 13 |
| 35–44 | 32 | 97 | 22 | 10 | 90 | 50 |
| 45–54 | 53 | 100 | 68 | — | — | — |
| 55–64 | 37 | 100 | 92 | — | — | — |
| 65–74 | 43 | 100 | 95 | — | — | — |
| | 221 | | | 80 | | |

[a] From Zielhuis (1975).

### 8.2.1.2 *Free erythrocyte porphyrins (FEP)*

The most recently identified biochemical correlate of blood lead concentration is the erythrocyte protoporphyrin concentration. Some of the analytical methods in use (see section 2.2.3) measure the protoporphyrin IX concentration in erythrocytes, while others measure the free erythrocyte porphyrins, more than 90% of which, however, consists of protoporphyrin IX (Baloh, 1974). A correlation between FEP and Pb-B levels has been reported for industrial workers (Haeger-Aronsen, 1971). The dose–effect relationship is linear if log FEP is plotted against Pb-B. Two reports have appeared showing this relationship (Piomelli, 1973; Sassa et al., 1973). In both cases, the subjects were young children with a wide range of blood lead values. For the data reported by Sassa et al. (1973)

the correlation of the logarithm of the protoporphyrin IX values and the blood lead concentrations was fairly good ($r = 0.72$). When only the data for children having had a constant blood lead level for three months or longer were used, the correlation was much better ($r = 0.91$). The point was made by the authors that the elevation of erythrocyte protoporphyrin IX reflected an inhibitory effect of lead on haem synthesis that occurs in erythroid cells in the bone marrow, whereas the absorption of lead by blood elements takes place both in circulating cells and in erythroid cells.

Table 25. Percentage of adult female subjects with FEP levels that exceeded those found in control subjects with Pb-B $< 20$ µg/100ml.

| Pb-B level (µg/100 ml) | No. | % with FEP level higher than normal |
|---|---|---|
| 11–20 | 28 | 4 |
| 21–30 | 9 | 33 |
| 31–40 | 8 | 90 |
| 41–50 ⎫ 51–60 ⎬ 61–70 ⎭ | 4 | 100 |
| | 49 | |

[a] From: Zielhuis, 1975.

Table 26. Percentage of adult male subjects with FEP levels that exceeded those found in control subjects with Pb-B $< 20$ µg/100 ml.

| Pb-B level (µg/100 ml) | No. | % with FEP level higher than normal |
|---|---|---|
| 11–20 | 26 | 0 |
| 21–30 | 43 | 7 |
| 31–40 | 32 | 19 |
| 41–50 | 4 | |
| 51–60 | 2 | 100 |
| 61–70 | 2 | |
| | 109 | |

[a] From: Zielhuis, 1975.

In recent years, it has become evident that the increase of FEP occurs at lower Pb-B levels than the increase in ALA in the urine (Stuik, 1974; Roels et al., 1975). In addition, the same authors observed that women were more sensitive than men with regard to the effect of lead on erythrocyte protoporphyrin IX. In women the effect was evident at a lower Pb-B level than in men, and the rate of increase in erythrocyte protoporphyrin IX with increasing Pb-B was greater than in men. From the results of a recent preliminary survey, children appear to display an FEP response to lead resembling that of women (Roels et al., 1975). Based on these limited data, for 109 men, 49 women, and for 219 children, Zielhuis (1975) calculated the dose–response relationship (see Tables 25, 26, and 27).

### 8.2.1.3 δ-aminolevulinic acid excretion in urine (ALA-U)

The rate of ALA excretion in urine has long been used as a measure of a biological effect of lead. The most recent studies of this relationship in industrially exposed subjects indicate that the logarithm of the ALA concentration in urine increases linearly with Pb-B levels from 40 µg/100 ml (Selander & Cramer, 1970; Haeger-Aronsen, 1971; Soliman et al., 1973). Chisolm (1973) reported a good correlation in children of log ALA excreted in urine per 24 hours per m² of body surface and Pb-B levels

Table 27. Percentage of children with FEP levels that exceeded those found in control subjects with Pb-B < 20 µg/100 ml.

| Pb-B level (µg/100 ml) | No. | % with FEP level higher than normal |
|---|---|---|
| 20 | 87 | 5 |
| 21–30 | 72 | 21 |
| 31–40 | 24 | 29 |
| 41–50 | 14 ⎫ | |
| 51–60 | 12 ⎬ | 64 |
| 61–70 | 10 ⎭ | |
| | 219 | |

ª From: Zielhuis, 1975.

Table 28. Percentage of male adults with ALA-U levels > 5 mg/litre and > 10 mg/litre according to Pb-B level

| Pb-B level (µg/100 ml) | No. | ALA-U level (mg/litre) | |
|---|---|---|---|
| | | > 5 | > 10 |
| 11–20 | 17 | 0 | 0 |
| 21–30 | 27 | 0 | 0 |
| 31–40 | 36 | 14 | 3 |
| 41–50 | 55 | 33 | 11 |
| 51–60 | 38 | 74 | 37 |
| 61–70 | 34 | 88 | 50 |
| | 207 | | |

ª From: Zielhuis, 1975.

over a wide range of blood lead values. In occupational exposure, the excretion of ALA in urine, at a given Pb-B level was higher in women than in men (Roels et al., 1975).

Using diagrams published by Haeger-Aronsen (1971) and by Selander & Cramer (1970) for 207 adult males, Zielhuis (1975) calculated the dose–response relationships for levels of ALA excretion greater than 5 mg/litre and greater than 10 mg/litre (see Table 28). Some of the dose–response relationships shown in Tables 24–28 are illustrated in Fig. 4.

#### 8.2.1.4 *Coproporphyrin excretion in urine (CP-U)*

Although there is some uncertainty, ALA-U is probably somewhat more sensitive to the effects of lead exposure than CP-U (Haeger-Aronsen, 1960; Djurić et al., 1966). ALA-U is also more lead-specific than CP-U. Data are insufficient for estimating dose–response relationships.

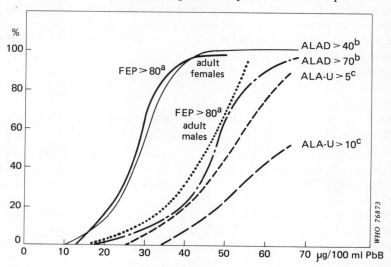

Fig. 4   Dose response relationships for some effects of lead. a = µg/100 ml of erythrocytes; b = % inhibition; c = mg/litre.

#### 8.2.1.5 *Effects of lead on cell morphology*

Punctate basophilia occurs in lead poisoning, but a quantitative relationship between the number of stippled cells and Pb-B levels is not to be expected (Zielhuis, 1971). Too many variables are involved in the preparation of smears. The same is probably true of reticulocyte counts.

#### 8.2.1.6 *Effects of lead on erythrocyte survival*

Increased rate of erythrocyte breakdown (decreased erythrocyte life) is often, but not consistently, seen in cases of anaemia due to lead poisoning. When erythrocytes are exposed to lead *in vitro*, they exhibit increased osmotic resistance and increased mechanical fragility (Waldron, 1966). They also show inhibition of Na-K-ATPase with increased loss of intracellular potassium (Hasan & Hernberg, 1966; Secchi et al., 1973). These effects have been cited to explain the fact that in many instances the anaemia in lead poisoning is accompanied by a shortening of the erythrocyte life span. It is presumed that one or more of these effects is responsible for the sensitivity of erythrocytes to spontaneous haemolysis. Erythrocyte survival time was reduced on the average by 20% in 17 occupationally-exposed workers, only 3 of whom showed clinical signs of poisoning (Hernberg, 1967). The author postulated that shortened cell life was due to the loss of membrane integrity secondary to Na-K-ATPase inhibition. Anaemia does not necessarily accompany a shortened red cell life span, and the correlation between blood haemoglobin and life span was not good in this particular study. The kinetics of disappearance of labelled cells indicated a shortening of life span by increased random destruction of cells of all ages. Leikin & Eng (1963) determined erythrocyte survival in 7 cases of lead poisoning in children. In 3 cases the erythrocyte survival time was shortened. All patients were mildly to moderately anaemic. It would seem from these and other studies that the anaemia in lead poisoning cannot be explained solely on the basis of reduced erythrocyte survival time.

### 8.2.1.7  *Effects of lead on haem synthesis*

The two general points of attack that have been identified are on haem synthesis and on globin synthesis. Of the two, the effects on haem synthesis are better understood. It is generally recognized, too, that manifestations of disturbed haem synthesis often occur in the absence of frank anaemia. These disturbances may also be significant for the numerous other haem-dependent enzymatic reactions essential for normal body functions. Thus, cytochromes, cytochrome *c* oxidase (EC 1.9.3.1), and hydroperoxidases are all part of electron transfer systems requiring haem.

Little is known about the effects of lead on the formation or activity of other haem-containing compounds. It has been reported that treatment with EDTA reversed the prolonged antipyrine half-life seen in two cases of clinical lead poisoning (Alvares et al., 1975). The authors suggested that in these cases, lead may have significantly inhibited the synthesis of cyto-chrome P-450.

### 8.2.1.7  *Relationship between lead exposure and anaemia*

It is well known that anaemia is a characteristic early toxic effect of lead in man. The Pb-B threshold level for this effect is still not certain. Williams (1966) reported that anaemia did not occur in industrial workers with Pb-B levels below 110 µg/100 ml. Cooper et al. (1973) reported that the average haemoglobin level (Hb) was not decreased at Pb-B levels of up to 100 µg/100 ml and Sakurai et al. (1974) did not observe any decrease of Hb or erythrocyte concentrations in workers at Pb-B levels of up to 50 µg/100 ml. On the other hand, Tola et al. (1973) reported a slight effect of lead on Hb at an average Pb-B level of about 50 µg/100 ml. This conclusion was drawn from analysis of the sequential change in Hb among workers newly introduced into an "industrial lead environment". This approach to the analysis of the effect of lead on Hb is certainly more sensitive for detecting an interaction between Pb-B levels and Hb than is a single Hb determination in a population of lead-exposed persons. Allowance must, however, be made for the possibility that sequential change in Hb may be due to seasonal effects independent of lead exposure (Coulthard, 1958).

Children appear to be more sensitive to lead anaemia than adults. Thus, Betts et al. (1973) found a significant negative correlation between Hb and Pb-B levels; a decrease in Hb was evident in 36% of children with Pb-B levels from 37 to 60 µg/100 ml, compared with only 14% in children with Pb-B levels less than 37 µg/100 ml. Pueschel et al. (1972) observed a curvilinear decrease in Hb between Pb-B levels of 40 and 130 µg/100 ml in children between 1 and 6 years old. On the other hand, McNeil & Ptaznik (1975) found no anaemia in children with Pb-B levels considerably higher than 40 µg/100 ml. Nutritional differences may explain the discrepancy. But this does not invalidate the proposition that for some groups of children a reduction in Hb may occur at a Pb-B level of approximately 40 µg/100 ml.

### 8.2.2 Nervous system

#### 8.2.2.1 *Central nervous system*

*Inorganic lead compounds.* The effects of lead on the nervous system vary with the duration and intensity of exposure. Distinction must also be made between the effects on the central nervous system and the effects on peripheral nerves. Further questions have been raised concerning the inherent differences in the sensitivity of the nervous system of adults and the nervous system of infants and young children. There is no doubt that lead effects on the brain are much more commonly associated with childhood lead poisoning than with poisoning as it is seen in adults. But it is also possible that these differences are related to the intensity of exposure at the time the cases are identified rather than to any difference in inherent sensitivity.

With chronic lead exposure, striking effects may occur referred to as lead encephalopathy. There are numerous detailed descriptions of adult lead encephalopathy (Crutcher, 1963; Whitfield et al., 1972; Teisinger & Styblova, 1961; Aub et al., 1926; Cantarow & Trumper, 1944). The major features are dullness, restlessness, irritability, headaches, muscular tremor, hallucinations, and loss of memory and ability to concentrate. These signs and symptoms may progress to delirium, mania, convulsions, paralysis, and coma. The signs and symptoms of encephalopathy in infants and young children are quite similar to those reported to occur in adults.

The brain lesions in fatal cases of lead poisoning are cerebral oedema and changes in cerebral blood vessels. The normal convolutions of the cerebral hemispheres are often obliterated. Capillary endothelial cells are usually swollen (Pentschew, 1965). Extravasation of red blood cells and perivascular haemorrhage occur rather commonly and patchy neuronal loss, serous exudate, glial proliferation, and occasional areas of demyelinization are all characteristic of lead poisoning (Blackman, 1937; Okazaki et al., 1963; Whitfield et al., 1972). But not all deaths due to lead encephalopathy are accompanied by histological lesions of the central nervous system (Pentschew, 1965).

Neurological sequelae can occur in severe or repeated episodes of lead encephalopathy. The sequelae are no different qualitatively from those that occur following traumatic or infectious cerebral injury. The occurrence of permanent sequelae seems to be much more common among young children than among adults. Approximately one-fourth of the children who survived an attack of acute lead encephalopathy sustained permanent sequelae (Byers, 1959; Chisolm & Harrison, 1956; Smith, 1964). At least this was true prior to the introduction of current therapeutic practices such as those described by Chisolm (1968a). The incidence of sequelae appears to have been substantially reduced in recent years, but central nervous system sequelae may still occur if therapy is initiated only after the onset of encephalopathy (Chisolm, 1973). The most severe sequelae are cortical atrophy, hydrocephalus, convulsive seizures, and idiocy. More commonly, the sequelae are of a more subtle nature. Learning ability may be impaired due to motor incoordination, lack of sensory perception, or inability to concentrate. Such subtle disturbances have also been claimed to occur in children with high lead exposure, but in the absence of a history of encephalopathy (Byers & Lord, 1943; Cohen & Ahrens, 1959).

The major concern today is that young children with elevated lead exposure, as reflected in Pb-B levels of 40–80 µg/100 ml, may be experiencing subtle neurological damage without ever exhibiting classical signs of lead encephalopathy. Studies have been reported of the neurological status of children with Pb-B values in this range. In view of the possible long-term effects of lead on the brain, association between Pb-B and neurological status at the time of evaluation may give a false impression concerning the level of lead exposure when the damage was initiated. Exposure levels at the time of examination may be lower than at the time toxic effects occurred. Thus, the Pb-B level–effect association may underestimate the dose responsible for the effect.

Burdé de la et al. (1972) and Peuschel et al. (1972) observed dysfunction of the central nervous system (irritability, clumsiness, fine motor dysfunction, impaired concept formation, etc.) in 70 and 58 children, respectively, whose Pb-B levels were always, in all cases, above 40 µg/100 ml. Albert et al. (1974) studied the psychological profiles and educational performances of children, 5–15 years of age, who had histories of lead exposure early in childhood. Those who had been treated for lead poisoning, with or without encephalopathy, exhibited a higher incidence of diagnosed mental disorders and of poor school performance than those who had no such history, even when their history showed elevated lead exposure early in childhood.

Kotok (1972) established that development deficiencies (using the Denver Development Screening test, which, according to the author is a somewhat insensitive measure of development) in a group of asymptomatic children with elevated lead levels (58–137 µg/100 ml) were identical to those of a control group similar in age, sex, ethnic group, environment, neonatal condition, and presence of pica, but whose Pb-B levels were lower (20–55 µg/100 ml). The deficiencies could be correlated with inadequacies in the children's environment. Klein et al. (1974) pointed out that in many studies, pica is not used as a controlled variable. In his view, pica may be part of a behavioural deficiency syndrome. In such a case the child would have the behavioural deficiency regardless of whether or not he ingested lead-containing objects. Indeed, there is evidence that among mentally subnormal children whose mental deficiency is unrelated to excessive lead absorption there is a high incidence of both pica and of moderately elevated Pb-B levels (Bicknell et al., 1968). In this study, 67 % of the children, whose subnormal state antedated pica, had Pb-B levels from 39 to 88 µg/100 ml, with a mean of 48 µg/100 ml. By contrast, among the subnormal group without pica all but one had a Pb-B level of less than 36 µg/100 ml. The study did not exclude the possibility that an excessive lead exposure could have aggravated the pre-existent subnormal state.

Recently McNeil & Ptasnik (1975) published an initial evaluation of the long-term effects of elevated Pb-B levels in asymptomatic children, living in El Paso, USA. In 138 out of 206 children aged from 21 months to 18 years (median 9 years), who volunteered (possibility of selection) to participate, the authors could not find any evidence of non-specific complaints, hyperactivity, or of abnormal psychometric testing values, if compared with a matched control group. There existed a significant difference in one personality test; however this was explained by geographic isolation and other factors and not by lead exposure. The average Pb-B levels were, respectively, 50 µg/100 ml (range 14–93) and 16 µg/100 ml (range 10–28).

More recently another psychological evaluation of the El Paso subjects was published by Landrigan et al. (1975a). Forty-six children, aged from 3 to 15 years, with Pb-B levels of 40–60 µg/100 ml were compared with 78 ethnically and socioeconomically similar controls with Pb-B levels below 40 µg/100 ml. The "Wechsler Intelligence Scale" showed that the age adjusted I.Q. was significantly lower in the first group. In addition, the lead exposed group also showed a significant slowing in the finger-wrist tapping test. The full-scale I.Q., verbal I.Q., and the behavioural and hyperactivity ratings did not differ. In this study, unfortunately, there were differences in age and sex between the study and control group which might account for the positive findings. It seems therefore that we have two studies of this situation that come to different conclusions regarding the possible effects of lead on neurological and psychological functions.

Another approach has been to identify children with neurological or behavioural disorders of obscure etiology and to determine whether they show evidence of current or past elevated lead exposure (David et al., 1972; Moncrieff et al., 1964; Gibson et al., 1967).

The work of David et al. (1972) is of particular interest because the neurological abnormality described was one that was reproduced experimentally in animals (see section 7.1.2). These workers reported occurrence of hyperactivity among children who had essentially normal blood lead concentrations, but who excreted abnormally large amounts of lead when treated with penicillamine. The children had no history of earlier lead encephalopathy. This study has been criticized because of statistical inadequacies (Bullpitt, 1972).

Lansdown et al. (1974) examined a population of schoolchildren in London (less than 17 years of age); there was no relationship between Pb-B levels and intelligence (Wechsler test), reading (Burt test), and behaviour (e.g. hyperactivity as rated by the teachers). The authors suggested that social factors were more important than exposure to lead in determining mental development. The design of the study has also been criticized. Neither Landsown's nor David's study are conclusive.

Morgan & Repko (1974) reported preliminary results of an extensive study of behavioural functions in 190 lead-exposed workers (Pb-B = 60.48 ± 16.96 µg/100 ml). In 68 % of the subjects the Pb-B level was less than 80 µg/100 ml. The majority of the subjects were exposed for between 5 and 20 years. The authors examined 36 non-independent measures of general performance. In addition, 44 measures of sensory, psychomotor, and psychological functions were obtained. Preliminary analysis suggested that Pb-B levels correlated with several reaction-time measures and ALAD correlated with measures from strength-endurance-recovery tasks. Both Pb-B levels and ALAD correlated with eye-hand co-ordination. This study, therefore, suggested that below a Pb-B level of 80 µg/100 ml some behavioural changes did occur in adult workers. In addition, variability of performance increased with increasing Pb-B levels. Only during periods of high-demand performance did a worker's capacity decrease due to lead exposure. The authors themselves stressed that this preliminary analysis still has to be confirmed by further work.

*Alkyllead compounds.* The encephalopathy of alkyllead intoxication is somewhat different from that due to inorganic lead exposure. In documented adult cases of poisoning the most frequent findings suggest a psychiatric problem. Hallucinations, tremor, delirium, insomnia, delusions, headaches, and violent mood swings are the most commonly reported symptoms (Boyd et al., 1957; Machle, 1935). The course of the intoxication runs from 1 to 10 weeks. Although alkyllead compounds are notorious for their high lethality, recovery is fairly complete among survivors (Akatsuka, 1973). Convulsions and coma apparently occur only in the most severe cases. There is insufficient information to establish dose–effect and dose–response relationships.

### 8.2.2.2 Peripheral nervous system

Inorganic lead has toxic effects on the peripheral nervous system. The older lead literature cites the frequent occurrence of lead palsy in occupational exposure to lead. The manifestations are mainly weakness of the extensor muscles, particularly those used most heavily. While motor function is mainly affected, hyperaesthesia, analgesia, and anaesthesia of affected areas have also been reported.

Catton et al. (1970) found evidence of reduced nerve conduction velocity in about one-third of a group of 19 occupationally-exposed men of whom only one showed any other overt signs of lead toxicity.

The most prominent finding of Seppäläinen & Hernberg (1972) in lead workers (Pb-B levels 80–120 µg/100 ml) without any clinical neurological signs was reduced motor conduction velocity of the slower fibres of the ulnar nerves; electromyographic changes included a diminished number of motor units on maximum contraction and fibrillations. Similar although less pronounced effects were reported by Seppäläinen et al. (1975) in 26 workers whose Pb-B levels had never exceeded 70 µg/100 ml (exposure time 13 months–17 years). Furthermore, in lead workers with Pb-B levels of 2–73 µg/100 ml, Araki & Honma (1976) reported statistically significant negative correlations between nerve conduction velocity and Pb-B, ALAD, and lead mobilization test values, respectively. More recently, Seppäläinen et al. (unpublished results[a]) reported a dose-response relationship between abnormally low conduction velocities, defined as values 2 standard deviations below the mean of an unexposed reference group, and the highest Pb-B recorded during employment (2–20 years). The results indicate that nerve conduction impairment is induced in some workers at Pb-B's exceeding 50 µg/100 ml.

---

[a] Reported at the Second International Workshop Permissible Levels for Occupational Exposure to Inorganic Lead, 21–23 September 1976. University of Amsterdam, The Netherlands. To be published shortly in *Int. Arch. Occup. Health.*

### 8.2.3 Renal system

The effects of lead on the kidney have been studied extensively. Two general types of effect have been described. The first is rather clear-cut renal tubular damage characterized by generalized aminoaciduria, hypophosphataemia with relative hyperphosphaturia, and glycosuria, which has been studied in some detail in children with clinical lead poisoning (Chisolm, 1962). The condition is characterized by decreased tubular reabsorption of glucose and α-amino acids and therefore reflects proximal tubular damage. Aminoaciduria was seen more consistently in Chisolm's studies than the other two manifestations of tubular damage. Thus, the amino acid transport system is probably more sensitive to the toxic actions of lead than the transport systems for glucose and phosphate. Limited data indicate that aminoaciduria is terminated by chelation (Chisolm, 1968b).

In a group of children with slight lead-related neurological signs, generalized aminoaciduria was found in 8/43 children with Pb-B levels of 40–120 µg/100 ml (Pueschel et al., 1972). A similar renal tubular syndrome has been reported to occur in industrially exposed adults (Clarkson & Kench, 1956; Goyer et al., 1972). In neither of these studies were Pb-B levels reported. However, Clarkson & Kench observed signs of lead poisoning (colic and punctate basophilia) in conjunction with aminoaciduria.

In a group of 7 carefully studied lead-exposed workers, aminoaciduria was not present. Inulin clearance and renal blood flow were also normal at the time of examination. For these cases, the average Pb-B level was 100 µg/100 ml and the minimum was 71 µg/100 ml. These workers had been exposed for up to 20 years (Cramer et al., 1974). All had markedly elevated urinary ALA excretion. Interestingly, some of these workers with prolonged exposure had diffuse interstitial and peritubular fibrosis as determined by renal biopsy. These pathological findings are associated with quite a different kind of renal effect which is seen with prolonged lead exposure. It is commonly referred to as chronic lead nephropathy. Chronic nephropathy is characterized by slow development of contracted kidneys with arteriosclerotic changes, interstitial fibrosis, glomerular atrophy, and hyaline degeneration of the vessels. This progressive disease sometimes ends in renal failure. There is evidence that it occurs in industrially exposed workers, in long-term drinkers of lead-contaminated whisky, and among middle-aged people who had developed clinical lead poisoning much earlier in life. Currently, it is only rarely encountered in occupational exposure.

This renal syndrome can develop and progress to renal failure long after abnormal lead exposure has terminated. As early as 1897, it was noted that deaths from chronic nephritis were much more frequent among people under 30 years of age in Queensland than in other sections of Australia. The first serious attempt to document a suspected relationship to earlier childhood lead poisoning was reported by Nye (1929). Further evidence of a causal relationship between chronic nephropathy and childhood lead exposure was provided later (Henderson, 1958). It was shown that people dying of chronic nephropathy in Queensland usually had a high concentration of lead in their bones (Henderson & Inglis, 1957). Emmerson (1963) later demonstrated abnormally elevated lead excretion in response to EDTA among surviving middle-aged cases of chronic nephropathy. Tepper (1963), however, was unable to find evidence of chronic nephropathy among young American adults with a history of childhood lead poisoning. The Americans had probably been exposed for a much shorter period of time than the Australians. Other unknown factors may also have played a role.

The Australian cases involved childhood exposure with an apparent latency of 10–30 years for the development of renal insufficiency. But there is evidence that the same effect can result from continuous, prolonged high lead exposure among adults (Lilis et al., 1968; Richet et al., 1966; Danilović, 1958; Morgan et al., 1966; Albahary et al., 1965; Albahary, 1964). In these cases, lead exposure was higher than is commonly encountered in industry today.

In a series of 102 cases of lead poisoning studied by Lilis et al. (1968), 18 cases of clinically verified chronic nephropathy were found. For the whole series, the mean Pb-B level was approximately 80 µg/100 ml with a range of 42–141 µg/100 ml. Nephropathy was more common among patients who had been exposed to lead for more than 10 years than among those who had been exposed for less than 10 years.

In the Danilović (1958) study 7/23 cases had Pb-B levels of about 100–200 µg/100 ml. In the studies of Albahary et al. (1965) Pb-B levels were not reported. But exposure levels must have been quite high since the mean ALA excretion was about 37 mg/24 h for 29 workers.

It seems likely, from all available evidence, that a prolonged high-level lead exposure is necessary, even in childhood, to produce this progressive chronic nephropathy.

One interesting feature of this syndrome of chronic renal insufficiency is the frequent association with gout (Emmerson, 1963; Morgan et al., 1966). Although uric acid excretion is largely dependent upon tubular secretion, it is not at all certain that tubular secretion is inhibited. As a matter of fact, a study by Emmerson et al. (1971) of 13 cases of renal insufficiency due to lead nephropathy failed to reveal any alteration in uric acid secretion. The authors suggested an increased tubular reabsorption to account for the observed decreased clearance of uric acid.

In summary, proximal tubular effects can occur in children and adults with subtle signs of lead poisoning.

Prolonged exposure to lead leading to a Pb-B level of more than 70 µg/100 ml may give rise to chronic irreversible nephropathy. However, little is known about dose–effect relationships or about time–effect relationships for lead-induced chronic interstitial nephritis.

### 8.2.4  Gastrointestinal tract

As a symptom of lead poisoning, colic is a fairly consistent early warning of potentially more serious effects likely to occur with prolonged periods of exposure. It is most commonly encountered in industrial exposure. But it is probably also common in lead-poisoned infants and young children. The occurrence of colic at relatively low exposure levels in industry is well-known. Although it has been reported that 13/64 industrially exposed men with presumably lead-related colic and constipation had blood lead levels from somewhat less than 40 µg to 80 µg/100 ml (Beritić, 1971), it was also reported that in every case the diagnosis of lead colic was confirmed by the findings of high CP-U, excessive basophilic stippling, reticulocytosis, and various degrees of anaemia. This is consistent with the general observation that lead colic seems to be accompanied by other signs of poisoning. There are not enough data available to establish a dose–response relationship for this lead effect.

### 8.2.5  Liver

There is no definite evidence for the effects of lead on the liver. Dodić et al. (1971) reported signs of impaired liver function in 11 out of 91 patients hospitalized for lead poisoning. Liver damage was more frequent in cases of severe lead poisoning in 7 out of 18 patients. However, the authors did not provide any information on Pb-B levels or on indices of disturbed porphyrin metabolism which would enable the assessment of the stage of lead poisoning. In a laboratory study of 301 workers in lead smelting and refining, Cooper et al. (1973) found 11.5 % increased aspartate aminotransferase (EC 2.6.1.1), (SGOT)[a] values (above 50 U/litre[b]) in subjects with a Pb-B level below 70 µg/100 ml, 20 % in those with a Pb-B level of about 70 µg/100 ml, and 50 % in workers with a Pb-B level above 100 µg/100 ml. The correlation between Pb-B levels and SGOT values was statistically significant. However, in the absence of information on the possible

---

[a] Formerly known as serum glutamic oxaloacetic transaminase.
[b] $= 50 \times 1.67 \times 10^{-5}$ mol/(m$^3$.s)

influence of diet, infections, or personal habits, the authors did not draw any definite conclusions concerning the etiology of these changes.

### 8.2.6 Cardiovascular system

Increased capillary permeability occurs in acute lead encephalopathy (section 8.2.2.1). Under conditions of long-term lead exposure at high levels, arteriosclerotic changes have been demonstrated in the kidney (section 8.2.3). Dingwell-Fordyce & Lane (1963) reported a marked increase in the cerebrovascular mortality rate as compared with the expected rate among heavily exposed lead workers (section 8.1.1). This observation applied to men exposed to lead during the first quarter of this century, when working conditions were quite bad. There was no similar increase in the mortality rate for men employed more recently. Hypertension is an important element in the etiology of cerebrovascular deaths. Cramer & Dahlberg (1966) studied the incidence of hypertension in a population of 364 industrially-exposed men, 273 of whom had a long-term exposure to lead. They subdivided these workers into "lead affected" and "non-lead-affected" groups, on the basis of the urinary coproporphyrin test. There was no statistically significant difference between the groups. Nor was the incidence higher than expected for non-exposed men in Sweden. This is contrary to the earlier findings of Vigdortchik (1935) and to the observations of Monaenkova & Glotova (1969). The disparity may have been due to differences in lead exposure. Other reports on the question do not show hypertension to be unduly prevalent among workers exposed to lead (Dressen et al., 1941; Lane, 1949). It is not clear whether vascular effects of lead in man are the result of an action on blood vessels directly, or whether the effects are secondary to renal effects.

There is a good evidence that signs of clinical lead poisoning sometimes include evidence of a toxic action on the heart. Cases have been described in adults and in children, always with clinical signs of poisoning. There is of course the possibility that the coexistence of lead poisoning and myocarditis is coincidental. But in many cases the electrocardiographic abnormalities disappeared with chelation therapy, suggesting that lead may have been the original etiological factor (Myerson & Eisenhauer, 1963; Silver & Rodriguez-Torres, 1968; Freeman, 1965). In a review of 5 fatal cases of lead poisoning in young children, heart failure was concluded to be the proximate cause of death in 2 cases (Kline, 1960). Kosmider & Petelenz (1962) examined 38 adults over 46 years of age with chronic lead poisoning. They found that 66 % had electrocardiographic changes, which was four times the expected rate for that age group. Orlova (1954) also reported electrocardiographic abnormalities in cases of lead poisoning. Dimitrova

(1972) reported cardiac abnormalities in workers with undefined degrees of lead intoxication. There was a correlation of urinary excretion of lead with duration of systolic contraction and with isometric tension. Lead mobilization by EDTA accentuated these effects on the heart. No dose–effect relationships are apparent from the limited data available.

## 8.2.7 Reproduction

There is no epidemiological evidence of an effect of lead on the fertility of women or on *in utero* fetal development, but there are numerous reports in the older literature of stillbirths and miscarriages among women working in the lead trade (Cantarow & Trumper, 1944; Oliver, 1914). These reports probably contributed to the promulgation of legislation forbidding the employment of women in the lead trades in many countries. Panova (1972) reported that women working in lead industries had a higher incidence, compared with a control group, of ovulatory dysfunction—mainly anovulatory cycles and cycles with luteal abnormality. A relationship was reported between ALA-U and the incidence of anovulatory cycles. The effect was seen at 8–10 mg ALA/litre of urine.

There are not any reliable data to indicate that infertility in women results from exposure of the male partner to lead.

Some of the early reports on lead poisoning (Oliver, 1914) suggested that reproductive failures such as sterility and miscarriages occurred even among the non-working wives of industrially-exposed men. The reproductive capability of 150 occupationally exposed men was recently studied by Lancranjan et al. (1975). The results indicated that both lead poisoning and moderately increased lead absorption decreased the fertility of men. An increased frequency of asthenospermia, hypospermia, and teratospermia was found. No interference with the hypothalamopituitary axis was demonstrated; thus, hypofertility was thought to be due to the toxic effect of lead on the gonads.

## 8.2.8 Endocrine organs

Impairment of thyroid function and of adrenal function has been reported in cases of lead poisoning (Monaenkova, 1957; Sandstead et al., 1969; Sandstead et al., 1970; Pines, 1965).

There is some evidence suggesting that lead may cause a derangement of tryptophan metabolism. This is based on the observation that urinary excretion of 5-hydroxyindoleacetic acid was increased in 227 children living near a lead smelter (Ghelberg, 1966). Unfortunately, the 5-hydroxyindoleacetic acid determinations were not quantitative. Furthermore, blood lead values or other indices of exposure were not determined. Urbanowicz et al. (1969) noted a rise in 5-hydroxyindoleacetic acid excretion in workers heavily exposed to lead (ALA-U—33.7 mg/litre of urine). The rise preceded the rise in ALA-U and CP-U. Dugandžić et al. (1973) also noted a rise in 5-hydroxyindoleacetic acid excretion in moderately exposed workers (ALA-U—$28.2 \pm 22.6$ mg/litre of urine). More recently Schiele et al. (1974a), using another analytical method, reported that they were unable to find any significant elevation in 5-hydroxyindoleacetic acid excretion in workers with relatively high blood lead levels ($88.5 \pm 16.1$ µg/100 ml).

### 8.2.9 Carcinogenicity

Dingwall-Fordyce & Lane (1963) did not find any evidence of an increased incidence of malignant diseases in their follow-up study of 267 workers (section 8.1.1).

In a more recent study of the causes of mortality among lead smelter and lead battery workers, it was concluded that while the incidence of malignant neoplasms was somewhat greater than expected, the difference was not statistically significant (Tabershaw & Cooper, 1974; Cooper & Gaffey, 1975). This seems to support the conclusion of a IARC Working Group that there is no evidence to suggest that exposure to lead salts causes cancer of any site in man (IARC, 1972).

### 8.2.10 Effects on chromosomes

The literature is controversial as regards chromosomal abnormalities induced by exposure to lead. On the one hand, chromosomal aberrations have been reported to result from lead exposure corresponding to mean Pb-B values of 38–75 µg/100 ml in various groups studied (Forni & Secchi, 1973; Schwanitz et al., 1970). Moreover, Deknudt et al. (1973) reported chromosomal aberrations in a group of 14 male workers with signs of lead poisoning. The authors concluded that, although the workers were exposed to zinc and cadmium as well as lead, the lead ought to be considered responsible for the aberrations. On the other hand, Schwanitz et al. (1975) were not able to corroborate their own findings among occupationally exposed workers and O'Riordan & Evans (1974) did not find any significant increase in chromosomal aberrations in shipbreakers with Pb-B values ranging from 40 to over 120 µg/100 ml. Schmid et al. (1972) did not find any evidence of lead-induced chromosome aberrations in a study on human peripheral lymphocytes *in vivo* and *in vitro*; furthermore, Bauchinger et al. (1972) did not find any abnormalities in the chromosomes of policemen with elevated Pb-B levels.

In a recent report, Bauchinger et al. (1976) found that chromosomal aberrations were significantly increased in a group of 24 male workers occupied in zinc electrolysis and exposed to zinc, lead, and cadmium. The workers had clearly elevated Pb-B and blood cadmium levels in comparison with a control group. The authors pointed out the similarity between this group and the group studied by Deknudt et al. (1973) as regards combined exposure. However, referring to studies indicating mutagenicity of cadmium (Oehlkers, 1953; Shiraishi et al., 1972; Shiraishi, 1975), Bauchinger and his colleagues were inclined to consider cadmium as being mainly responsible for the aberrations. They also emphasized the possibility of a synergistic effect of several metals on the chromosomes. Thus, the question as to whether chromosomal abnormalities occur as a result of lead exposure in man remains open. Furthermore, the human health significance of chromosomal abnormalities seen in lymphocyte cultures, as observed in some of these studies, is not yet known.

### 8.2.11   Teratogenicity

There is practically no information in the literature to suggest that lead is teratogenic for man (Wilson, 1973). Only one case has been reported of neuromuscular abnormalities and failure to grow in a child attributed to lead poisoning as a result of the consumption by the pregnant mother of illicit whisky (Palmisano et al., 1969).

### 8.3   Factors influencing Lead Toxicity

#### 8.3.1   Acquisition of tolerance to lead

Experience in industry does not suggest that, with continuous lead exposure, the human body becomes less reactive to lead. There have been two studies in which the biochemical parameters of lead exposure were followed for a long period after the initiation of industrial lead exposure. Tola et al. (1973) found that erythrocyte ALAD fell to a stable level in about 21 days, as the concentration of lead in the blood increased correspondingly. Then both blood lead and blood ALAD remained essentially stable for the next three months. There was no return toward normal values to suggest development of tolerance. Urbanowicz (1971) followed ALA-U and CP-U levels in 60 workers for 24 months after they first became industrially exposed. There was a build-up of both biochemical effects for several months. But the levels then stabilized for the remainder of the two-year period. These studies suggest that the toxicologically-active fraction of the body burden during steady, long-term exposure remains essentially unchanged.

### 8.3.2 Age

Young children absorb lead more readily than older people. It also seems that children are more susceptible than adults in the sense that toxic effects occur at lower blood lead concentrations. The susceptibility of old people in comparison with younger adults has not been studied.

### 8.3.3 Seasonal variations

It has long been recognized that the incidence of severe lead intoxication in children is highest during the summer months (Baetjer, 1959; NAS-NRC, 1972). The observation that urinary excretion of lead increases in late summer may have some bearing (Kehoe, 1961).

### 8.3.4 Nutrition

There are few reports of studies that point to nutritional variables as having a distinct effect on lead toxicity in man (NAS-NRC, 1972; Goyer & Rhyne, 1974). Iron deficiency and lead exposure both affect porphyrin metabolism at the point where protoporphyrin IX is converted to haem. An additive effect results.

### 8.3.5 Intercurrent disease, alcohol, and other metals

Little is known about the effects of intercurrent diseases on the toxicity of lead or about the effect of lead on the susceptibility of people to other diseases. People with haemoglobin and erythrocyte anomalies, such as sickle cell anaemia and thalassaemia, would probably be more sensitive to the effects of lead exposure, as would perhaps people with renal damage. It is also possible that an interaction may exist between lead exposure and infectious disease processes, although reliable human data are not available to prove the point.

The effect of ethanol on lead toxicity is of some interest because the encephalopathy of illicitly-distilled whisky drinkers could conceivably involve an interaction of lead and the alcohol consumed. Furthermore, it has been suggested that heavy drinkers among industrially-exposed men may be more prone to lead toxicity than non-drinkers (Cramer, 1966; Candani & Farina, 1972).

# 9. EVALUATION OF HEALTH RISKS TO MAN FROM EXPOSURE TO LEAD AND ITS COMPOUNDS

The evaluation of health risks to man from exposure to lead and its compounds involves the following considerations:

(1) the significance of different environmental sources of lead and of pathways of exposure;

(2) the probability of occurrence of biological effects at different levels and rates of lead intake;

(3) the significance for human health of the various known biological effects of lead;

(4) the validity and limitations of various indicators of lead exposure and of resultant effects.

These considerations have been used in arriving at the conclusions which are summarized in this chapter.

## 9.1 Relative Contributions of Air, Food, Water, and Other Exposures to Total Intake

### 9.1.1 Adult members of general population groups

For the general population, the major contribution of lead to the total daily intake is from food, but water and air may provide significant contributions under certain conditions. Separate consideration must be given to occupationally-exposed persons in whom both the total lead intake and the relative contributions of dietary and airborne lead are quite different.

The inhalation of airborne lead contributes comparatively little to the Pb-B level in the general population. This follows from the fact that the lead concentration in ambient air seldom exceeds 3 $\mu g/m^3$ when averaged over months and from the conclusions reached in section 6 that the contribution of airborne lead to Pb-B levels is probably within the range of 1.0 to about 2.0 $\mu g/100$ ml for every 1 $\mu g/m^3$ of air. Although deposition and retention of different forms of lead in air may vary, estimates of Pb-B levels from the concentrations of lead in air are similar for the ambient air and for the air in the work environment.

Even if we assume a concentration of 1 $\mu g$ of lead per cubic metre of air contributes as much as 2.0 $\mu g/100$ ml of blood, and that the ambient air concentration of lead is as high as 4.5 $\mu g/m^3$, the total contribution of airborne lead would not exceed 9.0 $\mu g/100$ ml. This is still less than two-thirds of the value estimated by a WHO Expert Committee (1973). The discrepancy arises from the different approaches used in making the estimate. The WHO Expert Committee's estimate was based on lung deposition figures for lead obtained using the ICRP model (Task Group on Lung Dynamics, 1966). However, the ICRP lung model probably overestimates deposition for particles smaller than 0.5 μm (aerodynamic diameter) (Mercer, 1975), and the assumption that all the lead that is deposited is absorbed is probably also incorrect.

Dietary intake of lead varies with eating habits and the lead content of water sources. The majority of estimates from various countries suggest that the daily oral lead intake from food by adults ranges from approximately 100 μg to more than 500 μg; most studies show lead intake from dietary sources to be 200–300 μg/day. Relating blood lead levels to known daily oral lead intake suggests that each 100 μg of oral lead intake contributes about 6–18 μg of lead/100 ml of blood. This source of lead therefore accounts for a very large fraction of the blood lead levels found in the general adult population with Pb-B values below 25 μg/100 ml.

The quantity of lead intake directly related to the lead content of drinking water is difficult to estimate. Assuming a lead concentration in drinking water of 50 μg/litre (which is the upper limit generally found in the absence of lead pipes or other lead contributing factors) and a daily intake of one litre of water, 50 μg of total dietary lead could be attributed to water. This may be regarded as an upper limit but it must also be pointed out that lead in water ingested independently of food may be more readily absorbed and may provide a relatively greater contribution to the blood lead level than lead in food.

In assessing the relative contributions of air and diet to Pb-B levels, attention is called to the possibility that air may be a significant source of dietary lead through fallout. However, there are no data to confirm this assumption.

Improperly glazed pottery and illicit whisky have been cited as potential sources of excessive lead exposure for members of the general population.

Smoking one packet of 20 cigarettes would result in the direct inhalation of about 1–5 μg of lead but this only indicates the order of magnitude.

### 9.1.2 Infants and children

Infants and preschool children are a high-risk group with regard to lead intake and absorption. Relative contributions from food, water, and air are difficult to estimate because of the different diet (e.g. milk) and more active metabolic rate of young children. Also, intestinal absorption of lead by young children and, in particular, by infants may be greater than by adults. Tolerable intake of lead for preschool children should be less than the 3 mg/week recommended provisionally for adults by a WHO Expert Committee on Food Additives (1972).

A special hazard for young children is the ingestion of non-food items, particularly lead-containing paint from surfaces in homes and lead-contaminated dust and soil.

### 9.1.3 Occupationally exposed population groups

Because of the variability of occupational exposure, no general conclusions are possible but precautions against excessive exposure must be exercised in view of the possibility of extremely high occupational lead exposures, as cited in section 5.

### 9.2 Evaluation of Haematological Effects

Based on information presented in section 8, the following conclusions have been reached concerning the significance of different effects on haematopoiesis.

*Inhibition of ALAD activity in erythrocytes.* The health significance of decreased ALAD activity is still open to discussion. Although inhibition of ALAD in erythrocytes is to a certain extent paralleled by a decrease in other organs, e.g. liver and brain, no effect on health of this decrease has ever been established. Inhibition of ALAD is generally regarded as a good indicator of lead absorption but not of health impairment.

*Increased excretion of ALA and CP in urine, and increase of FEP* are indicators of impaired haematopoiesis. Although at moderate levels of increase, no evidence has been brought forward to show that the vital functions of haematopoiesis are impaired, resulting, for example, in a reduced life-span of erythrocytes or anaemia, any increase should be regarded with suspicion and particularly so when it is more than twice the level found in non-exposed population groups. Because free erythrocyte protoporphyrins are also increased in the case of iron deficiency, this test may provide a better indication of impaired haematopoiesis in exposed iron deficient population groups (especially children) than the excretion of ALA and CP. Moreover, females and children appear to have an earlier and steeper increase of FEP than males for the same levels of Pb-B.

*Effects on erythrocyte membrane*, as evidenced by shortened life-span and a decrease of Na-K-ATPase clearly can result in adverse health effects since anaemia may occur. *Anaemia*, expressed by decreased haemoglobin level, may be regarded as a consequence of disturbed haem and globin synthesis and of the decreased life-span of erythrocytes and has clear adverse health consequences.

### 9.3  Dose–Effect Relationships

At present Pb-B levels are the best available indicator of the dose. It should, however, be recognized that Pb-B does not reflect the type of exposure. The dose–effect relationships based on Pb-B levels should generally be used for long-term exposure.

As stated in section 8 (page 99) a dose–effect relationship refers in this report to the relationship between the dose as estimated by Pb-B levels and the intensity of a specified effect in *individual* subjects. For most effects, not enough data are available to present adequate dose–effect curves; however, for some effects, some points on the dose–effect curve can be tentatively estimated; for other effects, the data available only permit a statement referring to the Pb-B level below which such an effect has not been reported. This level is referred to as the no-detected-effect level. The degree of confidence that can be placed on such estimates will vary depending on the sample size and the number of studies reporting no effect.

*ALAD activity in erythrocytes.* There is a negative linear relationship between the logarithm of ALAD and Pb-B levels. Increase in Pb-B levels is paralleled by a decrease in ALAD levels in the Pb-B range up to about 60 µg/100 ml. For higher Pb-B values, the ALAD activity levels off at a very low level of enzyme activity. The no-detected-effect level for Pb-B is probably about 10 µg/100 ml but may be even lower.

*ALA and CP in urine; PP in erythrocytes.* There is a positive linear relationship between the logarithm of ALA (CP, FEP) and Pb-B levels; the no-detected-effect level for ALA and CP is about 40 µg/100 ml; for FEP the no-detected-effect level in females is about 20–30 µg/100 ml, in males it is about 25–35 µg/100 ml; and in iron-deficient children in particular, it may be about 20–25 µg/100 ml.

*Effects on the erythrocyte membrane* start to occur at higher Pb-B levels, probably higher than 50–60 µg/100 ml; a study by Secchi et al. (1974), on Na-K-ATPase, however, reports a lower no-detected-effect level of between 30 and 40 µg/100 ml.

*Anaemia.* Some authors maintain that the no-effect level in workers is above a Pb-B level of 100 μg/100 ml; others, however, report a slight decrease in the haemoglobin level, at a mean level of Pb-B of about 50 μg/100 ml. In some population groups and particularly in iron-deficient children, the no-detected-effect level is at an approximate Pb-B level of 40 μg/100 ml.

*Nervous system effects.* The data on effects of lead compounds on the nervous system lead to the following tentative conclusions in regard to prolonged exposures:

(1) From Pb-B levels of approximately 40 μg/100 ml, the probability of the occurrence of subclinical peripheral electrophysiological changes increases.

(2) From approximately 50 μg/100 ml in children, the probability of noticeable brain dysfunction increases; in adults the level is probably somewhat higher (60–70 μg/100 ml).

(3) From approximately 60 μg/100 ml in children the probability of acute or chronic encephalopathy increases; in adults this level is higher, probably above 80 μg/100 ml.

(4) The potential effects of lead on the nervous system constitute one of the main concerns, particularly in children. More carefully considered prospective studies should be carried out taking into account various interacting variables such as nutrition, socioeconomic status, and parental care in order to establish better founded dose–effect and dose–response relationships.

(5) No dose–effect or dose–response relationships can be established for alkyllead exposure on the basis of currently available information.

The present no-detected-effect level for sub-clinical neuropathy appears to be a Pb-B value of 40–50 μg/100 ml. For minimal brain dysfunction it is probably 50–60 μg/100 ml in children and 60–70 μg/100 ml in adults, and for acute or chronic encephalopathy, 60–70 μg/100 ml in children, and over 80 μg/100 ml in adults. The establishment of relationships between Pb-B levels and effect is especially difficult in children because the effect may be detected months or years after the critical exposure occurred.

*Renal function.* Apparently prolonged exposure to Pb-B levels greater than 70 μg/100 ml is necessary to produce nephropathy; a no-detected-effect level cannot be given. The problem is non-correspondence in time between the determination of Pb-B level and the detection of effect.

Aminoaciduria, reflecting impaired amino acid transport through the renal tubules may occur in children and adults with increased lead absorption. The present data do not allow a no-detected-effect level to be estimated, but indicate that this effect is unlikely to be found in association with Pb-B levels below some 90–100 µg/100 ml (Chisolm, 1968b, Cramer et al., 1974).

*Changes in blood constituents* such as calcium, phosphorus, glucose, cholesterol, total proteins, serum albumins, alkaline phosphatase (EC 3.1.3.1), lactate acid dehydrogenase, and urea nitrogen, could not be found in male workers with a median Pb-B level of 63 µg/100 ml; 37 % showed an effect with a Pb-B level greater than 70 µg/100 ml (Cooper et al., 1973). There was an indication of increased bilirubin at a Pb-B level of about 70 µg/100 ml. An increased pyruvate level after glucose administration was reported in 50 % of children with Pb-B levels of 40–60 µg/100 ml (Moncrieff et al., 1964).

*The general pattern of morbidity and mortality* in workers does not appear to be affected if the Pb-B level never exceeds 70 µg/100 ml.

In assessing reported dose–effect relationships and no-detected-effect levels, one should take into account the fact that the available data are limited. Even from a theoretical viewpoint, the establishment of a definite no-effect level is not possible, because one can hardly ever expect to cover the whole range of susceptibility in human populations. Nevertheless, the available data suggest that the no-detected-effect levels given above are on the conservative side.

Table 29 summarizes the no-detected-effect levels discussed. For some of these effects, it is possible to elaborate dose–response relationships. These cases are considered in section 9.4.

Table 29. No-detected effect levels in terms of Pb-B (µg of lead per 100 ml of blood)

| No detected effect level | Effect | Population |
|---|---|---|
| <10 | Erythrocyte ALAD inhibition | adults, children |
| 20–25 | FEP | children |
| 20–30 | FEP | adult, female |
| 25–35 | FEP | adult, male |
| 30–40 | Erythrocyte ATPase inhibition | general |
| 40 | ALA excretion in urine | adults, children |
| 40 | CP excretion in urine | adults |
| 40 | Anaemia | children |
| 40–50 | Peripheral neuropathy | adults |
| 50 | Anaemia | adults |
| 50–60 | Minimal brain dysfunction | children |
| 60–70 | Minimal brain dysfunction | adults |
| 60–70 | Encephalopathy | children |
| >80 | Encephalopathy | adults |

## 9.4 Dose-response Relationships

A dose–response relationship considers the observed relative frequency of occurrence of a specified *effect*[a] in a group of subjects at a given dose level. As in the case of dose–effect relationships, the data available to evaluate a dose–response relationship are either limited or non-existent. The available information on dose–response relationships has been presented in section 8. In this section, attention is paid to the 5% response levels, i.e. that level of Pb-B at which not more than 5% of the group considered is expected to show the specified intensity of a specified effect. The 5% level has had to be stipulated, because not enough data are available to state the Pb-B levels for 0.5%, 1%, etc. Further investigations have to be carried out to enlarge the amount of data available. For further discussion see Zielhuis (1975). His review suggests the 5% response levels recorded in Table 30, which are in accordance with the data discussed in section 8. These response levels are also in agreement with those suggested by Hernberg (1975).

Table 30. Pb-B levels at which no more than 5% of the population will show the indicated intensity of effect

| Biochemical effect | Intensity of effect | Population | Pb-B (µg/100 ml) |
|---|---|---|---|
| ALAD inhibition | perceptible inhibition | adult, children | 10 |
| | >40% inhibition | adults | 15–20 |
| | >70% inhibition | adults | 30 |
| | >70% inhibition | children | 25–30 |
| ALA-U | perceptible increase | adults, children | 40 |
| | >10 mg/litre | adults, children | 50 |
| FEP | perceptible increase | adult males | 30 |
| | | adult females | 25 |
| | | children | 20 |

[a] From: Zielhuis, 1975.

## 9.5 Diagnosis of Lead Poisoning and Indices of Exposure and/or Effects for Epidemiological Studies

For epidemiological studies and for the detection of the early effects of lead in occupational exposure of individuals, the following tests have been used:

(1) Lead levels in blood.

---

[a] Graded effects may be specified in terms of their intensity.

(2) Excretion of lead in urine spontaneously or after administration of chelating agents.

(3) Lead levels in tissues (teeth, bones, hair, etc.).

(4) Activity of ALAD in blood.

(5) Indices of disturbed porphyrin metabolism:
ALA and/or CP in urine,
Protoporphyrin IX in erythrocytes.

(6) Haematological indices such as basophilic stippling and haemoglobin levels.

(7) Early (sub-clinical) symptoms and signs of other damage (e.g. to the nervous system or the kidneys).

(8) Clinical evidence of poisoning.

The criteria used by individual investigators correspond to the premises and purposes of their studies, for example, Pb-B for evaluating lead levels in the general population, and clinical signs of poisoning to assess morbidity caused by occupational exposure.

The following considerations should be kept in mind when using and interpreting the results.

### 9.5.1 Concentration of lead in blood (Pb-B)

Pb-B reflects the current state of the dynamic equilibrium between the amounts of lead entering the organism, transported in the blood, and deposited in the tissues (including the bones). To date, insufficient information has been collected about the quantitative aspects of these processes, but from the data available, it may be stated that:

(*a*) After a single inhalation of a soluble lead compound, the concentration of lead in the body will change in the same way as after an intravenous injection, i.e. there will be a rapid increase in Pb-B levels followed by a slower decrease; initially there will be a rapid elimination in the urine and a slow deposition in the tissues with subsequent redistribution according to the metabolism of lead in the various organs and systems.

(*b*) During long-term exposure at a constant rate, an equilibrium between the amount of lead absorbed, deposited, and excreted develops over a long period (weeks to months, according to the daily doses received), which can be considered as a steady state.

(*c*) There are only limited data as to how quickly this equilibrium (and Pb-B) changes when irregular variations in the dose of lead received (e.g. air lead concentrations) occur.

A long-term steady state probably exists normally in non-occupationally exposed general adult populations, at least in the Pb-B level. However, no direct evidence for this assumption is available.

In occupationally exposed persons, a steady state cannot be assumed because of the well known and marked variations of air lead concentrations in the working environment and of Pb-B levels in occupationally exposed individuals from one time to another and among the individuals in the same work place. Occasional exposure to a high lead concentration in the air could raise the Pb-B level for some time without contributing significantly to the body burden and to the biological effects.

If Pb-B is to be used as an indicator of the degree of environmental lead exposure the above-mentioned facts must be taken into account, as well as the analytical method used and the limitations (accuracy, precision, sensitivity, limits of detection).

### 9.5.2 Aminolevulinic acid dehydratase (ALAD)

For ALAD the same conditions can apply as for Pb-B. The behaviour of ALAD activity will follow closely the level of Pb-B up to 50–60 µg/100 ml.

### 9.5.3 Aminolevulinic acid (ALA) and coproporphyrin (CP) excretion in the urine

ALA and CP in urine are not so dependent on the current state of lead exposure and absorption as the Pb-B, although their excretion diminishes relatively quickly when exposure ceases; they reflect more the average short-term level of lead exposure and have proved useful in this way. ALA and CP estimates have found broad recognition as indices of lead absorption and as indicators of early effects they reflect individual susceptibility to lead.

### 9.5.4 Lead excretion in the urine

An elevated rate of spontaneous lead excretion in the urine is indicative of high lead absorbed, but a normal rate of excretion does not serve as a reliable means of excluding the possibility of excessive absorption. Lead excretion in urine is dependent on the Pb-B level but is also influenced by other—mostly unknown—factors, so that no direct conclusions about exposure and the extent of absorption can be derived from lead levels in urine (even in a 24-hour sample).

The excretion of lead provoked by chelating agents such as calcium disodium ethylenediamintetraacetate is thought to reflect the biologically active portion of the body burden. It is probably a more sensitive index of over exposure and excess absorption than the Pb-B level since clearly elevated values have been reported in cases of only marginally elevated Pb-B levels.

### 9.5.5 Haematological changes (stippled cells, anaemia)

These are not sensitive indices of over-exposure or excess absorption. They are not very useful for the early detection of possible health impairment.

### 9.5.6 Lead in tissues (teeth and hair)

These have been used as indicators of integrated long-term exposure and have the advantage that samples are easy to procure. As yet, the amount of information concerning the interpretation of the values obtained is inadequate for their evaluation as indices of exposure or dose.

### 9.5.7 Some practical aspects

9.5.7.1 *General population studies*

The Pb-B level is the epidemiological index of choice, assuming that a reasonable approximation of a steady state exposure exists. ALAD activity estimates are equally useful for such studies or as epidemiological indices of lead absorption. The decision to use Pb-B or ALAD depends on the laboratory facilities available. Signs of lead effects other than ALAD inhibition are not to be expected at Pb-B levels below 20 µg/100 ml. Lead in deciduous teeth and hair is potentially useful as an indicator of integrated exposure in infants but needs more study.

9.5.7.2 *Occupationally-exposed persons*

For screening the exposure of groups of workers, any method can be used that has the required sensitivity and specificity. Economic and time factors will determine the choice of test. When using the Pb-B level, the conditions of sampling must be well defined, taking into account the factors influencing the variations of the Pb-B concentrations. ALA and CP estimations in urine are widely used since they are simple, avoid the possibility of external contamination, and may provide a better picture of the integral exposure. ALAD activity is only useful at Pb-B levels below about 60 µg/100 ml. For early detection of the signs of lead effects in individuals, ALA-U or CP-U tests are the best established screening methods. When abnormal values are found, further tests (including clinical and laboratory investigations) will have to be applied to evaluate the kind of disturbance and the degree of health risk (WHO Study Group, 1975).

### 9.5.7.3 *Reliability of sampling and analytical methods*

The evaluation of the pollution of the environment by lead and of the health effects on man which might result, depends on the reliability of sampling procedures and analytical methods used.

The methods of sampling for different environmental media, and the possible exposure pathways of man have been discussed in section 3. The great spatial and temporal variability of these environmental media and their diversity make the accurate assessment of total exposure a difficult task. Unless elaborate schemes are set up and extreme precautions are taken, the total exposure of a population group cannot be evaluated with an error of less than about 50%, taking into account the analytical uncertainties.

In the determination of the dose received or the effects on haematopoiesis observed, the sampling problem is relatively minor but the accuracy and precision of analytical techniques play an important role. An evaluation with up to 20% relative precision is seldom achieved under normal operational conditions.

## 9.6 The Problem of Alkyllead Compounds

The principal risk of alkyllead compounds is in occupational exposure, either by inhalation or by absorption through the skin. Acute toxicity results in an encephalopathy that differs greatly from the effects of inorganic lead on the central nervous system. Some components of the toxic effects are probably due to the alkyl compound as a whole rather than its lead component. Workmen at greatest risk are those involved in mixing fuel additives, although other workmen engaged in related occupations such as the cleaning of storage tanks where inhalation is possible, are also at high risk. Over-exposure of the general population to alkyllead compounds has not been documented.

# REFERENCES

ABERNETHY, R. F., PETERSON, M. J., & GIBSON, F. H. (1969) Spectrochemical analyses of coal ash for trace elements. U.S. Bur. Mines, Rep. Inv. 7281.

AKATSUKA, K. (1973) Tetraalkyl lead poisoning. *Jpn. J. ind. Health*, **15**: 3–66.

ALBAHARY, C. (1964) Les troubles porphyriques dans le saturnisme. *Arch. Mal. prof.*, **25**: 495–507.

ALBAHARY, C., RICHET, G., GUILLAUME, J., & MOREL-MAROGER, L. (1965) Le rein dans le saturnisme professionnel. *Arch. Mal. prof.*, **26**: 5–19.

ALBERT, R. E., SHORE, R. E., SAYERS, A. J., STREHLOW, C., KNEIP, T. Y., PASTERNACK, B. S., FRIEDHOFF, A. J., COVAN, F., & CIMINO, J. A. (1974) Follow-up of children overexposed to lead. *Environ. Health. Perspect.*, Exp. Issue No. 7, 33–41.

ALEXANDER, F. W., & DELVES, H. T. (1972) Death from acute lead poisoning. *Arch. Dis. Child.*, **47**: 446–448.

ALEXANDER, F. W., DELVES, H. T., & CLAYTON, B. E. (1973) *The uptake and excretion by children of lead and other contaminants.* In: *Proceedings of the International Symposium; Environmental Health Aspects of Lead, Amsterdam, 2–6 October 1972.* Luxembourg, Commission of the European Communities, pp. 319–330.

ALLCROFT, R. (1951) Lead poisoning in cattle and sheep. *Vet. Rec.*, **63**: 583–590.

ALLCROFT, R., & BLAXTER, R. L. (1950) Lead as a nutritional hazard to farm livestock V. The toxicity of lead to cattle and sheep and an evaluation of the lead hazard under farm conditions. *J. Comp. Pathol.*, **60** (3): 209–218.

ALTMAN, P. U., & DETTMER, D. S., ed. (1968) *Nutritional standards in man.* In: *Metabolism.* Bethesda, MD, Fed. Am. Soc. Exp. Biol., pp. 95–96.

ALVARES, A. P., KAPELNER, S., SASSA, S., & KAPPAS, A. (1975) Drug metabolism in normal children, lead-poisoned children and normal adults. *Clin. Pharm. Ther.*, **17** (2): 179–183.

ALVARES, A. P., LEIGH, S., CONN, J., & KAPPAS, A. (1972) Lead and methyl mercury: effects of acute exposure of cytochrome P-450 and the mixed function oxidase system in the liver. *J. exp. Med.*, **135** (6): 1406–1409.

ARAKI, S. & HONMA, T. (1976) Relationships between lead absorption and peripheral nerve conduction velocities in lead workers. *Scand. J. Work environ. Health*, **4**: 225–231.

ASKEVOLD, R. (1951) Routine analysis of porphyrines in urine. *J. clin. Lab. Invest.*, **3**: 318–319.

ASTM (1970) *Book of standards. Tentative method of test for lead in the atmosphere.* Philadelphia, American Society for Testing and Materials, vol. 23, pp. 830–836.

ATKINS, P. R. (1969) Lead in suburban environment. *J. air Pollut. Contr. Ass.*, **19**: 591–594.

AUB, J. C., FAIRHALL, L. T., MINOT, A. S., & REZNIKOFF, P. (1926) *Lead poisoning.* Baltimore, Williams and Wilkins Co., pp. 206 (Medicine Monographs Vol. VII).

AZAR, A., SNEE, R., & HABIBI, K. (1973) *Relationship of community levels of air lead and indices of lead absorption.* In: *Proceedings of the International Symposium; Environmental Health Aspects of Lead, Amsterdam, 2–6 October 1972.* Luxembourg, Commission of the European Communities, pp. 581–594.

BAERNSTEIN, H. D., & GRAND, J. A. (1942) The relation of protein intake to lead poisoning in rats. *J. Pharmacol. exp. Ther.*, **74**: 18–24.

BAETJER, A. M. (1959) Effects of season and temperature on childhood plumbism. *Ind. Med. Surg.*, **28**: 137–144.

BAETJER, A. M., & HORIGUCHI, S. (1963) *Effects of environmental temperature and dehydration on lead poisoning in laboratory animals.* Amsterdam, Excerpta Medica, pp. 795–797 (International Congress Series, No. 62).

BAETJER, A. M., Joardar, S. N. D., & McQUARY, W. A. (1960) Effect of environmental temperature and humidity on lead poisoning in animals. *Arch. environ. Health*, **1**: 463–477.

BAGCHI, R. B., GANGULY, H. D., & SIRDAR, J. M. (1940) Lead in food. *Ind. J. med. Res.*, **28**: 441–445.

BAKER, R. W. R. (1950) Polarographic determination of lead in urine. *Biochem. J.*, **46**: 606–612.

BALOH, R. W. (1974) Laboratory diagnosis of increased lead absorption. *Arch. environ. Health*, **28**: 198–208.

BARLTROP, D. (1969) Transfer of lead to the human foetus. In: Barltrop, D. & Burland, W. L., ed., *Mineral metabolism in pediatrics*. Philadelphia, Davis Co., pp. 135–151.

BARLTROP, D., & SMITH, A. (1971) Interaction of lead with erythrocytes. *Experientia (Basel)* **27**: 92–95.

BARLTROP, D., & SMITH, A. (1972) Lead binding to haemoglobin. *Experientia (Basel)*, **28**: 76–77.

BARLTROP, D., BARRET, A. J., & DINGLE, J. T. (1971) Subcellular distribution of lead in the rat. *J. Lab. clin. Med.* **77**: 705–712.

BARLTROP, D., STREHLOW, C. D., THORNTON, J., & WEBB, J. S. (1974) Significance of high soil lead concentrations for childhood lead problem. *Environ. Health Perspect.*, Exp. Issue No. 7, 75–83.

BARRY, P. S. I. (1975) A comparison of concentrations of lead in human tissues. *Brit. J. ind. Med.*, **32**: 119–139.

BARRY, P. S., & MOSSMAN, D. B. (1970) Lead concentrations in human tissues. *Brit. J. ind. Med.*, **27**: 339–351.

BAUCHINGER, M., SCHMID, E., EINBRODT, H. J., & DRESP, J. (1976) Chromosome aberrations in lymphocytes after occupational exposure to lead and cadmium. *Mutat. Res.*, **40**: 57–62.

BAUCHINGER, M., SCHMID, E., & SCHMIDT, D. (1972) Chromosomenanalyse bei Verkehrspolizisten mit erhöhter Bleilast. *Mutat. Res.*, **16**: 407–412.

BEATTIE, A. D., MOORE, M. R., DEVENAY, W. I., MILLER, A. R., & GOLDBERG, A. (1972a) Environmental lead pollution in an urban soft-water area. *Brit. med. J.*, **2**: 491–493.

BEATTIE, A. D., MOORE, M. R., & GOLDBERG, A. (1972b) Tetraethyl-lead poisoning. *Lancet*, **2**: 12–15.

BECK, E. G., MANOJLOVIC, N., & FISHER, A. B. (1973) Die Zytotoxizität von Blei. In: *Proceedings of the International Symposium; Environmental Health Aspects of lead, Amsterdam, 2–6 October, 1972*. Luxembourg, Commission of the European Communities, pp. 451–461.

BECKMAN, K. (1925) Über die Beziehungen zwischen Blutdruck, Kapillardruck und Nierenveränderungen im Tierexperiment. *Arch. klin. Med.*, **149**, 177–188.

BENSON, F. (1972) Indoor-outdoor air pollution relationships. US EPA Publ. No. AP-112.

BERG, B., & ZENZ, C. (1967) Environmental and clinical control of lead exposure in a nonferrous foundry. *Am. ind. Hyg. Assoc. J.*, **28**: 175–178.

BERITIĆ, T. (1971) Lead concentration found in human blood in association with lead colic. *Arch. environ. Health*, **23**: 289–291.

BERITIĆ, T., & STAHULJAK, D. (1961) Lead poisoning from lead-glazed pottery. *Lancet*, **1**: 669.

BERK, P. D., TSCHUNDY, D. P., SHEPLEY, L. A., WAGGONER, J. G., & BERLIN, N. I. (1970) Hematologic and biochemical studies in a case of lead poisoning. *Am. J. Med.*, **48**: 137–144.

BERLIN, A., DEL CASTILHO, P., & SMEETS, J. (1973) *European intercomparison programmes*. In: *Proceedings of the International Symposium; Environmental Health Aspects of Lead, Amsterdam, 2–6 October 1972*. Luxembourg, Commission of the European Community, pp. 1033–1046.

BERLIN, A., SCHALLER, K. H., & SMEETS, J. (1975) *Standardization of ALAD activity determinations at the European level; Intercalibration and applications*. In: *Proceedings of CEC-EPA-WHO International Symposium; Recent Advances in the Assessment of the Health Effects of Environmental Pollution, Paris, 24–28 June 1974*. Luxembourg, Commission of the European Communities, pp. 1087–1101.

BERTINUSON, J. R., & CLARK, C. S. (1973) The contribution to lead content of soils from urban housing. *Interface*, **2**: 6.

BESSIS, M. C., & JENSEN, W. N. (1965) Sideroblastic anaemia, mitochondria and erythroblastic iron. *Brit. J. Haematol.*, **11**: 49–51.

BETTS, P. R., ASTLEY, R., & RAINE, D. N. (1973) Lead intoxication in children in Birmingham. *Brit. med. J.*, **1**: 402–406.

BICKNELL, J., CLAYTON, B. E., & DELVES, H. T. (1968) Lead in mentally retarded children. *J. Mental Def. Res.*, **12**: 282–293.

BINGHAM, E., PFITZER, E. A., BARKLEY, W., & RADFORD, E. P. (1968) Alveolar macrophages: reduced number in rats after prolonged inhalation of lead sesquioxide. *Science*, **162**: 1297–1299.

BLACKMAN, S. S. (1937) The lesions of lead encephalitis in children. *Bull. Johns Hopkins Hosp.*, **61** (1): 1–61.

BLANKSMA, L. A., SACHS, H. K., MURRAY, E. F., & O'CONNEL, M. J. (1969) Incidence of high blood lead levels in Chicago children. *Pediatrics*, **44**: 661–665.

BLAXTER, K. L., & COWIC, A. T. (1946) Excretion of lead in the bile. *Nature (Lond.)*, **157**: 588.

BOLANOWSKA, W. (1968) Distribution and excretion of triethyllead in rats. *Brit. J. ind. Med.*, **25**: 203–208.

BOLANOWSKA, W., PIOTROWSKI, J., & CARCZYNSKI, H. (1967) Triethyllead in the biological materia in cases of acute tetraethyllead poisoning. *Archiv J. Toxikol.* **22**: 278–282.

BOLANOWSKA, W., PIOTROWSKI, J., & TROJANOWSKA, B. (1964) *The kinetics of distribution and excretion of lead (PB²¹⁶) in rats.* In: *Proceedings of the 14th International Congress of Occupational Health, Madrid, 16–21 Sept.*, pp. 420–422.

BOLTER, E., BUTZ, T., & ARSENEAU, J. F. (1975) *Mobilization of heavy metals by organic acids in the soils of a lead mining and smelting district.* In: *Proceedings of the IXth Annual Conference on Trace Substances in Environmental Health, Columbia, MO, 10–12 June 1975*, pp. 107–112.

BONSIGNORE, D., CALISANO, P., & CARTASEGNA, C. (1965) Un semplice metodo per la determinazione della δ-amino-levulinico-deidratasi nel sangue. *Med. Lav.*, **56**: 199–205.

BOUDENE, C., ARSAC, F., & MEININGER, J. (1975) *Etude des taux de plomb dans l'air et dans la population en France.* In: *International Symposium on Environmental Lead Research, Dubrovnik, 14–15 May 1975. Arch. industr. Hyg. Toxicol.*, **26**: supplement, pp. 179–189.

BOYD, P. R., WALKER, G., & HENDERSON, J. N. (1957) The treatment of tetraethyl lead poisoning. *Lancet*, **1**: 181–185

BOYLAND, E., DUKES, C. E., GROVE, P. L., & MITCHLEY, B. C. V. (1962) The induction of renal tumours by feeding lead acetate to rats. *Brit. J. Cancer*, **16**: 283–288.

BRANDT, A. D., & REICHENBACH, G. S. (1943) Lead exposures at the government printing office. *J. ind. Hyg. Toxicol.*, **25** (10): 445–450.

BREZINA, M., & ZUMAN, P. (1958) *Polarography in medicine, biochemistry and pharmacy.* New York, Interscience Publishers Inc.

BRITISH MINISTRY OF AGRICULTURE, FISHERIES AND FOOD (1972) *Survey of lead in food. Working Party on the Monitoring of Foodstuffs for Heavy Metals: Second Report.* London, Her Majesty's Stationery Office, pp. 31.

BROWN S., DRAGAN, N., & VOGEL, W. (1971) Effects of lead acetate on learning and memory in rats. *Arch. environ. Health*, **22**: 370-372.

BRUCH, J., BROCKHAUS, A., & DEHNEN, W. (1973) *Elektronmikroskopische Beobachtungen an Rattenlungen nach Exposition mit partikelförmigem Blei.* In: *Proceedings of the International Symposium; Environmental Health Aspects of Lead, Amsterdam, 2–6 October 1972.* Luxembourg, Commission of the European Communities, pp. 221–229.

BRUCH, J., BROCKHAUS, A., & DEHNEN, W. (1975). *Local effects of inhaled lead compounds on the lung.* In: *Proceedings of CEC-EPA-WHO International Symposium; Recent Advances in the Assessment of the Health Effects of Environmental Pollution, Paris, 24–28 June 1974.* Luxembourg, Commission of the European Communities, pp. 781–793.

BULLOCK, J. D., WEY, R. J., ZAIA, J. A., ZAREMBOK, I., & SCHROEDER, H. A. (1966) Effect of tetraethyllead on learning and memory in rats. *Arch. environ. Health*, **13**: 21–22.

BULPITT, G. J. (1972) Lead and hyperactivity. *Lancet*, **II**: 1144.

BURDÉ, B. DE LA, & CHOATE, M. S.(1972) Does asymptomatic lead exposure in children have latent sequelae? *J. Pediar.*, **81** (6): 1088–1091.

BUTT, E. U., NUSBAUM, R. E., GILMOUR, T. C., DIDIO, S. L., & Sister MARIANO (1964) Trace metal levels in human serum and blood. *Arch. environ. Health*, **8**: 52–57.

BYERS, R. K. (1959) Lead poisoning. *Pediatrics*, **23**: 585–603.

BYERS, R. K., & LORD, E. E. (1943) Late effects of lead poisoning on mental development. *Am. J. Dis. Child.*, **66**: 471–494.

CANTAROW, A., & TRUMPER, M. (1944) *Lead poisoning*. Baltimore, Williams and Wilkins Co., p. 8.

CANDANI, A., & FARINA, G. (1972) Influence of alcoholic beverages consumption on lead-induced changes of haemebiosynthesis. *Med. Lav.*, **63**, 22–28.

CARPENTER, S. (1974) Placental permeability of lead. *Environ. Health, Perspect.* Exp. Issue, No. 7, 129–133.

CARSON, T. L., VAN GELDER, G. A., KARAS, G. G., & BUCK, W. B. (1974) Development of behavioural tests for the assessment of neurologic effects of lead in sheep. *Environ. Health Perspect.*, Exp. Issue No. 7, 233–239.

CASTELLINO, N., & ALOJ, S. (1964) Kinetics of the distribution and excretion of lead in the rat. *Brit. J. ind. Med.*, **21**: 308–314.

CASTELLINO, N., & ALOJ, S. (1969) Intracellular distribution of lead in the liver and kidney of the rat. *Brit. J. ind. Med.*, **26**: 139–143.

CASTELLINO, N., LIAMANNA, P., & CRIECO, B. (1966) Biliary excretion of lead in the rat. *Brit. J. ind. Med.*, **23**: 237–239.

CATTON, M. J., HARRISON, M. J. G., FULLERTON, P. M., & KAZANTZIS, G. (1970) Subclinical neuropathy in lead workers. *Brit. med. J.*, **2**: 80–82.

CERNIK, A. A. (1974) Determination of blood lead using a 4.0 mm paper punched disc carbon sampling cup technique. *Brit. J. ind. Med.*, **31**: 239–245.

CERNIK, A. A., & SAYERS, M. H. P. (1971) Determination of lead in capillary blood using a paper punched disc atomic absorption technique application to the supervision of lead workers. *Brit. J. ind. Med.*, **28** (4): 392–398.

CHANEY, R. L. (1973) *Crop and food chain effects of toxic elements in sludges and effluents. Recycling municipal sludges and effluents on land* July 9–13, 1973 Illinois, pp. 129–141.

CHEH, A., & NEILANDS, J. B. (1973) Zinc, an essential metal for beef liver δ-aminolevulinate dehydratase. *Biochem. biophys. Res. Comm.*, **55** (4): 1060–1063.

CHISOLM, J. J. (1962) Aminoaciduria as a manifestation of renal tubular injury in lead intoxication and a comparison with patterns of aminoaciduria seen in other diseases. *J. Pediatr.*, **60**: 1–17.

CHISOLM, J. J. (1968a) *Lead poisoning (plumbism)*. In: H. L. Barnett, ed., *Pediatrics* (14th Ed.) New York, Appelton-Century-Crofts, pp. 313–319.

CHISOLM, J. J. (1968b) The use of chelating agents in the treatment of acute and chronic lead intoxication in childhood. *J. Pediatr.*, **73**: 1–38.

CHISOLM, J. J. (1973) Management of increased lead absorption and lead poisoning in children. *New Engl. J. Med.*, **289**: 1016–1018.

CHISOLM, J. J. (1974) Chelation therapy in children with subclinical plumbism. *Pediatrics*, **53**: 441–443.

CHISOLM, J. J., & HARRISON, H. E. (1956) The exposure of children to lead. *Pediatrics*, **18**: 943–958.

CHOVIN, P., DUFFAUD, J., MERKEZ, F., FAVART, M., & TRUFFERT, L. (1973) *Resultats d'une année d'étude de la teneur en plomb particulaire et l'atmosphère de Paris*. In: *Proceedings of the International Symposium: Environmental Health Aspects of Lead, Amsterdam 2–6 October 1972*. Luxembourg, Commission of the European Communities, pp. 1003–1015.

CHOW, T. J. (1968) Isotope analysis of seawater by mass spectrometry. *J. Water Pollut. Control Fed.*, **40**: 399–411.

CHOW, T. J., & EARL, J. L. (1970) Lead aerosols in the atmosphere: Increasing concentrations. *Science*, **169**: 577–580.

CHOW, T. J., & BENNET, C. F. (1969) Lead aerosols in marine atmosphere. *Environ. Sci. Technol.*, **3**: 737–740.

CHOW, T. J., EARL, J. L., & SNYDER, C. B. (1972) Lead aerosol baseline: Concentration at White Mountain and Liaguna Mountain. *Science*, **178**: 401–402.

CLARKSON, T. W., & KENCH, J. E. (1956) Urinary excretion of amino acids by men absorbing heavy metals. *Biochem. J.*, **62**: 361–372.

CLARKSON, T. W., & KENCH, J. E. (1958) Uptake of lead by human erythrocytes in vitro. *Biochem. J.*, **69**: 432–439.

CLASEN, R. A., HARTMAN, J. F., STARR, A. J., COOGAN, PH. S., PANDOLFI, S., LAING, J., BECKER, R., & HASS, G. U. (1974) Electron microscopic and chemical studies of the vascular changes and edema of lead encephalopathy. *Am. J. Pathol.*, **74** (2): 215–233.

CLERCQ, E. DE, & MERIGAN, T. C. (1969) An active interferon inducer obtained from *Haemophilus influenzae* type B. *J. Immun.*, **103**: 899–906.

COGBILL, E. C., & HOBBS, M. E. (1957) Transfer of metallic constituents of cigarettes to the main-stream smoke. *Tob. Sci.*, **1**: 68–73.

COHEN, G. J., & AHRENS, W. E. (1959) Chronic lead poisoning. *J. Pediatr.*, **54** (2): 271–284.

COHEN, G. J., BOWERS, G. N., & LEPOW, M. (1973) Epidemiology of lead poisoning. *J. Am. Med. Ass.*, **266**: 1430–1433.

COHEN, N., KNEIP, T. J., GOLDSTEIN, D. H., & MUCHMORE, E. A. S. (1972) The juvenile baboon as a model for studies of lead poisoning in children. *J. Med. Primatol.*, **1**: 142–155.

COLE, L. J., & BACHHUBER, L. J. (1914) The effect of lead on the germ cells of the male rabbit and fowl as indicated by their progeny. *Proc. Soc. Exp. Biol. Med.*, **12**: 24–29.

COLLIER, H. B. (1971) A study of the determination of δ-aminolevulinate hydrolyase (δ-aminolevulinate dehydratase) activity in hemolysates of human erythrocytes. *Clin. Biochem.*, **4**: 222–232.

COLWILL, D. M., & HICKMAN, A. J. (1973) *The concentration of volatile and particulate lead compounds in the atmosphere: measurements at four road sites.* Transport and Road Research Laboratory Report LR 545 Crowthorne, Berkshire, pp. 5.

COMMISSION OF THE EUROPEAN COMMUNITIES (1973) *Air lead concentrations in the European Community. Yearly Report: April 1971–March 1972.* Luxembourg.

COOPER, W. C., & GAFFEY, W. R. (1975) Mortality of lead workers. *J. occup. Med.*, **17**: 100–107.

COOPER, W. C., TABERSHAW, I. R., & NELSON, K. W. (1973) *Laboratory studies of workers in lead smelting and refining.* In: *Proceedings of the International symposium: Environmental Health Aspects of lead, Amsterdam, 2–6 October 1972,* Luxembourg, Commission of the European Communities, pp. 517–529.

COULSTON, F., GOLDBERG, L., GRIFFIN, T. B., & RUSSELL, J. C. (1972a) *The effects of continuous exposure to airborne lead. 1. Exposure of rats and monkeys to particulate lead at a level of 21.5 $\mu g/m^3$.* Final Report to the US Environmental Protection Agency.

COULSTON, F., GOLDBERG, L., GRIFFIN, T. B., & RUSSELL, J. C. (1972b) *The effects of continuous exposure to airborne lead. 2. Exposure of man to particulate lead at a level of 10.9 $\mu g/m^3$.* Final Report to the US Environmental Protection Agency.

COULSTON, F., GOLDBERG, L., GRIFFIN, T. B., & RUSSELL, J. C. (1972c) *The effects of continuous exposure to airborne lead. 4. Exposure of man to particulate lead at a level of 3.2 $\mu g/m^3$.* Final Report to the US Environmental Protection Agency.

COULTHARD, A. J. (1958) Animal cycle of blood haemoglobin levels. *Clin. Chim. Acta*, **2**: 226–233.

CRAMER, K. (1966) Predisposing factors for lead poisoning. *Acta. Med. scand., (Suppl. 445):* **179**: 56–59.

CRAMER, K., & DAHLBERG, L. (1966) Incidence of hypertension among lead workers. *Brit. J. ind. Med.*, **23**: 101–104.

CRAMER, K., GOYER, R. A., JAGENBURG, R., & WILSON, M. H. (1974) Renal ultrastructure renal function, and parameters of lead toxicity in workers with different period of lead exposure. *Brit. J. ind. Med.*, **31**: 113–127.

CREMER, J. E. (1959) Biochemical studies on the toxicity of tetraethyl lead and other organo-lead compounds. *Brit. J. ind. Med.*, **16**: 191–199.

CREMER, J. E. (1965) Toxicology and biochemistry of alkyl lead compounds. *Occup. Health Rev.*, **17**: 14–19.

CREMER, J. E., & CALLAWAY, S. (1961) Further studies on the toxicity of some tetra- and trialkyl lead compounds. *Brit. J. ind. Med.*, **18**: 277–282.

CRUTCHER, J. C. (1963) Clinical manifestation and therapy of acute lead intoxication due to the ingestion of illicitly distilled alcohol. *Ann. intern. Med.*, **59**: 707–715.

Daines, R. H., Smith, D. W., Feliciano, A., & Trout, J. R. (1972) Air levels of lead inside and outside of homes. *Ind. Med. Surg.*, **41**: 26–28.

Dalldorf, G., & Williams, R. R. (1945) Impairment of reproduction in rats by ingestion of lead. *Science*, **102**: 668.

Danilović, V. (1958) Chronic nephritis due to ingestion of lead-contaminated flour. *Brit. med. J.*, **1**: 27–28.

David, O. J., Clark, J., & Voeller, K. (1972) Lead and hyperactivity. *Lancet*, **2**: 900–903.

Davidson, D. F., & Lakin, H. W. (1962) *Metal content of some black shales of the western United States*. U.S. Geological Survey Professional Paper 450C, p. C74.

Davis, J. R., & Andelman, S. L. (1967) Urinary delta aminolevulinic acid levels in lead poisoning. A modified method for the rapid determination of urinary delta-aminolevulinic acid using disposable ion-exchange chromatography columns. *Arch. environ. Health*, **15**: 53–59.

Davis, R. K., Horton, A. W., Larson, E. E., & Stemmer, K. L. (1963) Inhalation of tetramethyllead and tetraethyllead. *Arch. environ. Health*, **6**: 473–479.

Davis, W. E. (1973) *Emission study of industrial sources of lead air pollutants 1970*. US EPA, Document APTD-1543, pp. 1–123.

Dedolph, R. G., Ter Haaer, R., Holtzman, R., & Lucas, H. J. (1970) Sources of lead in perennial ryegrass and radishes. *Environ. Sci. Technol.*, **4**: 217–223.

Deknudt, G., Leonard, A., & Ivanov, B. (1973) Chromosome aberrations observed in male workers occupationally exposed to lead. *Environ. Physiol. Biochem.*, **3**: 132–138.

Delves, H. T. (1970) A micro sampling method for the rapid determination of lead in blood by atomic absorption spectrometry. *Analyst*, **95**: 431.

Dimitrova, M. (1972) Modifications of the contractile function of the myocard in chronic lead poisoning. *Arch. Mal. prof. Med. Trav.*, **33**: 383–387.

Dingwall-Fordyce, J., & Lane, R. E. (1963) A follow-up study of lead workers. *Brit. J. ind. Med.*, **20**: 313–315.

Dinischioutu, G. T., Nestorescu, B., Radielescu, J. C., Jonescu, C., Preda, N., & Hutza, G. (1960) Studies on the chemical forms of urinary lead. *Brit. J. ind. Med.*, **17**: 141–145.

Djurić, D. (1964) Fluorimetric determination of porphyrius. *Arch. environ. Health*, **9**: 742–744.

Djurić, D., Kerin, Z., Graovac-Leposavić, L., Novak, L., & Kop, M. (1971) Environmental contamination by lead from a mine and smelter; A preliminary Report. *Arch. environ. Health*, **23**: 275–279.

Djurić, D., Novak, L., Milic, S., & Kalic-Filipovic, D. (1966) Delta-ALA as an early sign of lead exposure. *Med. Lav.*, **57**: 161–166.

Dodić, S., Vidaković, A., Perišić, V., & Stefanović, S. (1971) *Stanje jetre u pojedinih profesionalnih intoksikacija*. In: *III Jugoslovanski Kongres Medicine Dela, Ljubljana, 1971*. Ljubljana, 20–24 September, 1971, Jzdange "Lek", pp. 285–293.

Donovan, D. T., Vought, V. M., & Rakow, A. B. (1971) Laboratories which conduct lead analysis on biologic specimens. *Arch. environ. Health*, **23**: 111–113.

Dresel, E. J. B., & Falk, J. E. (1954) Studies on the biosynthesis of blood pigments. Haem synthesis in hemolysed erythrocytes of chicken blood. *Biochem. J.*, **56**: 156–163.

Dressen, W. C., Edwards, T. J., Reinhart, W. H., Page, R. T., Webster, S. H., Armstrong, D. W., & Sayers, R. R. (1941) The control of the lead hazard in the storage battery industry. *Publ. Health Bull. (Wash.)*, No. 262.

Dugandžić, M., Stanković, B., Milovanović, Ly., & Koričanac, Z. (1973) Urinary excretion of 5-hydroxindolacetic acid in lead exposed persons. *Archiv. Hig. Rada.*, **24**: 37–43.

Durfor, C., & Becker, E. (1964) Selected data on public supplies of the 100 largest cities in the United States, 1962. *J. Am. Water Works Assoc.*, **56**: 237–246.

Ediger, R. D., & Coleman, R. L. (1973) An evaluation of anodic stripping voltamentry and non-flame atomic absorption as routine analytical tools. In: *Proceedings of the 6th Annual Conference on Trace Substances in the Environment, Columbia, MO., 13–15 June 1972*, pp. 279–283.

141

EGOROV, V. V., ZHIGALOVSKAJA, T. N., & MALAKHOV, S. G. (1970) Microelement content of surface air above the continent and the ocean. *J. geophys. Res.*, **75**: 18, 3650–3656.

EGOROVA, G. M., IVANOV, N. G., & SANOCSKIJ, J. V. (1966) Specificity of the effect of lead on spermatogenesis. *Toksikol., nov. prom. him., veščestv.*, **8**: 33–41.

EINBRODT, H. J., ROSMANITH, J., & SCHROEDER, A. (1974) Beim Spielen, Schulkind Vergiftet, *Umwelt*, 3: 726.

EISENBUD, M., & WRENN, M. E. (1970) *Radioactivity studies annual report NYO-3086-10 V1., 1970*. Springfield, VA, National Technical Information Service.

ELKINS, H. B. (1959) *The chemistry of industrial toxicology*. New York, J. Wiley & Sons (2nd edition), pp. 51–52.

EMMERSON, B. T. (1963) Chronic lead nephropathy: The diagnostic use of calcium EDTA and the association with gout. *Austr. Ann. Med.*, **12**: 310–324.

EMMERSON, B. T., MIROSH, W., & DOUGLAS, J. B. (1971) The relative contributions of tubular reabsorption and secretion to urate excretion in lead nephropathy. *Aust. N.Z. J. Med.*, **1**: 353–362.

ENGELS, L. H. VON, & KÜHNEN, G. (1973) Staubschutz in der Akkumulatoren-Industrie Bonn, Staubforschungsinstitut des Hauptverbandes der gewarblichen Berufsgenossenschaften, STF-Report No. 2, pp. 27.

ENGEL, R. E., HAMMER, D. J., HORTON, R. J. M., LANE, N. M., & PLUMLEE, L. A. (1971) *Environmental lead and public health*. Research Triangle, Park, NC Environmental Protection Agency. Air Pollution Control Office Publication No. AP 90, pp. 1–34.

ENVIRONMENT AGENCY, JAPAN (1975) *Results of air pollution survey*. Tokyo, pp. 148–153.

EPSTEIN, S. S., & MANTEL, N. (1968) Carcinogenicity of tetraethyllead. *Experientia Basel*, **24**: 580–581.

ERVE, P. R., & SCHUMER, W. (1972) Endotoxin sensitivity of adrenalectomized rats treated with lead acetate. *Res. J. Reticuloend. Soc.*, **11** (38): 427.

FAIREY, F. S., & GREY, J. W. (1970 Soil lead and pediatric lead poisoning in Charleston, S.C. *J. S. Carolina med. Assoc.*, **66**: 79–82.

FAIRHALL, L. T., & MILLER, J. W. (1941) A study of the relative toxicity of the molecular components of lead arsenate. *Public Health Rep. (Wash.)*, **56**: 1610–1625.

FEDERAL INSTITUTE FOR MINERALS RESEARCH AND GERMAN INSTITUTE FOR ECONOMIC RESEARCH (1972). Supply and demand for lead. (Translated into English by Lead Development Association, 34 Berkeley Square, London W1X 6AJ, April, 1972), pp. 1–47.

FERM, V. H. (1969) The synteratogenic effect of lead and cadmium. *Experientia Basel* **25**: 56–57.

FILKINS, J. P., & BUCHANEN, B. J. (1973) Effects of lead acetate on sensitivity to shock, intravascular carbon and endotoxin clearances, and hepatic endotoxin detoxification. *Proc. Soc. Exp. Biol. Med.*, **142**: 471–475.

FINE, P. R., THOMAS, C. W., SUHS, R. H., COHNBERG, R. E., & FLASHNER, B. A. (1972) Pediatric blood lead levels: A study in 14 Illinois cities of intermediate population. *J. Am. Med. Assoc.*, **221** (13): 1475–1479.

FORBES, G. B., & REINA, J. C. (1972) Effect of age on gastrointestinal absorption (Fe, Sr, Pb) in the rat. *J. Nutr.*, **102**: 647–652).

FORNI, A., & SECCHI, C. C. (1973) *Chromosome changes in preclinical and clinical lead poisoning and correlation with biochemical findings*. In: *Proceedings of the International Symposium, Environmental Aspects of lead, Amsterdam, 2–6 October 1972*. Luxembourg, Commission of the European Communities, pp. 473–483.

FOUTS, P. J., & PAGE, J. H. (1942) The effect of chronic lead poisoning on arterial blood pressure in dogs. *Am. Heart J.*, **24**: 329–331.

FREEMAN, R. (1965) Reversible myocarditis due to chronic lead poisoning in childhood. *Arch. Dis. Child.*, **40**: 389–393.

FRIBERG, L., PISCATOR, M., NORDBERG, G., & KJELLSTRÓM, T. (1974) *Cadmium in the environment*. 2nd ed. Ohio, Chemical Rubber Company.

FRIED, J. F., ROSENTHAL, N. W., & SCHUBERT, J. (1956) Induced accumulation of citrate in therapy of experimental lead poisoning. *Proc. Soc. Exp. Biol. Med.*, **92**: 331–333.

Fugaš, M., Wilder, B., Pauković, R., Hrsak, J., & Steiner-Škreb, D. (1973) *Concentration levels and particle size distribution of lead in the air of an urban and an industrial area as a basis for the calculation of population exposure.* In: *Proceedings of the International Symposium; Environmental Health Aspects of Lead, Amsterdam, 2–6 October 1972.* Luxembourg, Commission of the European Communities, pp. 961–968.

Fullerton, P. M. (1966) Chronic peripheral neuropathy produced by lead poisoning in guinea pigs. *J. Neuropathol. exp. Neurol.*, **25**: 214–236.

Gage, J. C., & Litchfield, M. H. (1968) The migration of lead from polymers in the rat gastro-intestinal tract. *Food Cosmet. Toxic.*, **6**: 329–338.

Gage, J. C., & Litchfield, M. H. (1969) The migration of lead from paint films in the rat gastro-intestinal tract. *J. Oil Col. Chem. Assoc.*, **52**: 236–243.

Gajdos, A., & Gajdos-Török, M. (1969 L'activité de l'acide $\delta$-aminolevulinique synthétase hépatique et médullaire au cours du saturnisme expérimental du lapin. *C.R. Soc. Biol.*, **163**: 60–63.

Gajdos, A., & Gajdos-Török, M. (1973) Erreur du diagnostique differentiel entre l'intoxication par le plomb et la porphyrie intermittente aiguë. In: Proceedings of the International Symposium; Environmental Health Aspects of Lead, Amsterdam, 2–6 October, 1972. Luxembourg, Commission of the European Communities, pp. 501–505.

Galle, P., & Morel-Marogen, L. (1965) Les lésions rénales du saturnisme humain et expérimental. *Nephron*, **2**: 273–286.

Galzigna, L., Brugnone, F., & Corsi, G. C. (1964) Excretion of 5-hydroxyindole acetic acid in experimental tetraethyllead poisoning. *Med. Lavoro.*, **55**: 102–106.

Garber, B. T., & Wei, E. (1974) Influence of dietary factors on the gastrointestinal absorption of lead. *Toxic. appl. Pharm.*, **27**: 685–691.

Georgii, H. W., & Jost, D. (1971) On the lead-concentration in an urban aerosol. *Atmosph. Environ.*, **5**: 725–727.

Ghelberg, N. W., Gorgan, J., & Checin, J. (1966) 5-hydroxyindoleacetic acid excretion in a population chronically exposed to low lead concentration in the atmosphere. *Igiena (Buc)* **15**: 87–92.

Gibson, K. G., Neuberger, A., & Scott, J. J. (1955) The purification and properties of $\delta$-aminolaevulic acid dehydrase. *Biochem. J.*, **61**: 618–629.

Gibson, S. L. M., Lam, C. N., McCrae, W. M., & Goldberg, A. (1967) Blood lead levels in normal and mentally retarded children. *Arch. Dis. Child.*, **42**: 573–578.

Gillet, J. A. (1955) An outbreak of lead poisoning in the Canklow district of Rotterdam. *Lancet*, **1**: 1118–1121.

Goldberg, A. (1972) Lead poisoning and haem biosynthesis. *Brit. J. Haematol.*, **23**: 521–523.

Goldberg, A. (1974) Drinking water as a source of lead pollution. *Environ. Health Perspect.*, Exp. Issue No. 7, pp. 103–107.

Goldberg, A., Ashenbrucker, H., Cartwright, G. E., & Wintrobe, M. M. (1956) Studies on the biosynthesis of heme *in vitro* by arian erythrocytes. *Blood*, **11**: 821–833.

Goldberg, E. D. (1976) *The health of the oceans. Chapter 5. Heavy metals: lead.* Paris, The Unesco Press, pp. 109–111.

Goldsmith, J. R., & Hexter, A. C. (1967) Respiratory exposure to lead epidemiological experimental dose-response relationships. *Science*, **158**: 132–134.

Golubović, E. Y., & Gevkovskaja, T. V. (1967) Effects of ethyleneimine and lead on several sides of nucleic acids changes in the spermary of rats. *Toksikol. nov. prom. him. veščestv*, **9**: 86–91.

Golubović, E. Y., Avhimenko, M. M., & Čirkova, E. M. (1968) Biochemical and morphological changes in testicles of rats under the effect of low doses of lead. *Toksikol. nov. prom. him. veščestv*, **10**: 64–72 (in Russian).

Goyer, R. A., & Krall, K. (1969) Ultrastructural transformation in mitochondria isolated from kidneys of normal and lead-intoxicated rats. *J. cell. Biol.*, **41**: 393–400.

Goyer, R. A., & Mahaffey, K. R. (1972) Susceptibility to lead toxicity. *Environ. Health Perspect.*, No. 2: 73–80.

Goyer, R. A., & Rhyne, B. C. (1973) Pathological effects of lead. *Intern. Rev. exp. Pathol.*, **12**: 1–77.

GOYER, R. A., LEONARD, D. L., MORRE, J. F., RHYNE, B., & KRIGMAN, M. R. (1970a) Lead dosage and the role of the intranuclear inclusion body. *Arch. environ. Health*, **20**: 705–711.

GOYER, R. A., MAY, P., CATES, M. M., & KRIGMAN, M. R. (1970b) Lead and protein content of isolated inclusion bodies from kidneys of lead poisoned rats. *Lab. Invest.*, **22**: 245–251.

GOYER, R. A., TSUCHIYA, K., LEONARD, D. L., & KAHYO, H. (1972) Aminoaciduria in Japanese workers in the lead and cadmium industries. *Am. J. clin. Pathol.*, **57**: 635–642.

GRABECKI, J., HADUCH, T., & URBANOWICZ, H. (1967) Die einfachen Bestimmungsmethoden der δ-Amino-lävulinsäure in Harn. *Int. Arch. Gewerbpath. Gewerbhyg.*, **23**: 226–240.

GRANICK, J. L., SASSA, S., GRANICK, S., LEVERE, R. D., & KAPPAS, A. (1973) Studies in lead poisoning II Correlation between the ratio of activated to inactivated aminolevulinic acid dehydratase of whole blood and the blood lead level. *Biochem. Med.*, **8**: 149–159.

GRANICK, S., SASSA, S., GRANICK, J. L., LEVERE, R. D., & KAPPAS, A. (1972) Assays for porphyrins. *Proc. Nat. Acad. Sci. USA*, **69**: 2382–2385.

GRAOVAC-LEPOSAVIĆ, L., DJURIĆ, D., VALJAREVIĆ, V., SENIČAR, L., MILIĆ, S., & DELIĆ, V. (1973) *Environmental lead contamination of Meža Valley—Study on lead exposure of populations.* In: *Proceedings of the International Symposium: Environmental Health Aspects of Lead, Amsterdam, 2–6 October 1972.* Luxembourg, Commission of the European Communities, pp. 685–703.

GREENFIELD, S. M., BRIDBORD, K., BARTH, D., & ENGEL, R. (1973) *The changing perspectives regarding lead as an environmental pollutant.* In: *Proceedings of the International Symposium: Environmental Health Aspects of lead, Amsterdam, 2–6 October 1972.* Luxembourg, Commission of the European Communities, pp. 19–27.

GRIFFITH, J. Q., & LANDAUER, M. A. (1944) The effect of chronic lead poisoning on arterial blood pressure in rats. *Am. Heart J.*, **28**: 295–297.

GRIGGS, R. C., SUNSHINE, J., NEWILL, V. NEWTON, B. W., BUCHANAN, S., & RASCH, C. A. (1964) Environmental factors in childhood lead poisoning. *J. Am. Med. Assoc.*, **187**: 703–707.

GRIMES, H., SAYERS, M. H. P., CERNIK, A. A., BERLIN, A., RECHT, P., & SMEETS, J. (1975) *Note on the lead exposure of children: determinations carried out on behalf of the Commission in Western Ireland.* Luxembourg, Commission of the European Communities (Report V/F/1491/75), pp. 7.

GUINEE, V. F. (1973) *Epidemiologic studies of lead exposure in New York city.* In: *Proceedings of the International Symposium; Environmental Health Aspects of Lead, Amsterdam, 2–6 October 1972.* Luxembourg, Commission of the European Communities, pp. 763–770.

GUTNIAK, O., KOZIOLOWA, H., & KOWALSKI, E. (1964) Free protoporphyrin content of erythrocytes in chronic tetraethyl lead poisoning. *Lancet*, **1**: 1137–1138.

HAAS, T., MACHE, K., SCHALLER, K.-H., MACHE, W., & VALENTIN, H. (1972a) Investigations into ecological lead levels in children.(Untersuchungen über die ökologische Bleibelastung im Kindesalter). *Zbl. Bakt. Hyg. I. Abt. Orig.*, B **156**: 353–360.

HAAS, T., WIEK, A. G., SCHALLER, K. H., MACHE, K., & VALENTIN, R. (1972b) Die usuelle Bleibelastung bei Neugeborenen und ihren Müttern. *Zbl. Bakt. Hyg., I. Abt. Orig.*, B **155**, 341–349.

HAČIROV, D. G. (1972) Data on electronmicroscopic investigations of erythrocytes and thrombocytes under saturnism. *Naučn. Tr. Sev.-Oset. Med. Inst.*, **36** (4): 59–63.

HAEGER-ARONSEN, B. (1960) Studies on urinary excretion of δ-aminolaevulic acid and other haem precursors in lead workers and lead-intoxicated rabbits. *Scand. J. clin. Lab. Invest.*, **12** (Suppl. 47) 1.

HAEGER-ARONSEN, B. (1971) An assessment of the laboratory tests used to monitor the exposure of lead workers. *Brit. J. ind. Med.*, **28**: 52–58.

HAEGER-ARONSEN, B., ABDULLA, M., & FRISTEDT, B. (1971) Effect of lead on δ-aminolevulinic acid dehydrase activity in red blood cells. *Arch. environ. Health*, **23**: 440–445.

HAMMOND, P. B. (1971) The effects of chelating agents on the tissue distribution and excretion of lead. *Toxicol. Pharmacol.*, **18**: 296–310.

HAMMOND, P. B., & ARONSON, A. L. (1964) Lead poisoning in cattle and horses in the vicinity of a smelter. *Ann. N.Y. Acad. Sci.*, **III**: 595–611.

HAMMOND, P. B., WRIGHT, H. N., & ROEPKE, M. H. (1956) A method for the detection of lead in bovine blood and liver. *University of Minnesota. Agric. Exp. Stat. Techn. Bull.*, **221**: 3–14.

HANKIN, L., HEICHEL, G. H., & BOTSFORD, R. A. (1973) Lead poisoning from coloured printing inks. *Clin. Pediatr.*, **12**: 654–655.

HAPKE, H. J. (1974) Effects and damage by lead, cadmium and zinc on useful animals. *Staub Reinhaltung der Luft*, **34** (1): 8–11.

HAPKE, H. J., & PRIGGE, E. (1973) Interactions of lead and glutathione with delta-aminolevulinic acid dehydratase. *Arch. Toxicol.*, **31**: 153–161.

HARRIS, R. W., & ELSEA, W. R. (1967) Ceramic glaze as a source of lead poisoning. *J. Am. med. Assoc.*, **202**: 544–549.

HARRISON, R. M., PERRY, R., & SLATER, D. H. (1974) An adsorption technique for the determination of organic lead in street air. *Atmos. Environ.*, **8**: 1187–1194.

HASAN, J., & HERNBERG, S. (1966) Interactions of inorganic lead with human blood cells. *Work environ. Health*, **2**: 26–44.

HELLER, A., & KETTNER, H. (1969) Probenahme und Bestimmung kleinster Bleimengen in der Luft. *Wasser-Boden- und Lufthygiene*, No. 29, 3–50.

HEMPHILL, F. E., KAEBERLE, M. L., & BUCK, W. B. (1971) Lead suppression of mouse resistance to salmonella typhimurium. *Science*, **172**: 1031–1032.

HENDERSON, D. A. (1958) The etiology of chronic nephritis in Queensland. *Med. J. Austr.*, **1**: 377–386.

HENDERSON, D. A., & INGLIS, J. A. (1957) The lead content of bone in chronic Bright's disease. *Austr. Ann. Med.*, **6**: 145–154.

HERNBERG, S. (1967) Lifespan, potassium fluxes and membrane ATPases of erythrocytes from subjects exposed to inorganic lead. *Work environ. Health*, **3**, suppl. 1, pp. 1–74.

HERNBERG, S. (1973) Prevention of occupational poisoning from inorganic lead. *Work environ. Health*, **10**: 53–61.

HERNBERG, S. (1975) *Lead* In: Zenz, C. (ed.) *Occupational Medicine Yearbook Pt 4: The Chemical Occupational Environment*, Chicago, pp. 715–769.

HERNBERG, S., LILIUS, H., MELLIN, G., & NIKKANEN, J. (1969) Lead exposure of workers in printing shops. *Work environ. Health*, **6** (2): 5–8.

HERNBERG, S., NIKKANEN, J., MELLIN, G., & LILIUS, H. (1970) $\delta$-aminolevulinic acid dehydrase as a measure of lead exposure. *Arch. environ. Health*, **21**: 140–145.

HEUSGEM, C., & deGRAEVE, J. (1973) *Importance de l'apport alimentaire en plomb dans l'est de la Belgique*. In: *Proceedings of the International Symposium; Environmental Health Aspects of Lead, Amsterdam, 2–6 October 1972*. Luxembourg, Commission of the European Communities, pp. 85–91.

HICKS, J. M., GUTIERREZ, A. N., & WORTHY, B. E. (1973) Evaluation of the Delves micro-system for blood lead analysis. *Clin. Chem.*, **19**: 322–325.

HILL, C. R. (1960) Lead-210 and polonium-210 in grass. *Nature (Lond.)*, **187**: 211–212.

HM CHIEF INSPECTOR OF FACTORIES (1972) H.M. Great Britain Department of Employment. Annual Report. London, Her Majesty's Stationery Office.

HOFFMAN, E. O., DI LUZIO, N. R., HOLPER, K., BRETTSCHNEIDER, L., & COOVER, J. (1974). Ultrastructural changes in the liver of baboons following lead and endotoxin administration. *Lab. Invest.*, **30**: 311–319.

HOFREUTER, D. H., CATCOTT, E. J., KEENAN, R. G., & XINTARAS, A. B. (1961) The public health significance of atmospheric lead. *Arch. environ. Health*, **3**: 568–574.

HOLMQVIST, J. (1966) Normal lead values in blood. 15th International Congress of Occupational Health. Vienna 19–24 Sept., A-III-1, pp. 179–183.

HOMMA, K. (1966) Experimental study of preparing metal fumes. *Ind. Health*, **4**: 129–137.

HOPKINS, A. P., & DAYAN, A. D. (1974) The pathology of experimental lead encephalopathy in the baboon *(Papio amibis)*. *Brit. J. ind. Med.*, **31**: 128–133.

HORIGUCHI, S., & UTSUNOMIYA, T. (1973) An estimate of the body burden of lead in the healthy Japanese population. An attempt to assume absorption and excretion of lead in the healthy Japanese population, Part 2. *Osaka City, med. J.*, **19**: 1–5.

HORIUCHI, K., (1970) Lead in the environment and its effect on man in Japan. *Osaka City, med. J.*, **16**: 1.

HORIUCHI, K., & TAKADA, J. (1954) Studies on the industrial lead poisoning. I. Absorption, transportation, deposition and excretion of lead, 1. Normal limits of lead in the blood, urine and feces among healthy Japanese urban inhabitants. *Osaka City, med. J.*, **1**: 117–125.

HORIUCHI, K., HORIGUCHI, T., KASAHARA, A., MORIOKA, T., UTSUNOMYA, T., & SHINAGAWA, K. (1964) Influences of temperature on manifestation of symptoms in industrial poisoning. 1. Influences of high temperature in lead poisoning. *Jpn. J. ind. Health*, **6**: 170–171.

HORIUCHI, K., HORIGUCHI, S., SHINAGAWA, K., TAKADA, F., & TERAMOTO, K. (1968) A polarographic method for the determination of a small amount of lead in biological materials. *Osaka City, med. J.*, **14**: 113–118.

HORIUCHI, K., HORIGUCHI, S., & SUEKANE, M. (1959) Studies on the industrial lead poisoning. I. Absorption, transportation, deposition and excretion of lead. 6. The lead contents in organ tissues of the normal Japanese. *Osaka City, med. J.*, **5**: 41–70.

HORIUCHI, K., YAMAMOTO, T., & TAMORI, E. (1956) Studies on the industrial lead poisoning. I. Absorption, transportation, deposition and excretion of lead. 2. A study on the lead content in daily food in Japan. *Osaka City, med. J.*, **3**: 84–113.

HOWER, J., PRINZ, B., GONO, E., & REUSMANN, G. (1975) *Untersuchungen zum Zusammenhang zwischen dem Blutbleispiegel bei Neugeboren en und der Bleiimmissionsbelastung der Mütter am Wohnort.* In: *Proceedings of CEC-EPA-WHO International Symposium; Recent Advances in the Assessment of the Health Effect of Environmental Pollution, Paris, 24–28 June 1974.* Luxembourg, Commission of the European Communities, pp. 591–603.

HUNT, W. F., PINKERTON, C., MCNULTY, O., & CREASON, J. (1971) *A study in trace element pollution in 77 midwestern cities.* In: ed. D. D. Hemphill *Proceedings of the University of Missouri's 4th Annual Conference on Trace Substances in Environmental Health, Columbia, Missouri, 23–25 June 1970.* University of Missouri, Columbia, 1971, pp. 56–68.

HUNTZICKER, J. J., FRIEDLANDER, S. K., & DAVIDSON, C. J. (1975) Material balance for automobile-emitted lead in Los Angeles basin. *Environ. Sci. Technol.*, **9**: 448–457.

HURSH, J. B., & MERCER, T. T. (1970) Measurement of $Pb^{212}$ loss rate from human lungs. *J. appl. Physiol.*, **28**: 268.

HWANG, J. Y., ULLUCCI, P. A., SMITH, S. B., & MALENFANT, A. L. (1971) Microdetermination of lead in blood by flameless atomic absorption spectrometry. *Anal. Chem.*, **43**: 1319–1321.

IARC (1972) Monographs. Evaluation of carcinogenic risk of chemicals to man. Vol. 1, Lyon, International Agency for Research on Cancer, pp. 184.

INTER-DEPARTMENT WORKING GROUP ON HEAVY METALS (1974) *Lead in the environment and its significance to man.* Report for the Department of the Environment Central Unit on Environmental Pollution. London, Her Majesty's Stationery Office, pp. 1–47, Pollution Paper No. 2.

INTERNATIONAL LEAD AND ZINC STUDY GROUP (1973) Lead in gasoline: A review of the current situation, International Lead and Zinc Study Group, United Nations, New York, pp. 1–37. First addendum 1974; Second addendum 1975; Third addendum 1976.

IUPAC-IUB COMMISSION ON BIOCHEMICAL NOMENCLATURE (1973) *Enzyme nomenclature.* Recommendations (1972) of the Commission on Biochemical Nomenclature on the nomenclature and classification of enzymes together with their units and the symbols of enzyme kinetics. Amsterdam, Elsevier.

JARVIE, A. W. P., MARKALL, R. N., & POTTER, H. R. (1975) Chemical alkylation of lead. *Nature (Lond.)*, **255**: 217–218.

JAWOROSKI, Z. (1968) Stable lead in fossil ice and bones. *Nature (Lond.)*, **217**: 152–153.

JERNIGAN, E. L., RAY, B. J., & DUCE, R. A. (1971) Lead and bromine in atmospheric particulate matter on Oahu, Hawaii. *Atmosph. Environ.*, **5**: 881–886.

JONES, R. D., COMMINS, B. T., & CERNIK, A. A. (1972) Blood lead in carboxyhemoglobin levels in London taxi drivers. *Lancet*, **2**: 302–303.

JOST, D., MULLER, J., & JENDRICKE, U. (1973) *Lead in atmospheric aerosol.* In: *Proceedings of the International Symposium; Environmental Health Aspects of Lead, Amsterdam, 2–6 October 1972. Luxembourg, Commission of the European Communities, pp. 941–948.*

KAMMHOLZ, L. P., THATCHER, L. G., BLODGETT, F. M., & GOOD, T. A. (1972) Rapid protoporphyrin quantitation for detection of lead poisoning. *Pediatrics*, **50**: 625–631.

KASSENAAR, A., MORELL, H., & LONDON, I. U. (1957) The incorporation of glycine into globin and the synthesis of heme *in vitro* in duck erythrocytes. *J. Biol. Chem.*, **229**: 423–435.

KATO, K. (1932) Lead meningitis in infants. *Am. J. Dis. Child.*, **44**: 569–591.

KEENAN, R. C., BYERS, D. H., SALTZMAN, B. E., & HYSLOP, F. L. (1963) The "USPHS" method for determining lead in air and in biological materials. *Am. Ind. Hyg. Assoc.*, **24**: 481–491.

KEHOE, R. A. (1961) The metabolism of lead in health and disease. The Harben Lectures, 1960. *J. Roy. Inst. Publ. Health Hyg.*, **24**: 81–96, 101–120, 129–143, 177–203.

KEHOE, R. A., CHOLAK, J., & LARGENT, E. J. (1944) The concentrations of certain trace metals in drinking water. *J. Am. Water Works. Assoc.*, **36**: 637–644.

KEHOE, R. A., THAMAN, F., & CHOLAK, J. (1933) Lead absorption and excretion in relation to the diagnosis of lead poisoning. *J. ind. Hyg.*, **15**: 320.

KELLO, D., & KOSTIAL, K. (1973) The effect of milk diet on lead metabolism in rats. *Environ. Res.*, **6**: 355–360.

KEMPF, TH., & SONNEBORN, M. (1973) Vergleich von Methoden zur Bestimmung von Spurenmetallen in Wasserkreislauf. *Z. anal. Chem.*, **267**: 267–270.

KENNEDY, V. C. (1960) Geochemical studies in the Coeur d'Alene district Shoshone County, Idaho. *U.S. geol Survey Bull.*, 1098-A.

KEPPLER, J. F., MAXFIELD, M. E., MOAS, W. D., TIETJEN, C., & LINCH, A. L. (1970) Interlaboratory evaluation of the reliability of blood lead analyses. *Am. Ind. Hyg. Assoc. J.*, **31**: 412–429.

KERIN, Z. (1972) Tägliche Bleiaufnahme mit der Bauernkost aus dem Emissionsgebiet einer Bleihütte. *Protectio vitae*, **71**: 22–23.

KERIN, Z. (1973) Lead in new-fallen snow near a lead smelter. *Arch. environ. Health*, **26**: 256–260.

KERIN, Z., KERIN, D., & DJURIĆ, D. (1972) Lead contamination of environment in Meza Valley. Lead content of the soil. *Int. Arch. Arbeitsmed.*, **29**: 129–138.

KLEIN, U. C., SAYRE, J. W., & KOTOK, D. (1974) *Am. J. Dis. Childh.*, **127**: 805–807. porphyrin in duck erythrocytes. *Am. J. Physiol.*, **203**: 971–974.

KLEIN, M., NANER, R., HARPUR, E., & CORBIN, R. (1970) Earthenware containers as a source of fatal lead poisoning. Case study and public-health considerations. *New Eng. J. Med.*, **283**: 669–672.

KLEIN, U. C., SAYRE, J. W., & KOTOK, D. (1974). *Am. J. Dis. Child.*, **127**: 805–807.

KLINE, T. S. (1960) Myocardial changes in lead poisoning. *A.M.A.J. Dis. Child.*, **99**: 48–54.

KNEIP, T. J., & LAUREN, G. R. (1972) Isotope excited X-ray fluorescence. *Anal. Chem.*, **44**: 57A.

KOBAYASHI, N., & OKAMOTO, T. (1974) Effects of lead oxide on the induction of lung tumors in Syrian hamsters. *J. Nat. Cancer Inst.*, **52** (5): 1605–1610.

KOLBYE, A. C., MAHAFFEY, R., FIORINO, A., CORNELIUSSEN, P. C., & JELINEK, C. F. (1974) Food exposures to lead. *Environ. Health Perspect.*, Exp. Issue No. 7, pp. 65–75.

KOPP, J. F., & KRONER, P. T. (1970) *Trace metals in water of the United States: a 5-year summary of trace metals in rivers and lakes of the US.* (Oct. 1962–Sept. 1967) Cincinnati, Ohio, US Department of the Interior.

KOSMIDER, S., & PETELNZ, T. (1962) Zmiany elektrokardiograficzne u starszych osob z przewleklym zawodowym zatruciem olowiem. *Pol. Arch. Med. Wewn*, **32**: 437–442.

KOSMIDER, S., & SROCZYNSKI, J. (1961) Zmiany elektrokardiograficzne w przewleklej doswiadczalney olowiej u krolikow (Electrocardiographic changes in chronic experimental plumbism in rabbits). *Postepy Hig. i Med. Dosw.*, **15**: 353–357.

147

KOSTIAL, K., & VOUK, V. B. (1957) Lead ions and synaptic transmission in the superior cervical ganglion of the cat. *Brit. J. Pharmacol., Chemother.*, **12**: 219–222.

KOSTIAL, K., MALJKOVIČ, T., & JUGO, S. (1974) Lead acetate toxicity in rats in relation to age and sex. *Arch. Toxicol.*, **31**: 265–269.

KOSTIAL, K., ŠIMONOVIĆ, I., & PISONIĆ, U. (1971) Lead absorption from the intestine in new born rats. *Nature (Lond.)*, **233**: 564.

KOTOK, D. (1972) Development of children with elevated blood lead levels; A controlled study. *J. Pediatr.*, **80**: 57–61.

KRIGMAN, U. R., DRUSE, U. J., TAYLOR, T. D., WILSON, U. H., NEWELL, L. R., & HOGAN, E. l (1974) Lead encephalopathy in the developing rat: Effect upon myelination. *Neuropath. exp. Neurol.*, **33** (1): 58–74.

KUBASIK, N. P., VALOSIN, U. T., & MURRAY, U. H. (1972) Carbon rod atomizer applied t measurement of lead in whole blood by atomic absorption spectrophotometry. *Cli. Chem.*, **18**: 410–412.

KUZ'MINSKAJA, T. N. (1964) Experimental atherosclerosis against a background of Pb intoxication. *Arch Patol.*, **26** (9): 21–24.

LAHMANN, E. (1969) Untersuchungen über Luftverunreinigungen durch den Kraftverkehr *Wasser-boden- und Lufthyg.*, No. 28, 3–80.

LAMM, S. H., & ROSEN, J. F. (1974) Lead contamination in milk fed to infants: 1972–197: *Pediatrics*, **53**: 137–141.

LAMOLA, A. A., & YAMANE, T. (1974) Zinc protoporphyrin in the erythocytes of patients wit lead intoxication and iron deficiency anaemia. *Science*, **186**: 936–938.

LAMOLA, A. A., JOSELOW, M., & YAMANE, T. (1975) Zinc protoporphyrin (ZPP): A simpl sensitive, fluorometric screening test for lead poisoning. *Clin. Chem.*, **21**: 93–97.

LAMPTER, P. W., & SCHOCHET, S. S. (1968) Demylination and remyelination in lea neuropathy. Electron microscopic studies. *J. Neuropath. exp. Neurol.*, **25**: 527 545.

LANCRANJAN, I., POPESCU, H. I., GĂVĂNESCU, O., KLEPSCH, I., & SERBĂNESCU, M. (1975) Reproductive ability of workmen occupationally exposed to lead. *Arch. environ. Health* **30**: 396–401.

LANE, R. E. (1949) The care of the lead worker. *Brit. J. ind. Med.*, **6**: 125–143.

LANE, R. E. (1964) Health control in inorganic lead industries. A follow-up of exposed workers. *Arch. environ. Health*, **8**: 55.

LANDING, B. H., & NAKAI, H. (1959) Histochemical properties of renal lead inclusions an: their demonstration in urinary sediment. *Am. J. clin. Path.*, **31**: 499–503.

LANDRIGAN, P. J., BALCH, R. W., BARTHEL, W. F., WHITWORTH, R. H., STAEHLING, N. W., & ROSENBLUM, B. F. (1975a) Neurophysiological dysfunction in children with chronic low level lead absorption. *Lancet*, **1**: 708–712.

LANDRIGAN, P. J., GEHLBACH, S. H., ROSENBLUM, B. F., SHOULTS, J. M., CANDELARIA, R. M BARTHEL, W. F., LIDDLE, J. A., SMREK, A. L., STAEHLING, N. W., & SANDERS, J. F. (1975b Epidemic lead absorption near an ore smelter; The role of particulate lead. *New Engl. J Med.*, **292**: 123–129.

LANSDOWN, R. G., CLAYTON, B. E., GRAHAM, P. J., SHEPHERD, J., DELVES, H. T., & TURNER W. C. (1974) Blood-lead levels, behaviour and intelligence: a population study. *Lancet* **1**: 538–541.

LAURS, A. J. (1976) *Review of European test methods for measuring lead and cadmium release* In: *Proceedings of the International Conference on Ceramic Foodware Safety, Geneva 1974.* New York, Lead Industries Association Inc., pp. 27–36.

LAVESKOG, A. (1971) *A method for determination of tetramethyl lead and tetraethyl lead in air* *Proceedings of the Second International Clean Air Congress*, H. H. England and W. T Beery, eds. New York, Acad. Press, pp. 549–557.

LAWTHER, P. J., COMMINS, B. T., ELLISON, J. MEK., & BILES, B. (1972) *Airborne lead and its uptake by inhalation.* In: Hepple, P. ed., *Lead in the environment.* Essex, UK, Applied Science Publishers Ltd., pp. 8–28.

LAZRUS, A. L., LORANGE, E., & LODGE, J. P. (1970) Lead and other metalions in US precipitation. *Environ. Sci. Technol.*, **4**: 55–58.

LEAD DEVELOPMENT ASSOCIATION (1976) Lead In Gasoline Bulletin No. 12, Lead Development Association, 34 Berkeley Square, London W1X 6AJ, 30 March 1976.

ᴇ, R. E., & GORANSON, S. (1972) National air surveillance cascade impactor network. I. Size distribution measurements of suspended particulate matter in air. *Environ. Sci. Technol.*, **6**: 1019–1024.

ᴇ, R. E., GORANSON, S. S., ENRIONE, R. E., & MORGAN, G. B. (1972) The NASH cascade impactor network, II Size distribution of trace-metal components. *Env. Sci. Technol.*, **6**: 1025–1030.

LEHNERT, G., MASTALL, H. SZADKOWSKI, D., & SCHALLER, K. H. (1970) Berufliche Bleibelastung durch Autoabgase in Grosstadtstrassen. *Dtsch. Med. Wschr.*, **95**: 1097–1099.

ᴇHNERT, G., SCHALLER, K. H., KÜHNER, A., & SZADKOWSKI, D. (1967) Auswirkungen des Zigarettenrauchens aus dem Blutbleisspiegel. *Int. Arch. Geweberpath. Geweberhyg.*, **23**: 358–363.

ᴇHNERT, G., SCHALLER, K. H., & SZADKOWSKI, D. (1969) Eine zuverlässige Schnellmethode zur Bleibestimmung in kleinen Blutmengen. *Z. Klin. Chem. Klin. Biochem.*, **7**: 310.

LᴇIKIN, S., & ENG, G. (1963) Erythrokinetic studies of the anemia of lead poisoning. *Pediatrics*, **31**: 996–1002.

ᵀILIS, R., GAVRILESCU, N., NESTORESCU, B., DURIMTIU, C., & ROVENTA, A. (1968) Nephropathy in chronic lead poisoning. *Brit. J. ind. Med.*, **25**: 196–202.

INCH, A. L., WIEST, E. G., & CARTER, U. D. (1970) Evaluation of tetraalkyl lead exposure by personnel monitor surveys. *Am. Ind. Hyg. Assoc. Y.*, **31**: 170–179.

ᴵVINGSTONE, D. A. (1963) Data of geochemistry. Sixth Edition. Chapter of chemical composition of rivers and lakes. US Geol. Survey Prof. Paper 440.

MACHLE, W. E. (1935) Tetraethyl lead intoxication and poisoning by related compounds of lead. *J. Am. Med. Assoc.*, **105**: 578–585.

MAHOFFEY, K. (1974) Nutritional factors and susceptibility to lead toxicity. *Environ. Health Perspect.*, Exp. Issue No. 7, pp. 107–113.

ᴵAKAŠEV, K. K., & KRIVDINA, L. V. (1972) Status of the interstitial tissues of vascular walls and their penetration under the lead poisoning. *Tr. Naučn-issl. In-ta Kraev. Pat. Kaz. SSR*, **23**: 11–13.

ᴵAKAŠEV, K. K., & VERBOLOVIĆ, V. P. (1967) Succinic dehydrogenase and cytochrome oxidase in the duodenum during lead poisoning. *Izv. Akad. Nauk Kazahst. SSR Ser. Biol.*, **5** (2): 59–64.

ᴹALCOLM, D. (1971) Prevention of long-term sequelae following the absorption of lead. *Arch. environ. Health*, **23**: 292–298.

ᴵALJKOVIĆ, J. (1971) A case of occupational poisoning with lead carbonate and stearate. *Sigurnost u pogonu.*, **13**: 123–124.

ᴵAMBEEVA, A. A., & AHMEDOVA, A. S. (1967) Changes of reactivity of the small intestine in healthy and lead-intoxicated homeotherms to histamine under the influence of solutions of lead acetate. *Biull. eksper. Biol.*, **32**: 34–37.

MAMBEEVA, A. A., & KOBKOVA, I. D. (1969) The concentration of catecholamines in tissues of the cardiovascular system in experimental lead intoxication. *Izv. Akad. Nauk Kazahst. SSR. Ser. Biol.*, No. 1: 77–82.

ᴵANALIS, R. S., & COOPER, G. I. (1973) Presynaptic and postsynaptic effects of lead at the frog neuromuscular junction. *Nature (Lond.)*, **243**: 354–355.

ᴵAO, P., & MOLNAR, J. J. (1967) The fine structure and histochemistry of lead-induced renal tumours in rats. *Am. J. Pathol.*, **50**: 571–603.

ᴹAPPES, R. (1972) Eine Fehlerquelle und ihre Kompensation bei der $\delta$-Aminolävulinsäurebestimmung im Harn nach Grabecki. *Int. Arch. Arbeitsmed.*, **30**: 81–86.

ᴵARTIN, A. E., FAIRWEATHER, F. A., BUXTON, R. S. J., & ROOTS, L. M. (1975) *Recent epidemiological studies of environmental lead of industrial origin.* In: *Proceedings of the CEP-EPA-WHO International Symposium; Recent Advances in the Assessment of the Health Effects of Environmental Pollution*, Paris, 24–28 June 1974. Luxembourg, Commission of the European Communities, pp. 1113–1120.

MARVER, H. S., TSCHUDY, D. P., PERLROTH, M. G., COLLINS, A., & HUNTER, G. JR. (1966) The determination of aminoketones in biological fluids. *Anal. Biochem.*, **14**: 53–60.

MATOUSEK, J. F., & STEVENS, B. J. (1971) Biological applications of the carbon rod atomizer in atomic absorption spectroscopy. I. Preliminary studies on magnesium, iron, copper, lead and zinc in blood in plasma. *Clin. Chem.*, **17**: 363–368.

MATSON, W. R. (1971) *Rapid sub-nanogram simultaneous analysis.* In: *Proceedings of the University of Missouri's 4th Annual Conference on Trace Substances in Environmental Health, Columbia, Missouri, 23–25 June 1970.* Columbia, ed. D. D. Hemphill, University of Missouri, pp. 396–406.

MATTSSON, R., & JAAKKOLA, T. (1974) Lead in the Helsinki air. *Ympäristö ja Terveys*, **5**: 721.

MAUZERALL, D., & GRANICK, S. (1956) The occurrence and determination of $\delta$- aminolevulinic acid and porphobilinogen in urine. *J. Biol. Chem.*, **219**: 435–446.

MAXFIELD, M. E., STOPPS, G. J., BARNES, J. P., SNEE, R. D., & AZAR, A. (1972) Effect of lead on blood regeneration following acute hemorrhage in dogs. *Am. Ind. Hyg. Assoc. J.*, **33**: 326–337.

MCCABE, L. J. (1970) Corrosion by soft water. Amer. Water Works Seminar on Corrosion by Soft Water. Wash. DC, pp. 9.

MCLAUGHLIN, M., & STOPPS, G. J.(1973) Smoking and Lead. *Arch. environ Health*, **26**:131–136.

MCLAUGHLIN, M., LINCH, A. L., & SNEE, R. D. (1973) Longitudinal studies of lead levels in US population. *Arch. environ. Health*, **27**: 305–312.

MCMULLEN, T. B., FAORO, R. B., & MORGAN, G. B. (1970) Profile of pollutant fractions in non-urban suspended particulate matter. *J. Air Pollut. Control Assoc.*, **20**: 369–372.

MCNEIL, J. L., & PTASNIK, J. A. (1975) *Evaluation of long-term effects of elevated blood lead concentrations in asymptomatic children.* In: *Proceedings of CEC-EPA-WHO International Symposium; Recent Advances in the Assessment of the Health Effects of Environmental Pollution, Paris, 24–28 June 1974.* Luxembourg, Commission of the European Communities, pp. 571–579.

MENDEN, R. E., ELIA, V. J., MICHAEL, L. W., & PETERING, H. C. (1972) Distribution of Cd and Ni of tobacco during cigarette smoking. *Env. Sci. Technol.*, **6**: 830–832.

MERCER, T. T. (1975) The deposition model of the Task Group on Lung Dynamics: A comparison with recent experimental data. *Health Phys.*, **29**: 673–680.

MERWIN, B. W. (1976) *Review of U.S. standards on test methods for lead and cadmium release.* In: *Proceedings of the International Conference on Ceramic Foodware Safety, Geneva, 1974.* New York, Lead Industries Association Inc., pp. 36–40.

MICHAELSON, J. A. (1973) Effect of inorganic lead on levels of RNA, DNA and protein on developing neonatal rats. *Toxicol. appl. Pharmacol.*, **26**: 539–548.

MICHAELSON, J. A., & SAUERHOFF, M. W. (1974) An improved model of lead induced brain dysfunction in the suckling rat. *Toxicol. appl. Pharmacol.*, **28**: 88–96.

MILLAR, J. A., BATTISTINI, V., CUMMING, R. L. C., CARSWELL, F., & GOLDBERG, A. (1970) Lead and $\delta$-aminolevulinic acid dehydratase levels in mentally retarded children and in lead-poisoned suckling rats. *Lancet*, **2**: 695–698.

MIRONČIK, L. M., & TIMOFEEVA, N. D. (1974) Lead effects on the experimental atherosclerosis process and several respiratory ferments of the myocard. *Probl. Gig. i organiz zdravoochr. v Uzbekistane*, **2**: 77–78.

MITCHELL, D. G., & ALDOUS, K. M. (1974) Lead content of foodstuffs. *Environ. Health Perspect.*, Exp. Issue No. 7, pp. 59–65.

MITCHELL, D. G., ALDOUS, K. M., & RYAN, F. J. (1974) Mass screening for lead poisoning, capillary blood sampling and automated Delves cup atomic absorption analysis. *N.Y. State, J. Med.*, **74**: 1599.

MITCHELL, R. L. (1963) Soil aspects of trace element problems in plants and animals. *J. Roy. Agric. Soc.*, **124**: 75–86.

MITCHELL, R. L., & REITH, J. W. S. (1966) The lead content of pasture herbage. *J. Sci. Food Agric.*, **17**: 437–440.

MÖLLER, B., AKSELSSON, R., BERLIN, M., FISCHBEIN, A., HAMMARSTRÖM, L. (1974) Sammanfatting av föredrag vid Läkaresällskapets Riksstämma 27–30 Nov., 1974 (Omgivningshygien No. 7).

Momčilović, B., & Kostial, K. (1974) Kinetics of lead retention and distribution in suckling and adult rats. *Environ. Res.*, **8**: 214–220.

Monaenkova, A. M. (1957) Functional state of the thyroid gland in chronic intoxication from several industrial poisons (Pb, Hg). *Gig. Trud. Prof. Zabol.*, **2**: 44–48.

Monaenkova, A. M., & Glotova, K. V. (1969) Cardiovascular changes in some chronic intoxications with industrial poisons (lead, carbon disulphide, benzene). *Gig. Trud. Prof. Zabol.*, **2**: 32–35.

Moncrieff, A. A., Koumides, O. P., Clayton, B. E., Patrick, A. D., Renwick, A. G. C., & Roberts, G. E. (1964) Lead poisoning in children. *Arch. Dis. Child.*, **39**: 1–13.

Moore, J. F., & Goyer, R. A. (1974) Lead-induced inclusion bodies; Composition and probable role in lead metabolism. *Environ. Health Perspect.*, Exp. Issue No. 7, pp. 121–129.

Moore, M. R. (1973) Plumbosolvency of waters. *Nature Lond.*, **243**: 222–223.

Morgan, B. B., & Repko, J. D. (1974) *Evaluation of behavioural functions in workers exposed to lead.* In: C. Xintaras *et al.*, eds., *Behavioural toxicology, early detection of occupational hazards.* Washington, U.S. Department of Health, Education and Welfare.

Morgan, J. M., Hartley, M. W., & Miller, R. E. (1966) Nephropathy in chronic lead poisoning. *Arch. intern. Med.*, **118**: 17–29.

Mueller, P. K. (1970) Discussion (characterization of particulate lead in vehicle exhaust experimental techniques). *Environ. Sci. Technol.*, **4**: 248–251.

Muir, D. C. F., & Davies, C. N. (1967) The deposition of 0.5 μ diameter aerosols in the lungs of man. *Ann. occup. Hyg.*, **10**: 161–174.

Muro, L. A., & Goyer, R. A. (1969) Chromosome damage in experimental lead poisoning. *Arch. Pathol.*, **87**: 660–663.

Murozumi, M., Chow, T. J., Patterson, C. C. (1969) Chemical concentrations of pollutant lead aerosols, terrestrial dusts and sea salts in Greenland and Antarctic snow strata. *Geochem. Cosmochim. Acta*, **33**: 1247–1294.

Murthy, G. M., & Rhea, U. S. (1971) Cadmium, copper, iron, lead, manganese and zinc in evaporated milk, infant products and human milk. *J. Dairy Sci.*, **54**: 1001–1005.

Myerson, R. M., & Eisenhauer, J. H. (1963) Atrioventricular conduction defects in lead poisoning. *Am. J. Cardiol.*, **11**: 409–412.

Nakao, K., Wada, O., & Yano, Y. (1968) Delta-aminolevulinic acid dehydratase activity in erythrocytes for the evaluation of lead poisoning. *Clin. Chim. Acta.*, **19**: 319–325.

NAS-NRC (1972) *Airborne lead in perspective.* Washington, DC, Nat. Acad. Sci.

Neal, P. A., Dressen, W. C., Edwards, T. J., Reinhart, W. H., Webster, S. H., Castberg, H. T., & Fairhall, L. T. (1941) A study of the effect of lead arsenate exposure on orchardists and consumers of sprayed fruit. *US publ. Health Bull.*, No. 267.

Needleman, H. L., & Shapiro, J. M. (1974) Dentine lead levels in asymptomatic Philadelphia school children: Subclinical exposure in high and low risk groups. *Environ. Health, Perspect.*, Exp. Issue. No. 7, pp. 27–33.

Nelson, W. C., Lykins, M. H., Mackey, J., Newill, V. A., Finklea, J. F., & Hammer, D. J. (1973) Mortality among orchard workers exposed to lead arsenate spray: A cohort study. *J. chron. Dis.*, **26**: 105–118.

Nikkanen, J., Hernberg, S., & Tola, S. (1972) Modifications of the δ-aminolevulinic acid dehydratase test and their significance for assessing different intensities of lead exposure. *Work-environ.-Health*, **9**: 46–52.

Niklowitz, W. J., & Yeager, D. W. (1973) Interference of Pb with essential brain tissue Cu, Fe and Zn as main determinant in experimental tetraethyllead encephalopathy. *Life Sci.*, **13**: 897–905.

Nordberg, G. F., ed. (1976) Effects and dose-response relationships of toxic metals. *Proceedings from an international meeting organized by the subcommittee on the Toxicology of Metals of the Permanent Commission and International Association on Occupational Health, Tokyo, 18–23 November 1974.* Amsterdam, Elsevier, 1976, p. 15.

Nordman, C. H. (1975) *Environment lead exposure in Finland. A study on selected population groups.* Doctoral Thesis, University of Helsinki, pp. 1–118.

151

NORDMAN, C. H., & HERNBERG, S. (1975) Blood lead levels and erythrocyte δ-aminolevulinic acid dehydratase activity of selected population groups in Helsinki. *Scand. J. Work-Environ.-Health*, **1**: 219–232.

NORDMAN, C. H., HERNBERG, S., NIKKANEN, J., & RYHÄNEN, A. (1973) Blood lead levels and erythrocyte-δ-aminolevulinic acid dehydratase activity in people living around a secondary lead smelter. *Work-Environ.-Health*, **10**: 19–25.

NOZAKI, K. (1966) Method for studies on inhaled particles in human respiratory system and retention of lead fume. *Ind. Health (Jpn)*, **4**: 118–128.

NRC, CANADA (1973) *Lead in the Canadian Environment.* Associate Committee on Scientific Criteria for Environmental Quality.

NYE, L. J. J. (1929) An investigation of the extraordinary incidence of chronic nephritis in young people in Queensland. *Med. J. Aust.*, **2**: 145–159.

OEHLKERS, F. (1953). Chromosome breaks influenced by chemicals. *Heredity*, **6** (suppl.) 95–105.

OKAZAKI, H., ARONSON, S. M., DiMAIO, D. J., & OLVERA, J. E. (1963) Acute lead encephalopathy of childhood. *Trans. Am. neurol. Assoc.*, **88**: 248–250.

OLIVER, T. (1914) *Lead Poisoning.* London, H. K. Lewis.

ORLOVA, A. A. (1954) Changes in cardiac activity of patients with lead and mercury intoxication. *Trud. Akad. Med. Nauk. SSSR.*, **31**: 102–112.

O'RIORDAN, M. L., & EVANS, H. J. (1974) Absence of significant chromosome damage in males occupationally exposed to lead. *Nature (Lond.)*, **247**: 50–53.

OTT, W., CLARKE, J. F., & OZOLINS, G. (1970) *Calculating future carbon monoxide emissions and concentrations from urban traffic data.* Durham, DHEW Publ. no. AP-41.

OYASU, R., BATTIFORA, H. A., CLASEN, R. A., McDONALD, J. H., & HASS, G. M. (1970) Induction of cerebral gliomas in rats with dietary lead subacutate and 2-acetylaminofluorene. *Cancer Res.*, **30**: 1248–1261.

PACKHAM, R. F. (1971) The leaching of toxic stabilizers for unplasticized PVC water pipe: P.1 A critical study of laboratory test procedures, P.2. A survey of lead levels in PVC distribution systems. *Water Treat. Exam.*, **20**: 108–124; 144–166.

PADILLA, F., SHAPIRO, A. P., & JENSEN, W. N. (1969) Effect of chronic lead intoxication on blood pressure in the rat. *Am. J. med. Sci.*, **258**: 359–365.

PALMISANO, P. A., SNEED, R. C., & CASSADY, G. (1969) Untaxed whisky and fetal lead exposure. *J. Pediatr.*, **75**: 869.

PANOVA, Z. (1972) Early changes in the ovarian function of women in occupational contact with inorganic lead. *Works United Res. Inst. Hyg. ind. Saf. (Sofia)*, **23**: 161–166.

PATEL, A. J., MICHAELSON, J. A., CREMER, J. E., & BALAZS, R. (1974a) The metabolism of ($C^{14}$) glucose by the brains of suckling rats intoxicated with inorganic lead. *J. Neurochem.*, **22**: 581–590.

PATEL, A. J., MICHAELSON, J. A., CREMER, J. E., & BALAZS, R. (1974b) Changes within metabolic compartments in the brains of young rats ingesting lead. *J. Neurochem.*, **22**: 591–598.

PATTERSON, C. C. (1965) Contaminated and natural lead environments of man. *Arch environ. Health*, **11**: 344–363.

PEGUES, W. L. (1960) Lead fume from welding on galvanized and zinc-silicate coated steels. *Ind. Hyg. J.*, **21**: 252–255.

PEKKARINEN, M. (1970) Methodology in the collection of food consumption data. *World Rev. Nut. Diet.*, **12**: 145–170.

PENTSCHEW, A. (1965) Morphology and morphogenesis of lead encephalopathy. *Acta Neuropathol.*, **5**: 133–160.

PENTSCHEW, A., & GARRO, F. (1966) Lead encephalo-myelopathy of the suckling rat and its implications on the porphyrinopathic nervous diseases. *Acta Neuropathol.*, **6**: 266–278.

PERNIS, B., DE PETRIS, S., BEARD, R. R., & KARLSBAD, G. (1964) The ultrastructure of red cells in experimental lead-poisoning. *Med. Lav.*, **55**: 81–101.

PINES, A. G. (1965) Indexes of general reactivity in saturnine. *Vrac. Delo*, **3**: 93–96.

PIOMELLI, S. (1973) A micromethod for free erythrocyte porphyrin: the FEP test. *J. Lab. clin. Med.*, **81**: 932–940.

152

POTT, F., & BROCKHAUS, A. (1971) Vergleich der enteralen und pulmonalen Resorptionsquote von Bleiverbindungen. *Zentralbl. Bakt. Hyg. J. Orig. B.*, **155**: 1–17.

PŘEROVSKÁ, I. (1973) *Einfluss von Blei auf biochemische Veränderungen im Serum und Veränderungen in der Aderwand im Hinblick auf Atherosklerose.* In: *Proceedings of the International Symposium. Environmental Health Aspects of Lead, Amsterdam, 2–6 October 1972.* Luxembourg, Commission of the European Communities, pp. 551–561.

PRPIĆ-MAJIC, D., MUELLER, P. K., LEW, V. C., & TWISS, S. (1973) δ-aminolevulinic acid dehydratase stability in human blood. *Am. ind. Hyg. Assoc. J.*, 315–319.

PUESCHEL, S. M., KOPITO, L., & SCHWACHMAN, H. (1972) A screening and follow up study of children with an increased lead burden. *J. Am. Med. Assoc.*, **333**: 462–466.

PUHAČ, J., HRGOVIĆ, N., STANKOVIĆ, M., & POPOVIĆ, S. (1963) Laboratory investigations of the possibility of application of lead nitrates compounds as a raticide means by decreasing reproductive capability of rats. *Acta Vet. (Beograd)*, **13**: 3–9.

PURDUE, L. J., ENRIONE, R. E., THOMPSON, R. J., & BONFIELD, B. A. (1973) Determination of organic and total lead in the atmosphere by atomic absorption spectrometry. *Anal. Chem.*, **45**: 527–530.

RABINOWITZ, M. B. (1974) Lead contamination of the biosphere by human activity. A stable isotope study. Ph. D. Thesis, University of California, Los Angeles.

RABINOWITZ, M. B., WETHERILL, G. W., & KOPPLE, J. D. (1973) Lead metabolism in the normal human: stable isotope studies. *Science*, **182**: 725–727.

RABINOWITZ, M. B., WETHERILL, G. W., & KOPPLE, J. D. (1974) Studies of human lead metabolism by use of stable isotope tracers. *Environ. Health. Perspect*, Exp. Issue No. 7, pp. 145–155.

RASBERRY, S. D. (1973) Investigation of portable X-ray Fluorescence analyzers for determining lead on painted surfaces. *Appl. Spectrosc.*, **27** (2): 102–108.

RICHET, G., ALBAHARY, C., MOREL-MAROGER, L., GUILLAUME, P., & GALLE, P. (1966) Les altérations rénales dans 23 cas de saturnisme professionnel. *Bull. Mem. Soc. Med. Hop. Paris*, **117**: 441–466.

RIEKE, F. E. (1969) Lead intoxication in shipbuilding and shipscrapping 1941–1968. *Arch. environ. Health*, **19**: 521–539.

RILEY, J. P., & SKIRROW, C. (1965) *Chemical oceanography.* New York, Academic Press, vol. 2.

RIMINGTON, C. (1938) An enzymic theory of haemopoiesis. *C.R. Lab Carlsberg. Ser. Chim.*, **22**: 454–464.

RIMINGTON, C., & SVEINSSON, S. L. (1950) The spectrophotometric determination of uroporphyrin. *Scand. J. clin. Lab. Invest.*, **2**: 209–216.

ROBERTS, T. M., HUTCHINSON, T. C., PACIGA, J., CHATTOPADHYAY, A., JERVIS, R. E., VAN-LOON, J., & PARKINSON, D. R. (1974) Lead contamination around secondary smelters: Estimation of dispersal and accumulation by humans. *Science*, **186**: 1120–1123.

ROBINSON, E., & LUDWIG, F. L. (1967) Particle size distribution of urban lead aerosols. *J. Air Pollut. Control Assoc.*, **17**: 664–669.

ROBINSON, E., LUDWIG, F. L., DEVries, J. E., & HOPKINS, T. E. (1963) *Variations of atmospheric lead concentrations and type with particle size. Final Report PA-4211.* Cal. Stanford Res. Inst. Menlo Park, pp. 1–80.

ROBINSON, T. R. (1974) Delta-aminolevulinic acid and lead in urine of lead antiknock workers. *Arch. Environ. Health*, **28**: 133–139.

ROE, F. Y., BOYLAND, C. E., DUKES, C. E., & MITCHLEY, B. V. C. (1965) Failure of testosterone or xanthopterin to influence the induction of renal neoplasms by lead in rats. *Brit. J. Cancer*, **19**: 860–866.

ROELS, H. A., BUCHET, J. P., & LAUWERYS, R. R. (1974) Inhibition of human erythrocyte δ-aminolevulinate dehydratase by lead. *Int. Arch. Arbeitsmed.*, **3**: 277–281.

ROELS, H. A., LAUWERYS, R. R., BUCHET, J. P., & VRELUST, M. (1975) Response of free erythrocyte porphyrin and urinary δ-aminolevulinic acid in men and women moderately exposed to lead. *Int. Arch. Arbeitsmed.*, **34**: 97–108.

ROSENBLUM, W. J., & JOHNSON, M. G. (1968) Neuropathologic changes produced in suckling mice by adding lead to the maternal diet. *Arch. Pathol.*, **85**: 640–648.

Rühling, A., & Tyler, G. (1968) An ecological approach to the lead problem. *Bot. Notiser,* **121**: 321–342.

Ruiter, N. de, Seemayer, N., & Manojlovic, N. (1977) Einfluss von Zink-Ionen auf die toxische Wirkung von Bleichlorid ($PbCl_2$) untersucht an Mäusemakrophagen *in vitro. Zentralbl. Bakt. Hyg., I. Abt. Orig. B.,* **164**: 90–98.

Sachs, H. K. (1974) Effect of a screening program on changing patterns of lead poisoning. *Environ. Health Perspect.,* Exp. Issue No. 7, pp. 41–47.

Sakurai, H., Sugita, M., & Tsuchiya, K. (1974) Biological response and subjective symptoms in low level lead exposure. *Arch. environ. Health,* **29**: 157–163.

Sandstead, H. H. (1967) Effect of chronic lead intoxication on *in vivo* I-131 uptake by the rat thyroid. *Proc. Soc. Exp. Biol. Med.* **124**: 18–20.

Sandstead, H. H., Orth, D. N., Abe, K., & Stiel, J. (1970) Lead intoxication: Effect on pituitary and adrenal function in man. *Clin. Res.,* **18**: 76 (abstract).

Sandstead, H. H., Stant, E. G., & Brill, H. B. (1969) Lead intoxication and the thyroid. *Arch. int. Med.,* **123**: 632–635.

Sano, S., & Rimington, C. (1963) Excretion of various porphyrins and their corresponding porphyrinogens by rabbits after intravenous injection. *Biochem. J.,* **86**: 203–212.

Sansoni, B., Kracke, W., Dietl, F., & Fisher, J. (1973) *Mikrospurenbestimmung von Blei in verschiedenartigen Umweltproben durch flammenlose Atomabsorption nach externer Nassveraschung mit $H_2O_2/Fe^2$.* In: *Proceedings of the International Symposium; Environmental Health Aspects of lead, Amsterdam, 2–6 October 1972.* Luxembourg, Commission of the European Communities, pp. 1107–1116.

Sassa, S., Granick, J. L., Granick, S., Kappas, A., & Levere, R. D. (1973) Studies in lead poisoning. I. Microanalysis of erythrocyte protoporphyrin levels by spectrofluorometry in the detection of chronic lead intoxication in the subclinical range. *Biochem. Med.,* **8**: 135–148.

Sayre, J. W., Charney, E., Vostal, J., & Pless, B. (1974) House and land dust as a potential source in childhood lead exposure. *Am. J. Dis. Child.,* **127**: 167–170.

Scarlato, G., Smirne, S., & Poloni, A. E. (1969) L'encefalopatia saturnina acuta dell'adulto. *Acta Neurol.,* **24**: 578–580.

SCEP (1970) *Study of critical environmental problems. Man's impact on the global environment.* Cambridge, Mass. London, England. Mit Press.

Schaller, K. H., Lindner, K., & Lehnert, G. (1968) Atomabsorptionsspektrometrische Schwermetallbestimmung im Trinkwasser. *Arch. Hyg. Bakteriol.,* **152**: 298–301.

Schiele, R., Schaller, K. H., & Valentin, H. (1974a) The influence of lead upon the tryptophan metabolism. *Klin. Wschr.,* **52**: 401–404.

Schiele, R., Schaller, K. H., & Wagner, H. M. (1974b) Die Bestimmung der freien Erythrozytenporphyrine als schneller Suchtest einer erhöhten Bleiexposition und seine Validität im vergleich zum Blutbleispiegel und zur Delta-Aminolävulinsäure-Dehydratase-Aktivität. SchrReihe Ven. *Wasser-Baden. Lufthyg. Berlin,* **41**: 231–240.

Schlaepfer, W. W. (1969) Experimental lead neuropathy: A disease of the supporting cells in the peripheral system. *J. Neuropathol. exp. Neurology,* **28**: 401–418.

Schlegel, H., Kufner, G., & Leinberger, H. (1972) *Die Praxis der Verhütung von Bleibeschädigungen in der metallverarbeitenden Industrie.* In: *Kommission für Umweltgefaren des Bundesgesundheitsamtes, Arbeitsgruppe Blei und Umwelt, Berlin,* pp. 67–69.

Schlegel, H., Kufner, G., & Leinberger, H. (1973) *Das Verhalten verschiedener Parameter der Hämsynthesestörung am Menschen bei experimenteller Aufnahme anorganischer Bleiverbindungen.* In: *Proceedings of the International Symposium; Environmental Health Aspects of Lead, Amsterdam, 2–6 October 1972.* Luxembourg, Commission of the European Communities, pp. 569–579.

Schlipköter, H. W., & Pott, F. (1973) *Die pulmonale Resorption von Bleistaub.* In: *Proceedings of the International Symposium; Environmental Health Aspects of Lead, Amsterdam, 2–6 October 1972.* Luxembourg, Commission of the European Communities, pp. 403–412.

Schlipköter, H. W., Ghelerter, L., & Ost, B. (1975) Untersuchungen zur Kombinationswirkung von Zink und Blei. *Zentralbl. Bakt. Hyg. I. Abt. Orig. B.,* **160**: 130–138.

SCHLIPKÖTER, H. W., IDEL, H., STILLER-WINKLER, R., & KRAUSE, G. H. M. (1977) Die Beeinflussung der Infektionsresistenz durch inhalierte Noxen. *Zentralbl. Bakt. Hyg., I. Abt. Orig. B.* (in press).

SCHMID, E., BAUCHINGER, M., PIETRUCK, S., & HALL, G. (1972) Die cytogenetische Wirkung von Blei in menschlichen peripheren Lymphocyten *in vitro* und *in vivo. Mutat. Res.* **16**: 401–406.

SCHRAMEL, P. (1973) Determination of eight metals in the international biological standard by flameless atomic-absorption spectrometry. *Analytica Chimica Acta.,* **67**: 69–77.

SCHRAMEL, P. (1974) The application of peak integration in flameless atomic-absorption spectrometry. *Analytica Chimica Acta.,* **72**: 414–418.

SCHROEDER, H. A., & TIPTON, I. H. (1968) The human body burden of lead. *Arch. environ. Health,* **17**: 965–978.

SCHROEDER, H. A., & BALASSA, J. J. (1961) Abnormal trace metals in man: lead. *J. chron. Dis.,* **14**: 408–425.

SCHÜTZ, A., DENCKER, I., & NORDÉN, A. (1971) Bly i kosten-undersökning med dubbelportionsteknik i Dalby. Svenska Läkeresällskapet, Medicinska Rikstämman (sammanfattningar), p. 424.

SCHWARTZ, S., & WIKOFF, H. (1952) The relation of erythrocyte coproporphyrin and protoporphyrin to erythropoiesis. *J. Biol. Chem.,* **194**: 563–573.

SCHWARTZ, S., ZIEVE, L., & WATSON, C. J. (1951) An improved method for the determination of urinary coproporphyrin and an evaluation of factors influencing the analysis. *J. Lab. Clin. Med.,* **37**: 843–859.

SCHWANITZ, G., GEBHART, E., ROTT, H. D., SCHALLER, K. H., ESSING, H. G., LAUER, O., & PRESTELE, H. (1975) Chromosome investigations in subjects with occupational lead exposure. *Dtsch. Med. Wschr.,* **100**: 1007–1011.

SCHWANITZ, G., LEHNERT, G., & GEBHART, E. (1970) Chromosomenschäden bei beruflicher Bleibelastung. *Dtsch. Med. Wschr.,* **95**: 1636–1641.

SECCHI, G. C., & ALESSIO, L. (1974) Laboratory results of some biological measures in workers exposed to lead. *Arch. environ. Health,* **29**: 351–355.

SECCHI, G. C., ALESSIO, L., & CAMBIOGGHI, G. (1971) Ricerche sull'attivita ALA-deidrasica eritrocitaria di soggetti non esposti a contatto professionale con plombo ed abitanti in zone rurali ed urbane. *Med. Lav.,* **62**: 435–450.

SECCHI, G. C., ALESSIO, L., & CAMBIOGGHI, G. (1973) Na/K-ATPase activity of erythrocyte membranes. *Arch. environ. Health,* **28**: 131–132.

SECCHI, G. C., ERBA, L., & CAMBIOGGHI, G. (1974) Delta-aminolevulinic acid dehydratase activity of erythrocytes and liver tissue in man. *Arch. environ. Health,* **28**: 130–133.

SELANDER, S., & CRAMER, K. (1970) Interrelationships between lead in blood, lead in urine and ALA in urine during lead work. *Brit. J. ind. Med.,* **27**: 28–39.

SELYE, H., TUCHWEBER, B., & BERTOK, L. (1966) Effect of lead acetate on the susceptibility of rats to bacterial endotoxins. *J. Bacteriol.,* **91**: 884–890.

SEPPÄLÄINEN, H. M., & HERNBERG, S. (1972) Sensitive technique for detecting subclinical lead neuropathy. *Brit. J. ind. Med.,* **29**: 443–449.

SEPPÄLÄINEN, A. M., TOLA, S., HERNBERG, S., & KOCK, B. (1975) Subclinical neuropathy at "safe" levels of lead exposure. *Arch. environ. Health,* **30**: 180–183.

SERVANT, J. (1973) Transport du plomb dans l'environnement. In: *Proceedings of the International Symposium; Environmental Health Aspects of Lead, Amsterdam, 2–6 October 1972.* Luxembourg, Commission of the European Communities, pp. 155–165.

SHELDON, R. P., WARNER, M. A., THOMPSON, M. E., & PEIRCE, H. W. (1953) Stratigraphic sections of the phosphoria formation in Idaho, 1949, Part I. *U.S. geol. Survey Cric.,* **304**: pt. I, p. 1.

SHIRAISHI, Y. (1975) Cytogenetic studies of 12 patients with Itai-Itai disease. *Humangentik,* **27**: 31–44.

SHIRAISHI, Y., KURAHASHI, H., & YOSIDA, T. H. (1972) Chromosomal aberrations in cultured human leucocytes induced by cadmium sulfide. *Proc. Jpn. Acad.,* **48**: 133–137.

SILBERGELD, E. K., & GOLDBERG, A. M. (1974) Lead-induced behavioral dysfunction: an animal model of hyperactivity. *Exp. Neurol.,* **42**: 146–157.

SILBERGELD, E. K., FALES, J. T., & GOLDBERG, A. M. (1974) Evidence for a functional effect of lead on neuromuscular function. *Nature (Lond.)*, **247**: 49–50.

SILVER, W., & RODRIQUEZ-TORRES, R. (1968) Electrocardiographic studies in children with lead poisoning. *Pediatrics*, **41**: 1124–1127.

SIX, K. M., & GOYER, R. A. (1970) Experimental enhancement of lead toxicity by low dietary calcium. *J. Lab. Clin. Med.* **76**: 933–942.

SIX, K. M., & GOYER, R. A. (1972) The influence of iron deficiency on tissue content and toxicity of ingested lead in the rat. *J. Lab. Clin. Med.*, **79**: 128–136.

SMITH, H. D. (1964). Pediatric lead poisoning. *Arch. environ. Health*, **8**: 256–261.

SNOWDON, C. T. (1973) Learning deficits in lead-injected rats. *Pharmacol. Biochem. Behavior*, **1**: 599–603.

SNYDER, L. J. (1967) Determination of trace amounts of organic lead in air. *Anal. Chem.*, **39**: 591–595.

SOBEL, A. E., GAWRON, O., & KRAMER, B. (1938a) Influence of vitamin D in experimental lead poisoning. *Proc. Soc. Exp. Biol. Med.*, **38**: 433–435.

SOBEL, A. E., WEXLER, I. B., PETROVSKY, D. D., & KRAMER, B. (1938b) Influence of dietary calcium and phosphorus upon action of vitamin D in experimental lead poisoning. *Proc. Soc. exp. Biol. Med.*, **38**: 435–437.

SOLIMAN, M., EL-SADIK, Y., & EL-WASSEF, A. (1973) *Evaluation of some parameters of lead exposure and possible correlation between them.* In: *Proceedings of the International Symposium; Environmental Health Aspects of lead, Amsterdam, 2–6 October 1972.* Luxembourg, Commission of the European Communities, pp. 531–543.

SOLOMINA, V. F. (1961) Effect of lead acetate and silica on the development of experimental skin cancer. *Isvest. Akad. Nauk Kazakh. SSR. Ser. Med. i Fiziol.*, **2** (16): 55–67.

SPURGEON, J. G. (1973) Response characteristics of a portable X-ray fluorescence lead detector; detection of lead in paint. Report to the Department of Housing and Urban Development by the National Bureau of Standards, N13 SIR, pp. 73–231.

SROCZYNSKI, J., ZAJUSZ, K., KOSSMANN, S., & WEGIEL, A. (1967) Effect of experimental lead poisoning on the development of arteriosclerosis. *Pol. Arch. Med. Wewn.*, **38** (5): 641–646.

STAPLES, E. L. J. (1955) Experimental lead-poisoning in dogs. *N.Z. vet. J.*, **3**: 39–46.

STOPPS, G. J. (1969) Discussion on epidemiological bases for possible air quality criteria for lead. *Air Pollut. Control Assoc.*, **19**: 719–721.

STOWE, H. D., & GOYER, R. A. (1971) The reproductive ability and progeny of F, lead-toxic rats. *Fertil. Steril.*, **22**: 755–760.

STOWE, H. D., GOYER, R. A., & KRIGMAN, M. R. (1973) Experimental oral lead toxicity in young dogs. *Arch. Pathol.*, **95**: 106–116.

STRAND, L. J., MANNING, J., & MARVER, H. S. (1972) The induction of $\delta$-aminolevulinic acid synthetase in cultured liver cells. *J. biol. Chem.*, **247**: 2820–2824.

STUBBS, R. L. (1975) Lead and zinc in 1975, Mining Annual Review—1976, pp. 45–49.

STUIK, E. J. (1974) Biological response of male and female volunteers to inorganic lead. *Int. Arch. Arbeitsmed.*, **33**: 89–97.

STUIK, E. J., & ZIELHUIS, R. L. (1975) *Increased susceptibility of females to inorganic lead.* In: *Proceedings of CEC-EPA-WHO International Symposium; Recent Advances in the Assessment of the Health Effects of Environmental Pollution, Paris, 24–28 June 1974.* Luxembourg, Commission of the European Communities, pp. 537–545.

SUN, M. W., STEIN, E., & GRUEN, F. W. (1969) A single column method for the determination of urinary $\delta$-aminolevulinic acid. *Clin. Chem.*, **15**: 183–189.

SWAINE, D. J. (1955) *The trace-element content of soils.* Commonwealth Bur. Soil. Sci. Technol. Comm. No. 48.

SZADKOWSKI, D., SCHULTZE, H., SCHALLER, K. H., & LEHNERT, G. (1969) Zur ökologischen Bedeutung des Schwermetallgehaltes von Zigaretten. *Arch. Hyg. Bakt.*, **153**: 1–8.

TABERSHAW, J. R., & COOPER, W. C. (1974) *Health study of lead workers.* A report prepared for the International Lead and Zinc Research Organization.

TABERSHAW, J. R., RUOTOLO, B. P. W., & GLAESON, R. P. (1943) Plumbism resulting from oxyacetylene cutting of painted structural steel. *J. ind. Hyg. Toxicol.*, **25**: 189–191.

TASK GROUP ON LUNG DYNAMICS (1966) Deposition and retention mod dosimetry of the human respiratory tract. *Health Phys.*, **12**: 173–207.

TATSUMOTO, M., & PATTERSON, C. C., (1963) The concentration of common lea Earth Sci. *Meteorics*, 74–89.

TEISINGER, J. (1935) Biochemical reactions of lead in blood. *Biochem. Zeitschr*, 178–185.

TEISINGER, J., & SRBOVA, J. (1959) The value of mobilization of lead by calcium t diamine-tetra-acetate in the diagnosis of lead poisoning. *Brit. J. ind. Med.*, **16**: 1

TEISINGER, J., & STYBLOVA, V. (1961) Neurological picture of chronic lead poisoning. *Universitatis Carolinae*—Med. Supp. 14, 199–206.

TEPPER, L. B. (1963) Renal function subsequent to childhood plumbism. *Arch. em Health*, **7**: 76–85.

TEPPER, L. B., & LEVIN, L. S. (1972) A survey of air and population lead levels in select American communities. Final report to the US EPA.

TER HAAR, G. L. (1970) Air as a source of lead in edible crops. *Environ. Sci. Technol.*, **4**: 226–229.

TER HAAR, G. L., & ARONOW, R. (1974) New information on lead in dirt and dust as related to the childhood lead problem. *Environ. Health Perspect.* Experim. Issue No. 7, 83–89.

TER HAAR, G. L., & BAYARD, M. A. (1971) Composition of airborne lead particles. *Nature (Lond.)*, **232**: 553–554.

TER HAAR, G. L., HOLTZMAN, R. B., & LUCAS, H. F. (1967) Lead and lead-210 in rainwater. *Nature (Lond.)*, **216**: 353–355.

THOMAS, J. A., DALLENBACH, F. D., & THOMAS, M. (1971) Considerations on the development of experimental lead encephalopathy. *Virchows Arch. Abt. A. Path. Anat.*, **352**: 61–74.

THOMPSON, J. A. (1971) Balance between intake and output of lead in normal individuals. *Brit. J. ind. Med.*, **28**: 189–194.

THORNTON, I. & WEBB, J. S. (1975) *Trace elements in soils and surface waters contaminated by plast metaliferous mining in parts of England.* In: *Proceedings of the IXth Annual Conference on Trace Substances in Environmental Health, Columbia, MO, 10–12 June 1975*, pp. 77–88.

TOLA, S. (1973) Effect of blood lead concentration, age, sex and exposure time on the erythrocyte δ-aminolevulinic acid dehydratase activity. *Work environ. Health*, **10**: 26–35.

TOLA, S. (1974) Occupational lead exposure in Finland. III: Lead scrap smelteries and scrap metal shops. *Work Environ. Health*, **11**: 114–118.

TOLA, S., HERNBERG, S., ASP, S., & NIKKANEN, J. (1973) Parameters indicative of absorption and biological effect in new lead exposure: A prospective study. *Brit. J. ind. Med.*, **30**: 134–141.

TOLA, S., HERNBERG, S., NIKKANEN, J., & VALKONEN, S. (1971) Occupational lead exposure in Finland: 1. Electric storage battery manufacturing and repair. *Work environ. Health*, **3**: 81–85.

TOMOKUNI, K., & OGATA, M. (1972) Simple method for determination of urinary δ-aminolevulinic acid as an index of lead exposure. *Clin. Chem.*, **18**: 1534–1536.

TRUHAUT, R., BOUDERIC, C., & ALBAHARY, C. (1964) Rôle possible de la consommation exagérée de vin dans l'étiologie du saturnisme. *Bull. WHO*, **31**: 127–129.

TSUCHIYA, K., & HARASHIMA, S. (1965) Lead exposure and the derivation of maximum allowable concentrations and threshold limit values. *Brit. J. ind. Med.*, **22**: 181–186.

TSUCHIYA, K., SUGITA, M., SEKI, Y., KOBAYASHI, Y., HORI, M., & BIN PARK, C. (1975) *Study of lead concentrations in atmosphere and population in Japan.* In: Coulston et al., ed., *Environmental quality and safety. Supplement Vol. II, Lead.* Stuttgart, Georg Thieme and NY, Academic Press, pp. 95–146.

TUREKIAN, N. K., & WEDEPOHL, K. H. (1961) Distribution of the elements in some major units of the earth's crust. *Geol. Soc. Am. Bull*, **72**: 175–191.

s, Statistical Office (1975) Statistical Yearbook 1974, Twenty-sixth issue,
ations Department of Economic and Social Affairs, Statistical Office, New
190.

& Granick, S. (1963) Biosynthesis of α-aminoketones and the metabolism o
acetone. J. Biol. Chem., **238**: 811–820.

icz, H. (1971) Occupational exposure to inorganic compounds of lead. Arch
iron. Health, **23**: 284–288.

owicz, H., Grabecki, J., & Kozielska, J. (1969) The urinary excretion of 5
hydroxyindoleacetic acid in industrial lead exposure. Med. Lav., **60**: 582–586.

Bureau of Mines (1969) Minerals Yearbook 1968, vol. I–II; Metals, minerals, and fuel:
Washington, pp. 631–660.

S DHEW (1965) *Survey of lead in the atmosphere of three urban communities.* USPHS Publ.
No. 999 AP 12.

US Environmental Protection Agency Office of Research and Monitoring (1972)
Water pollution aspects of street surface contaminants. Washington DC (EPA-R2-72-
081).

Van Esch, G. J., & Kroes, R. (1969) The induction of renal tumours by feeding basic lea
acetate to mice and hamsters. Brit. J. Cancer, **23**: 765–771.

Van Esch, G. J., van Gendeven, H., & Vink, H. H. (1962) The induction of renal tumours b
feeding of basic lead acetate to rats. Brit. J. Cancer, **16**: 289–297.

Verbolovič, V. P. (1965) Peculiarities of cytochrome oxidase activity and of the localizatio
of myoglobin during lead poisoning. T. Kazahsk. Inst. Kraevoi Pat. Akad. Med. Nau
SSSR, **14**: 74–80.

Vigdortchik, N. A. (1935) Lead intoxication in the etiology of hypertonia. J. ind. Hyg., **1**
(1): 1–6.

Vinogradov, A. P. (1956) Regularity of distribution of chemical elements in the earth's crus
Geohimija, p. 6 (English translation in Geochemistry, 1956, pp. 1–43).

Vinogradov, A. P. (1962) Average contents of chemical elements in the principal types
igneous rocks of the earth's crust. Geohimija, p. 555 (English translation in Geochemistr
1962, pp. 641–664).

Vostal, J. (1966) *Study of the renal excretory mechanisms of heavy metals.* 15th Internationa
Congress Occupational Health, Vienna, 19–24 September, v. 3., pp. 61–64.

Waldron, H. A. (1966) The anemia of lead poisoning. A review. Brit. J. ind. Med., **23**: 82
100.

Waldron, H. A. (1975) Lead levels in blood of residents near the M6-A38(M) interchange
Birmingham. Nature (Lond.), **253**: 345–346.

Warley, M. A., Blackledge, P., & O'Gorman, P. (1968) Lead poisoning from eye cosmetics
Brit. med. J., **1**: 117.

Warren, H. V., & Delavault, R. E. (1962) Lead in some food crops and trees. J. Sci. Foo
Agric., **13**: 96–98.

Weast, R. C., ed. (1974) *Handbook of Chemistry and Physics*, 55th ed., Cleveland, CRC Press
pp. B-100–B-101.

Webb, M. (1972) Binding of cadmium ions by rat liver and kidney. Biochem. Pharmacol., **21**
2751–2765.

Weber, H. J. (1947) Some experiences with polarographic methods in controlling a lea
hazard in brass foundries. J. ind. Hyg. Toxicol., **28**: 158–167.

Wedepohl, K. H. (1956) Untersuchungen zur Geochemie des Bleis. Geochim. Acta, **10**: 69
148.

Wedepohl, K. H. (1971) Zinc and lead in common sedimentary rocks. Econ. Geol., **66**: 240
242.

Weller, C. V. (1915) The blastophthoric effect of chronic lead poisoning. J. med. Res., **3**: 271
293.

White, J. M., & Harvey, D. R. (1972) Defective synthesis of α- and β-globin chains in lead
poisoning. Nature (Lond.), **236**: 71–73.

Whitehead, T. P., & Prior, A. P. (1960) Lead poisoning from homemade wine. Lancet, **2**:
1343–1344.

158

WHITFIELD, C. L., CHIEN, L. T., & WHITEHEAD, J. D. (1972) Lead encephalopathy in adults. *Am. J. Med.*, **52**: 289–298.

WHO EXPERT COMMITTEE (1969) Urban air pollution with particular reference to motor vehicles. *WHO Technical Report Series, No. 410*, p. 19.

WHO EXPERT COMMITTEE ON FOOD ADDITIVES (1972) Lead. Sixteenth Report, pp. 16–20.

WHO EXPERT COMMITTEE (1973) Trace elements in human nutrition. *WHO Technical Report Series No. 532*, p. 47.

WHO STUDY GROUP (1975) Early detection of health impairment in occupational exposure to health hazards. *WHO Technical Report Series No. 571*, pp. 55–61.

WHO WORKING GROUP (1973) Technical document on lead. In: The hazards to health of persistent substances in water. Annexes to the Report of a WHO Working Group, Copenhagen, WHO Regional Office for Europe, pp. 61–110 (Document EURO 3109 W(1)).

WILLIAMS, H., SCHULZE, W. H., ROTHCHILD, H. B., BROWN, A. S. & SMITH, F. R. (1933) Lead poisoning from the burning of battery casings. *J. Am. med. Assoc.*, **100**: 1485–1489.

WILLIAMS, U. K. (1966) Blood lead and haemoglobin in lead absorption. *Brit. J. ind. Med.*, **23**: 105–111.

WILLIAMS, M. K., & FEW, J. D. (1967) A simplified procedure for the determination of urinary δ-aminoloevulinic acid. *Brit. J. ind. Med.*, **24**: 294–296.

WILLIAMS, M. K., KING, E., & WALFORD, J. (1969) An investigation of lead absorption in an electric accumulator factory with the use of personal samplers. *Brit. J. ind. Med.*, **26**: 202–216.

WILLOUGHBY, R. A., MACDONALD, E., MCSHERRY, B. J., & BROWN, G. (1972) The interaction of toxic amounts of lead and zinc fed to young growing horses. *Vet. Rec.*, **91**: 382–383.

WILSON, J. G. (1973) *Environment and birth defects.* New York, Academic Press, p. 78.

WONG, C. S., & BERRANG, P. (1976) Contamination of tap water by lead pipe and solder. *Bull. environm. Contam. Toxicol.*, **15**: 530–534.

WONG, P. T. S., CHAU, Y. K., & LUXON, P. L. (1975) Methylation of lead in the environment. *Nature (Lond.)*, **253**: 263–264.

WRANNE, L., (1960) Free erythrocytes copro- and protoporphyrin: A methodological and clinical study. *Acta. Paediatrica.*, **49** (suppl. 124).

YAMAMOTO, Y., TANAKA, M., & ARAO, K. (1968) Hemispherical distribution of turbidity coefficient as estimated from direct solar radiation measurements. *J. Meteorol. Soc. Jpn*, **46**: 287–300.

YEAGER, D. W., CHOLAK, J., & HENDERSON, E. W. (1971) Determination of lead in biological and related material by atomic absorption spectrophotometry. *Environ. Sci. Technol.*, **5**: 1020–1022.

YOCOM, J. E., CLINK, W. L., & COTE, W. A. (1971) Indoor/outdoor air quality relationships. *J. Air Pollut. Control Assoc.*, **21**: 251–259.

ZAWIRSKA, B., & MEDRAS, K. (1968) Tumoren und Störungen des Porphyrinstoffwechsels bei Ratten mit chronischer experimenteller Bleiintoxikation. *Zentralbl. Allg. Path.*, **111** (1): 1–12.

ZEL'TSER, M. E. (1962) The functional state of thyroid gland in lead poisoning. *Tr. Just. kraevoi Patol. Akad. Nauk Kaz. SSR*, **10**: 116–120.

ZIEGFIELD, R. L. (1964) Importance and uses of lead. *Arch. environ. Health*, **8**: 202–212.

ZIELHUIS, R. L. (1971) Interrelationship of biochemical responses to the absorption of inorganic lead. *Arch. environ. Health*, **23**: 299–311.

ZIELHUIS, R. L. (1974) Dose response relationship for inorganic lead. Report to the Director, Health Protection EEC.

ZIELHUIS, R. L. (1975) Dose response relationships for inorganic lead. I. Biochemical and haematological responses. II. Subjective and functional responses. Chronic sequelae. No-response levels. *Int. Arch. occup. Health*, **35**: 1–18, 19–35.

ZOLLINGER, H. U. (1953) Durch chronische Bleivergiftung erzeugter Nierenadenome und Carcinome bei Ratten und ihre Beziehungen zu den entsprechenden Neubildungen des Menschen. *Virchows Arch. Bd.*, **323**: 694–710.

ZOOK, B. C., CARPENTER, J. L., & LEEDS, E. B. (1969) Lead poisoning in dogs. *J. Am. vet. Med. Assoc.*, **155**: 1329–1342.

ŽUKOVICKAJA, A. L., ZAMUATKINA A. A., & LUKAŠEV, K. I. (1966) Trace elements in the water of the upper Dnepr river. *Dokl. Akad. Nauk BSSR*, **10**: 891–893.

ZURLO, N., & GRIFFINI, A. M. (1973) *Le plomb dans les aliments et dans les boissons consommes à Milan.* In: *Proceedings of the International Symposium; Environmental Health Aspects of Lead, Amsterdam, 2–6 October 1972.* Luxembourg, Commission of the European Communities, pp. 93–99.

ZURLO, N., GRIFFINI, H. M., & VIGLIANI, E. C. (1970) The content of lead in blood and urine of adults, living in Milan, not occupationally exposed to lead. *Am. ind. Hyg. Assoc. J.* **31**: 92–95.

| | |
|---|---|
| ALGERIA | Société nationale d'Edition et de Diffusion, 3 bd Zirout Youcef, ALGIERS. |
| ARGENTINA | Libreria de las Naciones, Cooperativa Ltda, Alsina 500, BUENOS AIRES — Editorial Sudamericana S.A., Humberto 1° 545, BUENOS AIRES. |
| AUSTRALIA | Mail Order Sales, Australian Government Publishing Service P.O. Box 84, CANBERRA A.C.T. 2600; *or over the counter from* Australian Government Publications and Inquiry Centres at: 113 London Circuit, CANBERRA CITY; 347 Swanston Street, MELBOURNE; 309 Pitt Street, SYDNEY; Mr. Newman House, 200 St. George's Terrace, PERTH; Industry House, 12 Pirie Street, ADELAIDE; 156–162 Macquarie Street, HOBART — Hunter Publications, 58A Gipps Street, COLLINGWOOD, Vic. 3066. |
| AUSTRIA | Gerold & Co., I. Graben 31, VIENNA 1. |
| BANGLADESH | WHO Representative, G.P.O. Box 250, DACCA 5. |
| BELGIUM | Office international de Librairie, 30 avenue Marnix, BRUSSELS. |
| BRAZIL | Biblioteca Regional de Medicina OMS/OPS, Unidad de Venta de Publicaciones, Caixa Postal 20.381, Vila Clementino, 01000 SÃO PAULO, S.P. |
| BURMA | *see* India, WHO Regional Office. |
| CANADA | Information Canada Bookstore, 171 Slater Street, OTTAWA, Ontario K1A 0S9; Main Library, University of Calgary, CALGARY, Alberta; 1683 Barrington Street, HALIFAX, N.S. B3J 1Z9; 640 Ste Catherine West, MONTREAL, Quebec H3B 1B8; 221 Yonge Street, TORONTO, Ontario M5B 1N4; 800 Granville Street, VANCOUVER, B.C. V6Z 1K4; 393 Portage Avenue, WINNIPEG, Manitoba R3B 2C6. *Mail orders to* 171 Slater Street, OTTAWA, Ontario K1A 0S9. |
| CHINA | China National Publications Import Corporation, P.O. Box 88, PEKING. |
| COLOMBIA | Distrilibros Ltd, Pío Alfonso García, Carrera 4a, Nos 36-119, CARTAGENA. |
| COSTA RICA | Imprenta y Librería Trejos S.A., Apartado 1313, SAN JOSÉ. |
| CYPRUS | MAM, P.O. Box 1674, NICOSIA. |
| CZECHO-SLOVAKIA | Artia, Smecky 30, 111 27 PRAGUE 1. |
| DENMARK | Ejnar Munksgaard, Ltd, Nørregade 6, COPENHAGEN. |
| ECUADOR | Librería Científica S.A., P.O. Box 362, Luque 223, GUAYAQUIL. |
| EGYPT | Nabaa El Fikr Bookshop, 55 Saad Zaghloul Street, ALEXANDRIA — Anglo Egyptian Bookshop, 165 Mohamed Farid Street, CAIRO. |
| EL SALVADOR | Librería Estudiantil, Edificio Comercial B No 3, Avenida Libertad, SAN SALVADOR. |
| FIJI | The WHO Representative, P.O. Box 113, SUVA. |
| FINLAND | Akateeminen Kirjakauppa, Keskuskatu 2, HELSINKI 10. |
| FRANCE | Librairie Arnette, 2 rue Casimir-Delavigne, PARIS 6e. |
| GERMAN DEMOCRATIC REPUBLIC | Buchhaus Leipzig, Postfach 140, 701, LEIPZIG. |
| GERMANY FEDERAL REPUBLIC OF | Govi-Verlag GmbH, Ginnheimerstrasse 20, Postfach 5360, 6236 ESCHBORN — W. E. Saarbach, Postfach 1510, Follerstrasse 2, 5 COLOGNE 1 — Alex. Horn, Spiegelgasse 9, Postfach 3340, 62 WIESBADEN. |
| GREECE | G. C. Eleftheroudakis S.A., Librairie Internationale, rue Nikis 4, ATHENS (T. 126). |
| HAITI | Max Bouchereau, Librairie "A la Caravelle", Boîte postale 111-B, PORT-AU-PRINCE. |
| HUNGARY | Kultura, P.O.B. 149, BUDAPEST 62 — Akadémiai Könyvesbolt, Váci utca 22, BUDAPEST V. |
| ICELAND | Snaebjørn Jonsson & Co., P.O. Box 1131, Hafnarstraeti 9, REYKJAVIK. |
| INDIA | WHO Regional Office for South-East Asia, World Health House, Indraprastha Estate, Ring Road, NEW DELHI — Oxford Book & Stationery Co., Scindia House, NEW DELHI: 17 Park Street, CALCUTTA 16 (Sub-agent). |
| INDONESIA | *see* India, WHO Regional Office. |
| IRAN | Iranian Amalgamated Distribution Agency, 151 Khiaban Soraya, TEHERAN. |
| IRELAND | The Stationery Office, DUBLIN. |
| ISRAEL | Heiliger & Co., 3 Nathan Strauss Street, JERUSALEM. |
| ITALY | Edizioni Minerva Medica, Corso Bramante 83-85, TURIN; Via Lamarmora 3, MILAN. |
| JAPAN | Maruzen Co. Ltd. P.O. Box 5050, TOKYO International, 100-31. |
| KENYA | The Caxton Press Ltd, Head Office; Gathani House, Huddersfield Road, P.O. Box 1742, NAIROBI. |
| KUWAIT | The Kuwait Bookshops Co. Ltd. Thunayan Al-Ghanem Bldg, P.O. Box 2942, KUWAIT. |
| LAO PEOPLE'S DEMOCRATIC REPUBLIC | The WHO Representative, P.O. Box 343, VIENTIANE. |
| LEBANON | Documenta Scientifica/Redico, P.O. Box 5641, BEIRUT. |
| LUXEMBOURG | Librairie du Centre, 49 bd Royal, LUXEMBOURG. |

WHO publications may be obtained, direct or through booksellers, from :

| | |
|---|---|
| MALAYSIA | The WHO Representative, Room 1004, Fitzpatrick Building, Jalan Raja Chulan, KUALA LUMPUR 05-02 — Jubilee (Book) Store Ltd, 97 Jalan Tuanku Abdul Rahman, P.O. Box 629, KUALA LUMPUR — Parry's Book Center, K.L. Hilton Hotel, KUALA LUMPUR. |
| MEXICO | La Prensa Médica Mexicana, Ediciores Científicas, Paseo de las Facultades, 26, MEXICO CITY 20, D.F. |
| MONGOLIA | *see* India, WHO Regional Office. |
| MOROCCO | Editions La Porte, 281 avenue Mohammed V, RABAT. |
| NEPAL | *see* India, WHO Regional Office. |
| NETHERLANDS | N.V. Martinus Nijhoff's Boekhandel en Uitgevers Maatschappij, Lange Voorhout 9, THE HAGUE. |
| NEW ZEALAND | Government Printing Office, Government Bookshops at : Rutland Street, P.O. 5344, AUCKLAND; 130 Oxford Terrace, P.O. Box 1721, CHRISTCHURCH; Alma Street, P.O. Box 857, HAMILTON; Princes Street, P.O. Box 1104, DUNEDIN ; Mulgrave Street, Private Bag, WELLINGTON — R. Hill & Son Ltd, Ideal House, Cnr. Gilles Avenue & Eden Street, Newmarket, AUCKLAND S.E. 1 |
| NIGERIA | University Bookshop Nigeria, Ltd, University of Ibadan, IBADAN. |
| NORWAY | Johan Grundt Tanum Bokhandel, Karl Johansgt, 43, OSLO 1. |
| PAKISTAN | Mirza Book Agency, 65 Shahrah Quaid-E. Azam, P.O. Box 729, LAHORE 3. |
| PARAGUAY | Agencia de Librerías Nizza S.A., Estrella No. 721, ASUNCIÓN. |
| PERU | Distribuidora Inca S.A., Apartado 3115, Emilio Althaus 470, LIMA. |
| PHILIPPINES | World Health Organization, Regional Office for the Western Pacific, P.O. Box 2932, MANILA — The Modern Book Company Inc., P.O. Box 632, 926 Rizal Avenue, MANILA. |
| POLAND | Składnica Księgarska, ul. Mazowiecka 9, WARSAW (*except periodicals*). — BKWZ Ruch, ul. Wronia 23, WARSAW (*periodicals only*). |
| PORTUGAL | Livraria Rodrigues, 186 Rua Aurea, LISBON. |
| REPUBLIC OF KOREA | The WHO Representative, Central P.O. Box 540, SEOUL. |
| SINGAPORE | The WHO Representative, 144 Moulmein Road, G.P.O. Box 3457, SINGAPORE 1. |
| SOUTH AFRICA | Van Schaik's Bookstore (Pty) Ltd, P.O. Box 724, PRETORIA. |
| SPAIN | Comercial Atheneum S.A., Consejo de Ciento 130-136, BARCELONA 15 ; General Moscardó 29, MADRID 20 — Librería Díaz de Santos, Lagasca 95, MADRID 6; Balmes 417 y 419, BARCELONA 6. |
| SRI LANKA | *see* India, WHO Regional Office. |
| SWEDEN | Aktiebolaget C.E. Fritzes Kungl. Hovbokhandel, Fredsgatan 2, STOCKHOLM 16. |
| SWITZERLAND | Medizinischer Verlag Hans Huber, Länggass Strasse 76, 3012 BERNE 9. |
| THAILAND | *see* India, WHO Regional Office. |
| TUNISIA | Société Tunisienne de Diffusion, 5 avenue de Carthage, TUNIS. |
| TURKEY | Librairie Hachette, 469 avenue de l'Indépendance, ISTANBUL. |
| UGANDA | *see address under* KENYA. |
| UNITED KINGDOM | H.M. Stationery Office : 49 High Holborn, LONDON WC1V 6HB ; 13a Castle Street, EDINBURGH EH2 3AR ; 41 The Hayes, CARDIFF CF1 1JW ; 80 Chichester Street, BELFAST BT1 4JY ; Brazennose Street, MANCHESTER M60 8AS ; 258 Broad Street, BIRMINGHAM B1 2HE ; Southey House, Wine Street, BRISTOL BS1 2BQ. *All mail orders should be sent to* P.O. Box 569, London SE1 9NH. |
| UNITED REP. OF TANZANIA | *see address under* KENYA. |
| UNITED STATES OF AMERICA | *Single and bulk copies of individual publications (not subscriptions):* Q Corporation, 49 Sheridan Avenue, ALBANY, NY 12210. *Subscriptions:* Subscription orders, accompanied by check made out to the Chemical Bank, New York, Account World Health Organization, should be sent to the World Health Organization, P.O. Box 5284, Church Street Station, NEW YORK, NY 10249. Correspondence concerning subscriptions should be forwarded to the World Health Organization, Distribution and Sales Service, 1211 GENEVA 27, Switzerland. *Publications are also available from the* United Nations Bookshop, NEW YORK, NY 10017 (*retail only*). |
| USSR | *For readers in the USSR requiring Russian editions:* Komsomolskij prospekt 18, Medicinskaja Kniga, Moscow — *For readers outside the USSR requiring Russian editions:* Kuzneckij most 18, Meždunarodnaja Kniga, Moscow G-200. |
| VENEZUELA | Editorial Interamericana de Venezuela C.A., Apartado 50785, CARACAS — Librería del Este, Av. Francisco de Miranda 52, Edificio Galipán, CARACAS. |
| YUGOSLAVIA | Jugoslovenska Knjiga, Terazije 27/II, BELGRADE. |

Orders from countries where sales agents have not yet been appointed may be addressed to :
World Health Organization, Distribution and Sales Service, 1211 Geneva 27, Switzerland,
but must be paid for in pounds sterling, US dollars, or Swiss francs.